The Agricultural Field Experiment

A Statistical Examination of Theory and Practice

The Agricultural Field Experiment

A Statistical Examination of Theory and Practice

S. C. PEARCE

Mathematical Institute
University of Kent at Canterbury U.K.

A Wiley–Interscience Publication

JOHN WILEY & SONS

Chichester · New York · Brisbane · Toronto · Singapore

Library of Congress Cataloging in Publication Data:

Pearce, S. C. (Stanley Clifford)
The agricultural field experiment.
'A Wiley–Interscience publication.'
Bibliography: p.
Includes index.
1. Field experiments. 2. Field experiments—Statistical methods. I. Title.
S540.F5P4 1983 630′.724 82-13711

ISBN 0 471 10511 2

British Library Cataloguing in Publication Data:

Pearce, S. C.
The agricultural field experiment.
1. Agriculture—Statistical methods
2. Mathematical statistics
I. Title
519.5′02463 HD1425

ISBN 0 471 10511 2

Photo Typeset by Macmillan India Ltd, Bangalore
Printed by Page Bros (Norwich) Ltd.

From *Gulliver's Travels* by Jonathan Swift, 1667–1745.
The learning of this people is very defective, consisting only in morality, history, poetry, and mathematics, wherein they must be allowed to excel. But the last of these is wholly applied to what may be useful in life, to the improvement of agriculture and all mechanical arts; so that among us it would be little esteemed.

A voyage to Brobdingnag, Chapter VII

In these colleges the professors contrive new rules and methods of agriculture and building, and new instruments and tools for all trades and manufactures, whereby, as they undertake, one man shall do the work of ten; a palace may be built in a week, of materials so durable as to last for ever without repairing; all the fruits of the earth shall come to maturity at whatever season we think fit to choose, and increase a hundredfold more than they do at present, with innumerable other happy proposals. The only inconvenience is that none of these projects are yet brought to perfection . . .

A voyage to Laputa, etc. Chapter IV

Contents

(Sections marked * are for the introduction of new ideas that are to be examined in more detail later.)

Preface

At the beginning of his *Discourse on Method* Descartes explained how he came to write his book. It was not, he said, that he supposed himself to be more talented than other people but the course of his life had led him to considerations that might be of wider interest. This book has been written in much the same spirit.

I explain myself with some reluctance because I do not like mixing personal and technical matters, but the potential reader will reasonably want to know what the book is about and I can best tell him by setting out the experiences that led to its being written. I began by taking a degree in mathematics, but I quickly found that I had missed my vocation. There are people who can 'do' mathematics without being much interested in the subject for its own sake and clearly I was one of them. I was fascinated rather by the uses to which it could be put, so I enrolled for another degree course, this time in statistics under Professor E. S. Pearson. His approach was a revelation to me and deeply satisfying, because to him there was no distinction between theory and practice; the one illuminated and justified the other. (Later, at his eightieth birthday party, I was to learn that his experiences with mathematics had been very like my own.) I can assure my reader that I have expressed myself in mathematical terms only because they provide the best way of saying some of the things that need to be said. Further, I have mostly used matrix algebra rather than the usual kind, again because it gives the best means of expression. In statistics the subject is often an aggregation of numbers, e.g. the data or the block sizes, rather than a single number, so the language must be of a kind that deals readily with such ideas.

If that sounds forbidding, let me emphasize that the mathematics is as straightforward as I have been able to make it. It is there as a bridge, not as an obstacle. If anyone is unfamiliar with matrix algebra, there is an appendix to help. If there is still difficulty, crucial points are illustrated by worked examples that show the form of any necessary calculations. As far as possible concepts are explained in words as well as in mathematical notation.

Although this text is not intended primarily for them, I hope that the mathematical statisticians will find something of interest in it. There is not a lot that is new, but a single notation has been used throughout, beginning with Chapter 2. As a consequence, a number of topics that had previously appeared to be unrelated can be seen as part of a whole. To quite an extent the text unifies much of the theory of the comparative experiment.

So much for the mathematics. Now let me continue with the background. When I had graduated a second time I needed a job and here I was fortunate because Pearson had a cousin, Dr R. G. Hatton, who was engaged on a rather unusual venture. Dissatisfaction with government policy had led to the founding of an unofficial research institute at East Malling in Kent to study the problems of English fruit growing and Hatton was the director. Among others he had the help of Mr T. N. Hoblyn, who had gone to East Malling as a recorder, in which capacity he had complained repeatedly that the data he was producing were so variable as to be useless. As a consequence he had been packed off to Dr R. A. Fisher at the Rothamsted Experimental Station to learn better and had come back to start a Statistics Section. He now needed an assistant and Hatton applied to his cousin for help in finding one. As a result I was sent along and, it appears, I won everyone's heart with the remark that statistical methods are a tool and no more. (To be truthful, I had no thoughts of my own on the subject. I was just repeating what I had been told, but now I believe it because my experience has amply confirmed it.) Anyhow, I was appointed at some ludicrously low salary—East Malling had no money—and later I was put in charge of the design of experiments and the analysis of data, which remain my chief centres of interest. Some who have seen the scheme of this book have commented that there is rather a lot on those two topics, but they provide my own window on to field experiments and I make no apology.

Looking back I can see that East Malling in the early days was rather unusual, being more like a religious community than a scientific institute. It paid such low salaries that one must wonder why anyone ever stayed there, but the comradeship was wonderful. I have often pondered how it came about that a team, recruited as that one was, achieved so much and produced so many scientists of high distinction. I can attribute its success only to the respect with which each member of staff regarded the diverse skills of colleagues, perceiving them to be complementary to his own and therefore assets to be used, not alien influences to be repelled. In that spirit everyone's work took on a depth and perspective that transcended anything he could have achieved unaided. In this book there is quite a lot about cooperation and the way in which people trained in various disciplines need one another. There are many institutes with expensive buildings and equipment. Some do good work and others not. Others are lamentably short of funds. Again some are good and others bad. Everything depends upon the staff, who as individuals must value one another as possessing different but necessary skills. Agricultural research requires people of various backgrounds and they must work together as a team. It has few rewards for a collection of non-cooperating ignoramuses, however high their individual qualifications may be.

With the passage of time Hatton was succeeded by Dr F. R. Tubbs and Hoblyn by myself. (Meanwhile East Malling had become completely 'official', a change that neither of us liked very much.) As Head of Statistics I was involved in a lot of planning, i.e. considering objectives and allocating resources to achieve them. It is a branch of statistics that is strangely neglected in the literature and, while my experience of it is not extensive, I have thought it worthwhile to put down some ideas on the subject. Certainly no one can design a good experiment without a clear appreciation of what has to be done and a knowledge of what is available in the way of land, labour and everything else.

When Tubbs retired I realized that I no longer had any distinctive role at East Malling, so I took advantage of the readiness of the Ministry of Overseas Development, as it then was, to fund a post for me at the University of Kent at Canterbury on the understanding that I would devote myself to the problems of the Third World. For a long time I had been going overseas under the auspices of various agencies and had become interested in experiments conducted in conditions very different from those I had known before. Some principles applied as much in tropical conditions as in temperate, but others did not transport very well from the areas where they had been evolved. In fact, too many statistical practices have been developed in the conditions of Europe or North America and then applied without further thought on coral islands, in areas of very low and very high rainfall, such as do occur in equatorial regions, and on the sides of mountains. In this book I have tried to produce a text that would not be restricted to meeting the needs of the so-called developed world but would be of use elsewhere. In fact, many of my examples concern tropical crops.

Such then is the scope of the book. It remains for me to thank the many people who have helped me along the way. They are too numerous for me to name them all, but a few stand out. For example, Chapter 3 is a summary of what I have learned from Professor T. Caliński of the Academy of Agriculture, Poznan, Poland. It is a pity that he did not write it himself because he would have done it better. Also, Dr G. H. Freeman, who worked with me at East Malling and is now at the National Vegetable Research Station, Wellesbourne, England, set me on the path that led to Chapter 5, but I had better be careful here because I am not sure that nowadays he would approve of it. Not least, there are two colleagues at the University of Kent to whom I am much indebted. One is Dr D. A. Preece, now returned to the Rothamsted Experimental Station, whose meticulous scholarship has corrected many of my own impetuous conjectures. The other is Dr Byron Jones, who has checked every piece of mathematics and every worked example to prevent anyone being misled.

However, books, like field experiments, are not just the outcome of getting the technicalities right and here again I am indebted. First, I thank my secretary, Mrs Margaret Wells, for the help she has given me throughout and especially for the extreme care with which she has prepared the typescript. Then there is my wife, who has supported me and provided a base from which to work. She tells me that there is no need to say anything. Much as I value her judgement in most matters, I think she is mistaken in that and I thank her very much.

1982 S. C. Pearce

List of symbols used consistently throughout the text

Note: A circumflex (^) over a symbol indicates a value that has been estimated from data subject to error, so it is not necessarily exact.

$\mathbf{1}_p$	A vector ($p \times 1$) with all elements equal to one.
n	The number of plots in an experiment.
b	The number of blocks in an experiment. (In a few instances b has been used in other senses to preserve the sequence a, b, c, etc. but not in a way that need cause confusion.)
v	The number of treatments in an experiment.
$\mathbf{y}$	A vector ($n \times 1$) of data, one value for each plot.
$\boldsymbol{\eta}$	A vector ($n \times 1$) of random, independent residuals, each with an expected value of zero and a variance of σ^2.
$\mathbf{D}$	The design matrix for blocks ($b \times n$), which has a row for each block and a column for each plot. An element equals one if the plot lies in that block and zero otherwise. Note that $\mathbf{D}'\mathbf{1}_b = \mathbf{1}_n$.
$\boldsymbol{\Delta}$	The design matrix for treatments ($v \times n$) which has a row for each treatment and a column for each plot. An element equals one if the plot receives that treatment and zero otherwise. Note that $\boldsymbol{\Delta}'\mathbf{1}_v = \mathbf{1}_n$.
$\mathbf{N} = \boldsymbol{\Delta}\mathbf{D}'$	The incidence matrix ($v \times b$). It has a row for each treatment and a column for each block. The elements show the number of times each treatment occurs in each block.
$\mathbf{B} = \mathbf{D}\mathbf{y}$	A vector ($b \times 1$) formed by summing the data in each block.
$\mathbf{T} = \boldsymbol{\Delta}\mathbf{y}$	A vector ($v \times 1$) formed by summing the data for each treatment.
G	The grand total of data. Note that $G = \mathbf{1}'_n\mathbf{y} = \mathbf{1}'_b\mathbf{B} = \mathbf{1}'_v\mathbf{T}$.
σ	See entry for $\boldsymbol{\eta}$ above.
$\mathbf{k} = \mathbf{D}\mathbf{1}_n$	The vector ($b \times 1$) of block sizes.
$\mathbf{r} = \boldsymbol{\Delta}\mathbf{1}_n$	The vector ($v \times 1$) of treatment replications.
$\boldsymbol{\phi}$	The residuals matrix ($n \times n$) (thus $\hat{\boldsymbol{\eta}} = \boldsymbol{\phi}\mathbf{y}$) when no allowance has been made for the effects of treatments.

$\boldsymbol{\psi}$ The residuals matrix $(n \times n)$ (thus $\hat{\boldsymbol{\eta}} = \boldsymbol{\psi}\mathbf{y}$) when the effects of treatments have been taken into account.

$\mathbf{Q} = \boldsymbol{\Delta\phi}\mathbf{y}$ The vector $(v \times 1)$ of adjusted treatment totals. (cf. $\mathbf{T} = \boldsymbol{\Delta}\mathbf{y}$.)

$\mathbf{C} = \boldsymbol{\Delta\phi\Delta}'$ The coefficient matrix $(v \times v)$, such that $\mathbf{Q} = \mathbf{C}\hat{\boldsymbol{\tau}}$.

$\mathbf{C}^-$ Any generalized inverse of $\mathbf{C}$, i.e., any matrix $(v \times v)$ such that $\mathbf{CC}^-\mathbf{C} = \mathbf{C}$.

$\boldsymbol{\Omega}$, $\boldsymbol{\Xi}$, $\boldsymbol{\Upsilon}$, and $\mathbf{C}^+$ are possible solutions for $\mathbf{C}^-$.

THE DELTA NOTATION

If $\mathbf{x}$ is a vector, $\mathbf{x}^{p\delta}$ is a diagonal matrix. Its elements correspond respectively to those of $\mathbf{x}$ but each is raised to the power, p. Thus, if

$$\mathbf{x} = \begin{bmatrix} 1 \\ 4 \\ 9 \end{bmatrix}, \quad \mathbf{x}^{\delta} = \begin{bmatrix} 1 & 0 & 0 \\ 0 & 4 & 0 \\ 0 & 0 & 9 \end{bmatrix}, \quad \mathbf{x}^{2\delta} = \begin{bmatrix} 1 & 0 & 0 \\ 0 & 16 & 0 \\ 0 & 0 & 81 \end{bmatrix},$$

$$\mathbf{x}^{\frac{1}{2}\delta} = \begin{bmatrix} 1 & 0 & 0 \\ 0 & 2 & 0 \\ 0 & 0 & 3 \end{bmatrix}, \quad \mathbf{x}^{-\frac{1}{2}\delta} = \begin{bmatrix} 1 & 0 & 0 \\ 0 & \frac{1}{2} & 0 \\ 0 & 0 & \frac{1}{3} \end{bmatrix}.$$

Note that $\mathbf{x}^{p\delta}\mathbf{x}^{q\delta} = \mathbf{x}^{(p+q)\delta}$, so $\mathbf{x}^{p\delta}\mathbf{x}^{-p\delta} = \mathbf{I}$, the identity matrix. Also, if $\mathbf{x}$ has m elements, $\mathbf{x}^{\delta}\mathbf{1}_m = \mathbf{x}$. Note also that $\mathbf{DD}' = \mathbf{k}^{\delta}$ and $\boldsymbol{\Delta\Delta}' = \mathbf{r}^{\delta}$.

THE ASTERISK NOTATION

*A indicates the final figure arrived at for the yield given by Treatment A, all needful adjustments having been made, e.g. for non-orthogonality, covariance, missing plots, etc.

Chapter 1

The Experiment in Context

1.1 THE HISTORICAL CONTEXT

Mankind must have been conducting agricultural experiments for a very long time, because the transition from wild grasses to high-yielding wheats—to take no other example—must have been a protracted process. Most of the time improvement was not consciously planned; the chief gains were probably made when the best grain was used for sowing in the following year. Perhaps too it was helped by famine. In societies based on tradition, in which son learns from father in one generation from another, families that practice poor methods are wiped out, leaving the more successful to people the earth.

In ancient times progress was slow, but the first decades of the nineteenth century saw a quickening of pace with the founding of properly constituted research institutes in which the new sciences of plant physiology and chemistry could be applied to agricultural needs. The story of one country has been told in some detail (Russell, 1966). At first methods were simple but the realization that fields are not uniform, which appears obvious now but had to be learnt (Mercer and Hall, 1911; Harris, 1920), led to consideration how best to take comparable samples of land. As a result 'replication' was introduced; it was no longer thought good enough to divide the experimental area into parts, one for each treatment, and to proceed as if all were the same. Instead, each treatment had to be assigned to a set of small areas (or 'plots') chosen to be representative of the whole. The author recalls clearly the classical apple rootstock trials at the East Malling Research Station in England, started in 1917. Each rootstock had been planted once in the shelter of the railway embankment and each had its share of the land at the other end, which was exposed to wind. Apart from that they were disposed so that as far as possible each treatment had the centre of mass of its plots on the mid-point. In that way everything was balanced spatially.

Along with these developments people were turning their attention to field technique, e.g. the sowing and recording of small areas and the meticulous routines needed for the reliable conduct of an experiment. The names of Student (W. S. Gosset) and E. S. Beaven especially come to mind in this connection. Also, the biometrical techniques being developed by Karl Pearson and his school began to influence thinking. Indeed, Gosset's *t*-test (Student, 1908) is a landmark in statistical history.

The modern approach is usually credited to R. A. Fisher, whose appointment to the Rothamsted Experimental Station in England in 1919 led to developments of great importance. A man of extraordinary width of mind, he was later to find his fulfilment rather in genetics and in the formalization of the process of inference, but his thoughts ranged over many topics and there were few that they did not illuminate. His early work was in experimental design and there he perceived that replication, if the plots are assigned treatments at random, provides a way of estimating the extent of error from the experiment itself. That was a great step forward. However, to suggest randomization to a generation accustomed to the careful balancing described above was to ask for trouble. Controversy centred on the relative advantages of minimizing experimental error and measuring it. Here Fisher made a good point. If plots of a treatment are dispersed systematically to cover all parts of the field, it may well be true that the mean value found will approximate closely to that for the field as a whole. The difficulty he saw was this: although the error might be minimized, the estimate obtained by comparing similarly treated plots would be increased by their dispersal. Consequently, although a systematic allocation of plots might indeed reduce the variability of means, it would in fact increase the estimate made of it. That paradoxical conclusion took some time to absorb.

Fisher's own approach was to divide the experimental area into 'blocks', chosen to represent areas of similar characteristics. Thus, to take the rootstock trials described above, he would no doubt have recognized the special nature of the land along the railway and of the slope exposed to the wind and he would have made a block of each. He would then have divided the remaining land into blocks that were as compact as possible, i.e. squares rather than strips. He required also that each block should contain as many plots as there were treatments. He then allocated the treatments at random within each block rather than within the area as a whole. In that way, he obtained a good level of 'local control' yet kept an unbiassed estimate of experimental error, since all data could be referred to the block means instead of the general mean, thus removing differences due to local variation. He called the design 'randomized blocks'; the method of dealing with the data was named the 'analysis of variance'. Both remain today much as Fisher left them.

At this point it may be useful to describe briefly what is meant by the analysis of variance. For a long time statisticians had been measuring the variability of figures by working out an expectation for each based on means and noting the deviation of each actual value from its expected value. They then used the sum of squared deviations as a measure of variability. What Fisher did was to 'partition' that quantity into independent components, one for treatments, one for error and so on. That, however, would have done little good had he not also obtained estimates of their relative values on the 'null hypothesis' that there are no genuine treatment effects, i.e. that any apparent differences between treatments arise solely from the operation of error. He found such estimates in the concept of 'degrees of freedom', which may be explained thus: If there are two quantities, a and b, it is sufficient to know $(a-b)$ and the variability is completely determined. Similarly, if there are three quantities, a, b and c, any two pieces of information will show the variability, provided they are independent. Thus, if $(a-b)$ is known and also $(a-c)$ there is no need to enquire about $(b-c)$ because its value has already been implied. (The requirement of independence raises questions that need not be discussed immediately.) In general there are $(p-1)$ degrees of freedom between p quantities. The basis of the analysis of variance lies in an ability to partition the $(n-1)$ degrees of freedom among the n data so that the expectations of the corresponding sums of squared deviations are proportionate. Then each sum of squared deviations, i.e. that for treatments and that for error, can be divided by its appropriate degrees of freedom to give two quantities called 'mean squares'. If the effect of treatments does indeed arise solely from the operation of error, the two mean squares should be about the same. The equality will not be exact, but tables can be prepared to show how far they can differ. If, however, there are real treatment effects, the treatment mean square will be inflated and F (= treatment mean square/error mean square) will be more than one and tables will show the probability of so large a value arising solely by chance. If that probability is found to be less than 0.05, by convention the effects of treatment are said to be 'significant'. Other levels of significance are, of course, possible and are sometimes used. (As a matter of nomenclature, it may be added that the error mean square is often called the 'error variance' and its square root the 'standard error of an observation'. The last quantity expressed as a percentage of the mean value of the data is called the 'coefficient of variation'.)

Many consider that the concept of significance has been pushed too far. In some instances no one doubts that the treatments must have had an effect. In that case a statement that they do not differ significantly sounds more like an indictment of the experimenter's skill than a scientific assertion. On the other hand, there are occasions when there is genuine doubt whether treatments of a certain kind have anything to do with the problem and it is then helpful to know that phenomena have appeared that can scarcely have arisen by chance. Quite apart from facilitating an F-test the analysis of variance is useful because the error mean square gives the internal estimate of variability that is an essential feature of the modern approach.

To return to the design considerations, Fisher's other suggestion was the use of a double blocking system, in which two sets of linear blocks cross, the one set being called 'rows' and the other 'columns', the plots being formed by their intersection. He then contrived to assign treatments so that each occurred once in each row and once in each column, e.g.

C	A	D	B
D	B	C	A
B	D	A	C
A	C	B	D

That design he called a 'Latin square'. The name is a tribute as much as anything to Fisher's width of reading, because it relates to Euler's (1782) famous problem of the 36 officers. It appears that the Emperor was to visit a garrison town in which six regiments were quartered and the commandant took it into his head to arrange 36 officers in a square, one of each rank from each regiment, so that, whichever row or column the Emperor walked along, he would meet one officer of each of the six ranks and one from each of the six regiments. Designating rank by Roman (or Latin) letters and regiments by Greek ones, Euler proceeded to show, using an argument that is fallacious in method but correct in its conclusion, that the commandant had set himself an impossible task. A Latin square is always possible. So is a Graeco-Latin square for most orders other than six, though it is necessary to choose the Latin square carefully, e.g.

Cα	Aγ	Dβ	Bδ
Dγ	Bα	Cδ	Aβ
Bβ	Dδ	Aα	Cγ
Aδ	Cβ	Bγ	Dα

The Greek letters are added to the design so that each occurs once in each row and in each column and once on each Latin letter. Truth to tell, Graeco-Latin squares are not very useful in designing experiments, but they have crept in on account of their historical association.

More useful, at least for theoretical purposes, are complete 'orthogonal sets' in which there is a $p \times p$ array and $(p+1)$ classifications, such that each level of each classification occurs with any level of any other classification once and once only, e.g.

Cα4	Aγ2	Dβ3	Bδ1
Dγ1	Bα3	Cδ2	Aβ4
Bβ2	Dδ4	Aα1	Cγ3
Aδ3	Cβ1	Bγ4	Dα2

Here the five classifications are rows, columns, Latin letters, Greek letters and numbers. The analysis of variance was readily extended to cover these new designs.

A number of other developments followed, among them the 'analysis of covariance'. Here a second quantity is measured on each plot, not on account of its own importance but for the light it might shed on the variability of the quantity under study. First suggested by Fisher (1934), the technique was put on to a sound mathematical basis by Wishart (1936).

Another vital contribution was that of Yates (1936b, 1937), who recognized the difficulties of requiring all blocks to be of such a size that each treatment must occur once and only once in each. As he perceived, there could be appreciable advantages in using smaller blocks that might fit the fertility patterns better. In particular, he suggested the use of 'balanced incomplete blocks', in which all treatments 'occur' with equal frequency and all pairs of treatments 'concur', i.e. come together in blocks, equally often, e.g. as in the design on the next page.

Block I	D	E	F	G	Block IV	A	C	E	F
II	B	C	F	G	V	A	B	D	F
III	B	C	D	E	VI	A	C	D	G

Block VII A B E G

Here all treatments occur four times and all pairs concur in two blocks. The effect of blocks and treatment can no longer be estimated independently, because blocks are not made up the same way in respect to treatments and treatments are not dispersed the same way among blocks. Accordingly, such designs are said to be 'non-orthogonal', the term being taken from mechanics, where forces that act at right angles, i.e. orthogonal forces, have effects that can be calculated for each without reference to the action of the other. The analysis of variance was again extended, as it has been for the many other forms of non-orthogonality that have subsequently been introduced.

Also Yates (1935) suggested a refinement that was to prove of the greatest importance. At that stage of knowledge the analysis of variance was being used to partition the total variability of the data into three parts. First, there was a component due to blocks, rows, columns etc., which represented the effects of local variation. It was identified only so that it could be banished. Secondly, there was a component for the effects of the various treatments and, thirdly, another that represented the residual variation for which no causes could be assigned. The last was regarded as error. What Yates proposed now was to partition the treatment component itself. Suppose, for example, that the twelve 'treatment combinations' have been made up by associating each of four varieties with each of three fertilizer schedules. There are eleven degrees of freedom in all for treatments; three of them represent the 'main effect' of the varieties and two the fertilizer treatments. What do the other six represent? As Yates saw, they arise from a possible 'interaction' between the two effects, i.e. it may well be that the various varieties respond to fertilizer in different ways. (Expressed differently, it may well be that the best fertilizer schedule depends upon the variety. The two statements come to the same thing.) Further, the interaction is not a mere ornamental addition. If varieties and fertilizers do interact, any assertion about main effects will need qualification. It may be true that a certain programme of fertilization is best for all the varieties in the experiment, but the extent of its improvement will need to be determined for each. Further, the fact that different varieties give different results raises the question whether there are not perhaps varieties that would do better under some other regime. 'Factorial' sets of treatments, as they are called, pose questions that sometimes call for quite subtle answers, but that is not an objection to their use. Rather they should be welcomed as taking an experimenter to the heart of what he is doing. Also, there is no need to stop at two factors. Many more are possible, but too many can give rise to interactions so abstruse that no one is really interested in them. It is then that 'confounding' becomes important, i.e. designing experiments with smaller blocks so that unimportant interactions are lost. That possibility was examined by Yates (1937) in a publication that remains of immense value. Incidentally factorial design was foreshadowed much earlier (Fisher and Mackenzie, 1923). Indeed, there were signs of it even before that, but it was Yates who saw its potential for situations in which a thorough survey of some problem is called for.

By 1934 the subject had become established in its present form and the formation of the Industrial and Agricultural Research Section of the Royal Statistical Society in that year led to a succession of papers of the highest quality (Wishart, 1934; Irwin, 1934; Neyman, 1935; Yates, 1935; Gosset, 1936; Cochran, 1937; Bartlett, 1937 and several others). The author, who was a student of the University of London at the time, recalls those discussions, often acrimonious and disputatious as they were, as a wonderful introduction to agricultural experimentation; they remain so still. Few later writers have attained the breadth of outlook that was taken for granted when the subject was young. For the same reason some of the early books remain unsurpassed (Fisher and Wishart, 1930; Hoblyn, 1931; Wishart and Sanders, 1936; Stewart, 1947, and Crowther, 1947, the last two being bound in one volume). Pride of place, however, must go to the two great works of Fisher, namely, *Statistical Methods for Research Workers* (1925) and *The Design of Experiments* (1935), which likewise retain their value even today.

So much for the past, but what of the future? Since the work of the pioneers the subject of statistics has developed enormously and the number of techniques is now so great that few can appreciate them all, even in outline. If the width of interest of the early investigators has been lost, everything is now more soundly established and there are few gaps. It is not clear, however, that statistical methods are keeping pace with the problems. In so far as questions have become more complex, there need be no fear. Indeed some might think that techniques have run ahead of requirements. The difficulty is rather that there is an ever-pressing need for greater precision. Once upon a time diseases devastated and it needed little skill at experimentation to show the efficacy of a proposed control measure. Nowadays the problems concern rather endemic diseases that insidiously reduce crops over a wide area by a small amount. To take another example, the larger effects of fertilizer appear to have been discovered. Present studies do not relate to the need for essential elements, the absence of which seriously diminishes growth and cropping, but to matters like fertilizer placement, where differences are much smaller, though not negligible. Indeed, a gain of five per cent in crop can spell the difference between bankruptcy and sufficiency for a small farmer or it could establish an industry that would otherwise collapse. Modern developments of statistical technique have done much, but in the main they have not been directed to improving precision. A return to the preoccupations of the pioneers would perhaps be of some value.

To write thus is not to belittle subsequent work, e.g., Goulden (1952), Kempthorne (1952), Quenouille (1953), Federer (1955), Finney (1955, 1956), Snedecor (1956), Cox (1958), Bailey (1959), Finney (1960, 1962), Pearce (1965), Campbell (1967), Rayner (1969), Gomez and Gomez (1976), and Dagnelie (1981). The first two volumes of Bliss (1967, 1970) are notable for the excellence of the examples. A third volume was promised but the death of its distinguished author has made its appearance uncertain. Mention must also be made of the classic work of Cochran and Cox (1950) with its encyclopaedic catalogue of available designs. Among more specialized works are those of the present writer (Pearce, 1953, 1976a) on long-term experiments and of Dyke (1974), who deals with those practicalities in the field that are so often ignored.

Modern experiments can be of many kinds and can use a bewildering range of designs proposed by the biometricians. With any one, however, there are at least two areas that have to be considered, namely, the largest and the smallest. The largest is the 'tract', i.e. the area for which results are required; the smallest is the 'plot', i.e. the units to which the separate treatments are to be applied. Both need consideration before anything else can be decided, so this monograph will deal with them first.

1.2 THE GENERALIZATION OF RESULTS

To take the tract first, it is a weakness of some research institutes that they conduct experiments on their own farms and then announce the results without giving much thought to generalization. They may defend their attitude by saying that their local conditions are fairly representative of those in the region, but that surely needs qualification. No region is uniform and the differences can take many forms. A recommendation about fertilization, for example, may not apply on heavier, richer soils, while another about the control of a fungal disease may not apply where rainfall is higher. The recipient of such advice is entitled to ask how far it applies to his local conditions. There are others, e.g. administrators in capital cities, who are chiefly interested in the general effect over a tract, but they do not want to sanction expenditure, say, on the bulk purchase of imported fertilizers or spray chemicals, without knowing of limitations on their use. Accordingly recommendations relating to farming practices need to be widely based. One way is to supplement the main experiment, which may well have tested a range of treatments, by others in varying conditions to test only those that appear to be promising; another is to try out practical recommendations at a number of sub-stations in the first place. Also, there are variants, like the use of farmers' fields.

If an experiment can be dispersed over many centres in the tract, it takes on many features of a survey. If, for example, there are two plots at each site, one treated conventionally and the other according to some new idea, the whole can be regarded as one big design in randomized blocks. Equally, the difference between the two treatments can be evaluated for each site and the differences related to rainfall, type of soil, etc. in the hope of finding the conditions in which the new treatment is most advantageous—or perhaps damaging.

The advocacy of experiments at local centres and on farms is not meant to decry a main research centre, which is indeed indispensable when studies are more fundamental and concern phenomena which notionally do not depend much on local conditions, i.e., the tract is virtually the range of the species. For example, someone might want to know the symptoms that arise when a certain mineral nutrient is in short supply, or perhaps the life cycle of a pest is being studied with a view to suggesting possible control measures, or an effort is being made to gauge the seriousness of partial defoliation at a time of year when it might occur accidentally from spray damage. Such more fundamental studies are essential if those who carry out experiments of more immediate application are to keep up a supply of ideas. They do not constitute the whole of useful research but they form a necessary part of it. A research worker should always know where he stands in the spectrum that reaches from studies in pure biology at one end to the trial of empirical expedients at the other because it must affect how and where he does his experiment.

There is another reason why he should know it. If the investigation is botanical in nature, the experimenter will want to continue with a treatment even if it does have deleterious effects. For example, if symptoms of mineral deficiency do appear, he will want to know how they develop and the extent to which the plants will become stunted or unfruitful and whether or not they will die as a result. In an agronomic experiment, on the other hand, the appearance of deficiency symptoms might justify the prompt application of some remedial measure lest other results be invalidated. Even if the purpose of the experiment is to study fertilizer requirements, the appearance of marked deficiency symptoms can be evidence enough that the treatment concerned is not for general adoption. To argue thus is not to set botanical and agronomic considerations in opposition. Indeed, they are complementary. Every experiment should be sound botanically and it should also be directed to agronomic application, whether immediately or by guiding further research. If the two kinds of consideration ever appear to be in conflict, the experimenter should be quite clear what sort of investigation he is conducting. Further, all such decisions should be consistent with one another. It is no good resolving a conflict this week so as to clarify an underlying biological mechanism and then upsetting everything next week by adapting a treatment to please practical farmers. Equally, in an experiment to find the best of several cultural practices, each must be carried out as it would be on a farm. It would be foolish to do something differently for fear of disguising a physiological phenomenon that was not under investigation. Here, as in many other instances, the first need is for a clear appreciation of aims.

Try as the experimenter will, there is one problem of generalization that must be accepted. Given a range of treatments, e.g. a batch of new seedlings, those that appear to be outstandingly good in the trial will be those that are both good in themselves and have the luck on that occasion to excel their usual standard. When they go out into commerce or proceed to a further trial, they may well retain their good characteristics but in the long run they will not have the same luck and somebody may be disappointed. Similarly the worst in the trial will not really be as bad as all that. Altogether the selection of varieties on the basis of field trials is a more complex business than appears at first sight. A good survey is that of Curnow (1961).

The subject of generalization has been raised at this early stage because it is one that should always be in an experimenter's mind. Little more will be said about it.

Perhaps the answer the experimenter seeks is so fundamental that supportive evidence is not called for; perhaps the experiment needs to be repeated at many sites in many seasons; perhaps it is enough to disperse its blocks; perhaps . . . The possibilities are endless. The warning about the over-enthusiastic acceptance of one result can only be emphasized.

1.3 PLOT TECHNIQUE

As has been said, the plots of an experiment are its smallest unit, being parcels of land each of which receives a designated treatment, that treatment probably being different from those assigned to its neighbours. It should be emphasized that some of the treatments at least, and usually all of them, must have two or more plots. The reason is the need to gauge the reliability of results in the absence of any prior estimate of general applicability. The experimental error has therefore to be derived from a comparison of values that would be the same.

This absence of prior information about errors is a feature of agricultural experimentation, but reflection shows that it is inevitable. Broadly speaking there are two main sources, environmental variation in an area apparently uniform and differences between the plants themselves, though that may be less important when a plot consists of many small plants rather than a few large ones. (Error also arises for technical reasons, like loss of grain at reaping or inaccuracy of weighings, but for the moment they can be ignored.) To take environmental variation first, a piece of land that appears to be quite uniform may have unexpected patches of good or bad soil, but they may show up only in certain conditions, such as a spell of dry weather. Losses from pests may cause a lot of variability, but again the extent and the pattern of incidence may be unforeseeable. Things are no better with plant variation. Plants are most uniform when all are dead. Growth and cropping are then exactly zero for each. Apart from that the nearest approach to uniformity comes when all are growing freely. However, in the middle of the range, when plants are under stress, differences are emphasized and variability is at a maximum. It is no more possible to forecast stress than the incidence of environmental factors. A dry season or one with little sun can increase variability as can any other adverse circumstance. The only thing to do is to hope for the best but to assume nothing, deriving an estimate of the error from the data themselves by the comparison of replicate values. In statistical parlance, there must be degrees of freedom for error. Further, there must be enough, say, eight or more, to make the estimation of error effective.

The primary property required of a plot is to serve as a unit for the application of treatments and it must be of a size suitable for that purpose. For example, if the treatments are different methods of land preparation, the plots must be large enough to allow the use of the customary farm implements. Hand digging is not the same as ploughing. Also, tractors have difficulty in turning, so plots may have to extend across the field. If not, special provision must be made for implements to turn somewhere in the middle. If that makes plots sound awkwardly large, it may be countered by the comment that they sometimes need to be very small. For example, the treatments may involve a lot of manual work like the emasculation of blossoms, which is time-consuming and a job only for those few who know what they are doing.

The next quality required in plots concerns the practicability of making the desired measurements. In an experiment on propagation by cuttings, for example, the percentage to survive can be found only from plots in which there are many plants, but if the enquiry requires intricate measurements on those that do survive, the experimenter might be well advised to make his plot a single cutting in a pot large enough for the resultant plant to achieve its full potential. There are, of course, a number of intermediate possibilities.

A third requirement is that the plots shall not interfere with one another. Sometimes that is nearly impossible to achieve as, for example, when treatments are intended to control the incidence of a very mobile insect. There can be two difficulties. One is that of confining the application of the treatments to the plot intended. Thus, in a spraying experiment wind can deposit a pesticide outside the area to which it is assigned, while fertilizer applied to one plot can be robbed by roots from a neighbouring plot that is being starved. The other is that of confining the effects of the treatments. Thus, a high concentration of insects on an unsprayed plot can lead to migration; also, well-developed plants can cast sun and wind shadows on others that have done less well. The usual remedy is to introduce discard areas between plots. (Strictly they are called 'guard areas' when the intention is to confine the treatments themselves and 'buffer areas' when it is the effects of treatments that are to be confined, but the distinction is not always easily made.) The solution is, however, only partial because something has to be done about the discard areas themselves. They cannot be left unsprayed because that would make them foci of infection or infestation. On the other hand, if they are sprayed to an unusually high degree to eliminate pests and diseases, problems can arise from drift. Deciding how to fertilize a discard area can be equally puzzling. The mechanical separation of plots to prevent the spread of fertilizer has been considered by Goode and Marchant (1962), and by Marchant and Boa (1962).

An interesting study is that of Lockwood and Martin (1976) on cocoa. They enquired whether the bias introduced into progeny trials by using all trees was compensated by a gain in precision as compared with using discard areas and a smaller number of trees in the middle of each plot. Later (Martin and Lockwood, 1979) they studied interactions.

Nevertheless, those are practical problems, not statistical ones. People sometimes ask about the best plot size for a certain species as if it were a constant of nature. Really they should be asking themselves questions that only they can answer. How am I to apply my treatments? What do I want to measure and how am I going to do it? Shall I need guards or buffers and, if I do, how shall I treat any resulting discard areas? If the answers give a clear indication about plot size and shape, the experimenter is lucky. He can then take into account the statistical considerations that result from the size chosen, e.g. the number of replicates required and the best means of controlling local variation.

There is another property required in plots, namely, that they shall give data that are independent. Since plots are disposed spatially, the answer is that their yields etc. will necessarily show a pattern of associations, some plots being close together and therefore probably similar, while others are separated and therefore possibly dissimilar. There is, in fact, no way of satisfying the statistical requirement of independence of data except by randomization procedures, which will be considered later as they arise.

There is one point at which practical and statistical considerations cannot be separated. If plots have to be very small, the coefficient of variation may be very large. If it exceeds, say, 30 per cent, the distribution of the data must be so far from normal that statistical difficulties must result. Perhaps a transformation of data will save the situation (see Section 9.12). If not, some compromise must be sought.

It should be remembered that the assessment of a treatment may depend upon several quantities. Although yield is usually of prime importance, it is not always wise to achieve bulk at the expense of quality. In other kinds of experiment, e.g. those to study the control of a disease, yield may not enter at all, but there will nonetheless usually be several quantities to be measured, e.g. the persistence of spray deposits as well as the efficacy of the spray. Difficulties sometimes arise if the various measurements ideally call for plots of widely different size. Thus spray deposit can hardly be measured over a wide area, but disease incidence cannot be assessed from a few leaves. Again, compromise may be the answer, or perhaps two experiments are needed on different scales. Sometimes it is possible to resolve the conflict by sampling one of the quantities in an experiment designed on a large scale to suit the other.

1.4 PATTERNS OF FIELD VARIABILITY

One of the chief features of an experiment is the field in which it takes place but its characteristics are often taken for granted. In fact surprisingly little is known about the spatial patterns with which investigators have to contend. An attempt will here be made to present some general considerations.

First of all, in any field there may well be features that are permanent or nearly so. Thus, soil may be deeper at one end than at the other; texture also may not be uniform. Then there could be streams and springs near enough to give damper patches, their effect depending upon the wetness of the season. Not least, the field may slope. In that case the top end will probably have shallower soil, less moisture and greater exposure to wind. Again, woods and areas of heathland can lead to damage from birds, small animals and insects, but only at certain times of the year and not necessarily then. Among the most troublesome features of a site can be obstructions like buildings that set up wind eddies, sometimes to a damaging extent. Some may even have been introduced deliberately, like rows of trees planted as windbreaks. Great discretion is needed. A partial obstruction that lets the wind through the gaps, e.g. between the trunks, slows it, but a solid one lifts it. It may then return to ground level with a damaging eddy at a distance that depends upon the height of the windbreak. That is the trouble with planted rows of trees. When young they slow the wind, but later they lift it, the eddies getting further away as the trees get taller. It is not easy to design an experiment in such conditions.

However, such features are at least capable of identification and precise mapping. The difficulty is rather that their relative importance varies from one season to another. The fertility pattern of a field can be quite different in a dry season from a wet one or in a windy season from a still one. Also, on account of the incidence of pests and the existence of alternative food supplies for them, it can be different when the season is early and when it is late. In short, permanent though such site features are, their effects can be transient and unpredictable.

There are other patterns that the experimenter or his predecessors may have introduced into the land. For example, there may in the past have been differential fertilizer applications or depths of ploughing or herbicide use. It is true that residual effects can sometimes be reduced and even eliminated by blanket treatments like an overall high fertilizer application or general deep ploughing, possibly preceded by a reversal of the differential treatments. At least, such effects occur on determinate areas with sharp boundaries, so the design of a future experiment can allow for them, if need arises, provided that the precise position of the former experiment has been surveyed and related to fixed points. Compaction patterns are somewhat similar. It has already been suggested that a good experimental plan provides sensible paths for tractors and their implements. It should also facilitate the collection of the crop by trailers and trucks, but vehicles compact the soil and their effect is not easy to remove. (When it is recalled that aerial photographs sometimes show prehistoric roads that have been disused for a thousand years or more, it is apparent that compaction patterns can last a long time despite continual cultivation. Further, such roads show up most clearly when crops are growing over them.)

In addition, any experiment may exhibit fertility patterns for reasons that reside in itself, even if only because it has an inside and an outside, not to mention corners. It is a wise precaution to surround each experiment with a discard area of its own species. If that is not done the outside plots will be more exposed to wind while their roots will be able to forage further for nutrients. If no such external guard areas are provided, not only will those plots be different but the effect of being outside may well interact with the effects of treatments and lead to misleading results. If no external guards are feasible, the blocking system, which will be discussed in a later chapter, should at least keep inside and outside plots in different blocks. Corner plots raise even more problems. The effects of microclimate and greater root run are more serious, but even if external guards are provided, those are the plots to which unthinking experimenters take visitors, who compact the soil underfoot and brush against the plants. Also, a corner plot is usually the last to be fertilized or sprayed. Sometimes it will go short if the operator does not go back for the small amount needed to finish the job or perhaps it will get too much because he thinks it would be a pity to waste what remains. (It is true that good experimenters avoid such errors but some never think about them.) Where feasible it can be wise to keep the corner plots to demonstrate selected treatments but to disregard them for purposes of comparison.

It is not suggested that the designer of an experiment can take all these effects into account. In general he can give them emphasis in the reverse order of their presentation here. Differences between inside and outside plots, if they are not eliminated by external guards, are for him to deal with. Given adequate surveying of previous experiments he ought to locate the new one so that residual effects are either avoided or can be taken into account. As to permanent features he must do his best, but he does not know what weather the future will bring and he cannot judge therefore which feature will be of greatest importance. Also their effects are not delimited by definite boundaries; they taper off and often they overlap with the effects of other features. Nevertheless, they can be important and he must judge from experience how to act. He will be greatly helped if he makes a point of looking back on each experiment as it comes to an end to see how far there were differences he should have done something about and how far he wasted effort controlling factors that mattered very little. Good experimenters pick up a lot in that way. That point will come up again in Section 10.2.

1.5 NON-PATTERNED FIELD VARIABILITY

The sources of variation considered in the last section may be considered as imposing patterns of fertility. Other sources may give variation in which there is a large random element.

Indeed, some variation is effectively completely random. Errors of measurement are of that sort and for some variates they dominate. Nematode counts provide a good example. Again, when plots contain only a few plants the variability arising from the experimental material can be so high that other sources are relatively unimportant. For example, the genetic variation of coconut palms is such that environmental differences are mostly negligible in comparison. In such cases, where a source of random variation dominates all else, the error variance may be awkwardly large but at least the biometrician is relieved of the obligation of trying to control the variation due to the environment.

In general, when plots consist of only a few plants, it is sound practice to ensure that each receives a comparable set, e.g. one large, two medium and one small. If that is not feasible, a random allocation of plants to plots is desirable. Whether the resulting random variation dominates or not it serves to dilute the effects of patterned variation and to make measures to control them less necessary.

Non-patterned sources of variation are usually approached by way of the Law of Fairfield Smith (Smith, 1938). It states that if yields from plots of unit size give a variance of V_1, then yields per unit area derived from plots of area, x, will have a variance, V_x, such that

$$V_x = V_1/x^b, \qquad (1.5.1)$$

where b is some constant between 0 and 1. Consequently if $\log V_x$ is plotted against $\log x$ the result should be a straight line of slope $-b$. Smith showed also that the consequence of taking plots from within a finite area is to give the line slight downward concavity. The derivation of the law was entirely empirical, being based on 44 sets of data from various species, and it is indeed true that most curves did exhibit the forecast downward concavity, though some did not and one or two were distinctly sigmoidal. The law has been applied by many subsequent investigators and has not been found wanting.

Later, Freeman (1963) pointed out that the effect of adding a term for random variation to the expression in (1.5.1) leads to the relationship

$$V_x = \frac{V'_1}{x^b} + \frac{V''_1}{x}, \tag{1.5.2}$$

and that implies upward concavity. His approach finds confirmation in Smith's own curves. Eight of them refer to experiments on trees, i.e. cases where plots would contain only a few plants, and with six of them there is the upward concavity expected. Rives (1969) took Freeman's work further. Working with the prunings of grape vines he found the adjacent plants in a row exhibited a negative correlation, the result no doubt of competition, and so b was negative.

Meanwhile others had been thinking in terms of the correlations between crops from adjacent units, e.g. Li and Keller (1951). The approach was systematized by Matérn (1972). Let ρ_d be the correlation coefficient between the performance of two plants a distance, d, apart, then it is to be expected that ρ_d will equal some monotonic decreasing function, $f(d)$, such that $f(0) = 1$ and $f(\infty) = 0$. Further, if two unit plots are amalgamated to give a plot double the size ($x = 2$), then it should be possible from a knowledge of $f(d)$ to determine b. Matérn noted further that a function that gave the Law of Fairfield Smith would be so complicated and arbitrary as to defy belief. The matter was taken further by computer simulation (Pearce, 1976b) with conclusions so obvious that it is a wonder that no one foresaw them. When plots are miniscule the correlation coefficient between performances at any two points will approximate to 1. Hence one plot will perform almost exactly like its neighbour and amalgamating them will have little effect. In such circumstances b tends to zero. On the other hand, when plots are enormous the only effective correlation is between plants near the common boundary. Most plants in one plot will be cropping independently of most plants in the other, so, as the plot size becomes greater, b tends to 1. Consequently, working over this very wide range of plot sizes, the graph of log (V_x) against log (x) must start horizontal with $b = 0$ and, as x becomes larger, the slope must change steadily towards an asymptotic value of -1, i.e. $b = 1$. An empirical investigation can study only a small range over which b will be nearly constant and between 0 and 1, as the law states. Also, there will in general be downward concavity quite apart from considerations of finiteness of total area. The shape of the curve was found to depend upon the particular $f(d)$ used. Thus, for $f(d) = 1/(1+d)$ the change in slope was fairly steady, but for $f(d) = 1 - 2 \arctan (d^2)/\pi$ there was a rather abrupt change. It comes approximately when the side, D, of the square unit plot is such that $f(D) = \rho_D = 0.5$.

It thus appears that Fairfield Smith's Law is not exact, but neither is it arbitrary.

Figure 1. Much local variation derives from unnoticed irregularities in the land itself. This picture shows the large differences in depth of top soil that result from the layer of chalk underneath. If an experiment were planted on the site it could well be unsatisfactory on account of a large error variance, but to casual examination there would be nothing wrong. (Copyright: G. V. Dyke)

Figure 2. Even where precautions have been taken to achieve uniformity plants can be very variable. Not only can there be disguised differences in the soil but plants themselves can be variable and the two sources of error can interact on one another and make things worse. The picture shows some cassava in Nigeria. They should be uniform but clearly something has gone wrong. (Copyright: S. C. Pearce)

Further, the reason why it holds with crop yields would apply in other contexts, so it could well be of wider usefulness.

The above argument does, however, start a line of thought, which for the moment can be only speculative. The error of block experiments arises not only from the variance of the plot yields but also from that of the blocks in which they find themselves. Smith studied what happens when b is constant. What happens if it changes fairly abruptly and the point for the plots lies above the break and that for the blocks lies below it? Would there not be an increase in error variance? If so, the startling conclusion is reached that an experiment can be of wrong size; it would be better if it were scaled differently, whether upwards or downwards. Reflection will show that the idea is not absurd. If land is patchy, in an experiment on a large scale a plot could be so large that it would make little difference if it did contain a patch or two. On the other hand, if the scale were much smaller, the whole experiment might be on a patch or off it, but the worst scale would be one in which a plot was of about the same size as one of the patches. That argument is vague and non-mathematical, but the idea finds confirmation in some other work of Matérn's (1972) in which he made computer simulations for various designs using $\rho_d = e^{-hd}$ where h is a scaling constant. He also used an expression based on Bessel functions. Interestingly he did find an increase in error about the middle of the range. Using $\rho_d = e^{-hd}$, Duby *et al.* (1977) reached a similar conclusion. Of course, simulated data prove nothing, especially when no account has been taken of the patterned variation discussed in the last section, but there is clearly something here that deserves exploration.

Attempts to express the fertility pattern of a field in mathematical terms are welcome, but so far they have not progressed to the point where they can be accepted as empirically justified. One obvious approach is to think in one-dimensional terms (R. M. Williams, 1952; Atkinson, 1969) and to write

$$x_i = \rho x_{i-1} + \eta_i, \qquad (1.5.3)$$

where x_i is the datum from the ith plot along a row and η_i is a random residual, uncorrelated with x_{i-1}, η_{i-1} or η_{i+1}. The relationship may be made symmetric by writing

$$x_i = \beta(x_{i-1} + x_{i+1}) + \varepsilon_i, \qquad (1.5.4)$$

where

$$\beta = \frac{\rho}{1+\rho^2} \quad \text{and} \quad \varepsilon_i = \frac{\eta_i - \rho\eta_{i+1}}{1+\rho^2}. \qquad (1.5.5)$$

The difficulty with this relationship is that the ε_i are no longer independent of their neighbours, as the η_i were, the correlation coefficient between ε_{i-1} and ε_i being $-\rho\sigma_\eta^2/(1+\rho^2) = -\beta\sigma_\varepsilon^2$. This negative correlation has little intuitive appeal, while the difficulties increase in two dimensions. In general it is probably better to use the 'auto-logistic' model proposed by Besag (1972), namely,

$$x_{ij} = \xi + \beta_1(x_{i-1,j} + x_{i+1,j}) + \beta_2(x_{i,j-1} + x_{i,j+1}) + \eta_{ij}, \qquad (1.5.6)$$

though that too raises problems.

1.6 CONTRASTS

An experiment should be seen not only in its physical context, i.e. the field in which it takes place and the plant material available, but also in the context of what its initiator is hoping to achieve. It exists because someone has asked a question that no one could answer and the matter was thought to be important enough to justify the labour and expense of carrying out a field trial. In such circumstances it is necessary at every stage to keep in mind the originating question to ensure that it receives a definite answer, yet somehow the whole subject of how to adapt the design of an experiment to purpose is sadly neglected.

The first need is to select the right treatments. There can be pressure to adopt some that are unrealistic because that will ease practical problems, but it should be resisted at all costs. Similar pressures arise, almost unregarded, when a required treatment is objectionable to scientific thinking. For example, an experiment to convince local farmers that there are better ways of deciding sowing date than reliance upon superstition or astrology must necessarily include at least one treatment that the investigator himself regards as absurd. It must nonetheless be included. No less important is the specification of the differences between treatments that need to be studied. There seems to be an unstated assumption that all treatments are of equal status and that all differences between them need to be estimated with equal precision. It is not clear how this extraordinary misconception arose or how it came to be so firmly grafted into statistical theory, but it is there. Nevertheless, it is quite mistaken.

To take an example, it may have been decided that an experiment is needed for four treatments. That of itself gives little indication of the design that should be adopted. Perhaps one of the four is a standard and the other three are to be compared with it, or perhaps one of them is an untreated control introduced only in order to establish a base-line for the others, or perhaps the four form a factorial set with two factors, each at two levels. In the last event there are still several possibilities, e.g. the intention may be to take two factors of known importance to see if they interact or it may be that an interaction is considered to be very unlikely and attention is to be concentrated on the two main effects. The various approaches could all lead to different designs.

The desiderata of an experiment are best summed up by setting out the 'contrasts' of chief interest. A contrast is expressed by a vector **c**, say, such that its elements, of which there is one for each treatment, sum to zero. To take an example, let an asterisk before the designation of a treatment represent the mean value assigned to it after the experiment has taken place, allowing for any adjustments required. Then in the first possibility suggested above, i.e. the case where one treatment, S, is an established standard and the others, i.e. A, B and C, are new treatments to be compared with S, the contrasts under study are (*S – *A), (*S – *B) and (*S – *C), which can be represented by the three vectors

$$\begin{array}{l} (1, \; -1, \; 0, \; 0)'/\sqrt{2} \\ (1, \; 0, \; -1, \; 0)'/\sqrt{2} \\ (1, \; 0, \; 0, \; -1)'/\sqrt{2} \end{array} \qquad (1.6.1)$$

For many purposes, though not for all, it is convenient to scale them so that $\mathbf{c}'\mathbf{c}$ equals 1, which explains the divisor, $\sqrt{2}$, in (1.6.1). It is usual to write contrast vectors in columns. Here, where it is more convenient to write them in rows, they are transposed.

If, however, the treatments had been an untreated control, O, with three others, A, B and C, and if there had been no reason to distinguish those others, the contrasts primarily in need of study are

$$\begin{array}{l} (0,\ 1,\ -1,\ \ \ 0)'/\sqrt{2} \\ (0,\ 1,\ \ \ 1,\ -2)'/\sqrt{6} \end{array} \qquad (1.6.2)$$

Here the first studies (*A − *B) and the second studies the mean of *A and *B with *C. (The order of the treatments is immaterial.) Of course, if the contrasts (1.6.2) were really the only ones of interest, there would be no need of Treatment O, so there must be others to be considered, perhaps those in (1.6.1).

In the factorial cases the treatments may be termed 1, A, B and AB. The main effect of A is found from

$$-{}^*1 + {}^*A - {}^*B + {}^*AB$$

and that of B from

$$-{}^*1 - {}^*A + {}^*B + {}^*AB$$

The interaction concerns the extent to which (*AB − *B) differs from (*A − *1), i.e. the extent to which

$${}^*1 - {}^*A - {}^*B + {}^*AB$$

differs from zero. Hence the contrasts involved are

$$\begin{array}{l} (-1,\ \ \ 1,\ -1,\ \ 1)'/2 \\ (-1,\ -1,\ \ \ 1,\ \ 1)'/2 \\ (\ \ 1,\ -1,\ -1,\ \ 1)'/2 \end{array} \qquad (1.6.3)$$

though, according to the purposes of the experiment, there could be different emphases between them.

If a design is proposed for the experiment, it can be assessed by first inquiring what the covariance matrix of the adjusted treatment means would be. (The methods of deriving such a matrix will be considered later.) If it is going to be $\mathbf{V}\sigma^2$, where σ^2 is the error variance, then the variance of estimation of a contrast $\mathbf{c}$ will be $\mathbf{c}'\mathbf{V}\mathbf{c}\sigma^2$. If $\mathbf{c}$ represents an important contrast, it is clearly desirable that $\mathbf{c}'\mathbf{V}\mathbf{c}$ should be as small as possible and a design is needed that makes it so. If, on the other hand, no one is much interested in the contrast, it matters little how large $\mathbf{c}'\mathbf{V}\mathbf{c}$ turns out to be. The skill lies in devising an experiment such that its $\mathbf{V}$ fits the desired contrasts as well as possible. At the practical level, that is what matters.

A further set of contrasts can be suggested. If the four treatments represented equally spaced levels of some quantity, e.g. 200, 400, 600 and 800 kilograms of fertilizer per hectare, it would be quite usual to consider the following:

$$\begin{array}{l} (-3\ \ -1\ \ \ \ 1\ \ \ 3)' \\ (\ \ 1\ \ -1\ \ -1\ \ \ 1)' \\ (-1\ \ \ \ \ 3\ \ -3\ \ \ 1)' \end{array} \qquad (1.6.4)$$

They are known respectively as the 'linear', 'quadratic' and 'cubic' effects. The reason is this: If the response to fertilizer were linear, i.e. represented by a straight line, all treatment differences would be comprised in the first contrast, the other two being equal to zero. If, however, there were also a quadratic component, the response curve being then a parabola, the first two contrasts would show the fact, the third still being equal to zero. Only if there were a cubic component also would the last show anything. Similar sets of contrasts can be found for other numbers of treatments and the table of Fisher and Yates (1938) provides a convenient way of writing them down. Thus, in Examples 2D and 2E it is supposed that there are three levels, so there are only two degrees of freedom and only two contrasts, namely $(-1 \quad 0 \quad 1)'$ and $(1 \quad -2 \quad 1)'$, the linear and quadratic effects respectively.

There is, however, one thing more; it is often desirable that the contrasts should be estimated independently. Thus, the existence of an interaction should not imply that there must be a main effect also. If there are two contrasts with vectors, $\mathbf{c}_1$ and $\mathbf{c}_2$, independence is shown if $\mathbf{c}_1' \mathbf{V} \mathbf{c}_2 = 0$. Nevertheless, although this property is desirable, it cannot always be achieved. For example, it is not possible in the case of the standard, S, and the three new treatments A, B and C, though that is quite a realistic example, because the evaluation of each contrast involves *S. Hence, they are necessarily correlated unless *S can be determined without error, which of course it cannot.

The standardization of contrasts should not become a fetish. An experimenter may think of the contrast, $(1 \quad -2 \quad 1)'$, to show how the effect of a median dose differs from what would be expected from a straight-line relationship. Then he thinks it would be better as $(\frac{1}{2} \quad -1 \quad \frac{1}{2})'$. Then, he transforms kilograms per plot to tonnes per hectare and gets $\alpha(1 \quad -2 \quad 1)'$, where α looks very arbitrary. For mathematical purposes he becomes very correct and writes $(1 \quad -2 \quad 1)'/\sqrt{6}$. So long as he acts reasonably at each point, he need not strive too hard for consistency.

From what has been said it will be seen that listing the contrasts of interest is the statistician's way of saying what he is interested in. In what follows contrasts of interest will play a large part since it is their effective estimation that makes an experimental design good. In fact the whole scheme of this book, which is set out in the next section, depends upon that approach.

Equations (1.6.3) exhibit a feature that will be noted on a number of future occasions, namely, the structure of interactions. If there are two contrasts,

$$(a_1, a_2, \ldots, a_v)'$$
$$(b_1, b_2, \ldots, b_v)'$$

such that $\sum_i a_i b_i = 0$, then

$$(a_1 b_1, a_2 b_2, \ldots, a_v b_v)'$$

is also a contrast and is said to be the interaction of the first two. In (1.6.3) the third contrast was obtained in that way, but formally any of the three contrasts is the interaction of the others, a property that plays a large part in the theory of factorial designs. The topic will be examined in some detail in Section 8.1.

Also, in (1.6.3) there would be little point in studying the first two contrasts if the effect of the third were appreciable. For example, the first measures the effect of A averaged over both levels of B. If A and B do not interact, that is correct, but if the effect of A does depend upon the level of B, it is necessary to look at the particular effects, i.e.

$$(-1,\ 1,\ \ 0,\ \ 0)'/\sqrt{2}$$
$$(\ \ 0,\ 0,\ -1,\ \ 1)'/\sqrt{2}$$

Clearly, since they are different an average would be misleading.

It follows that there should be a defined order of testing—first the interaction, then if that is not significant, the two main effects. Such a situation arises fairly often. Thus, given three evenly spaced doses, a natural pair of contrasts is given by

$$(-1,\ \ 0,\ \ 1)'/\sqrt{2}$$
$$(\ \ 1,\ -2,\ \ 1)'/\sqrt{6}$$

The second contrast measures the extent to which the response to the middle dose differs from the mean of the responses to the outer doses, i.e., it measures the extent to which the dosage-response curve departs from a straight line. If the line can be regarded as straight, the first contrast measures its constant slope; if it is not straight, there is no constant slope to measure. Where the treatment effects can be broken down into isolated components, each with its own interpretation, care is often needed as to the order in which they should be studied.

To sum up, the designer of an experiment should not think of all possible contrasts as being equally important or equally valid. He should know which ones are of interest and he should concentrate on them. Perhaps he wants to know whether a contrast is significant or not or perhaps he wants to estimate its magnitude; either way he needs to know its variance. Perhaps the contrast is 'elementary', i.e. it concerns only the difference between two treatments, or perhaps it involves several treatments, as in the case of an interaction; so long as it corresponds to some question in the mind of the experimenter it can be valuable.

The worst way to conduct an experiment is to insist on giving all treatments equal status, so far as that is possible, and then to apply some technique like multiple comparisons to sort them out. It is wrong for several reasons. One is the concentration upon elementary contrasts, which are not necessarily of chief importance. Another is the emphasis on significance, which is not necessarily relevant. The main objection, however, is that such an approach corresponds to the tactics of the bad scientist, who makes an experiment by putting together a lot of treatments, more or less related, which look as if they might have something to do with his problem, in the hope that the data, when obtained, will suggest something. By contrast, the good scientist has thought about what he is doing and builds a set of treatments to contain the contrasts about which he needs information. It is the latter approach that will be considered in this monograph.

1.7 THE CULTURAL CONTEXT

It should not be supposed from what has been said that the choice of contrasts for study is an easy exercise. Sometimes it can be difficult on account of genuine uncertainty, the way forward being far from obvious. Sometimes too there can be a deep division of opinion. If so, a biometrician involved must insist that the experts settle their differences first, because an experiment designed with a dual purpose will be a bad one. Anyhow, someone will have to decide eventually. Occasionally some powerful person will insist upon the inclusion of a treatment that no one else wants; nor are powerful persons always wrong. So long as his reasons are clear, it can probably be accommodated without tension. For example, the experts may want a factorial design but someone may insist upon the inclusion of a supernumerary treatment to represent current commercial practice. There is no need to resist such a proposal fiercely on scientific grounds. Statistically, for purposes of the analysis of variance, it is enough to introduce an additional contrast to pick up the difference between the supernumerary treatment and those in the factorial set taken as a group, leaving other contrasts as the scientists want them. It is true that the additional treatment may lead to reduced replication or larger blocks or some other undesirable feature, but the tactful will seize the opportunity to convince the powerful person that more resources are needed and they may well succeed and even end with a better experiment for their own purposes than they had expected.

The example, imaginary though it is, will serve to remind the reader that a group of scientists, highly trained in their own ways and accustomed to discuss matters in isolation, form a sub-culture in the society that supports them. Administrators in capital cities and the farmers round about may have different ideas, which are not necessarily wrong for being 'unscientific'.

When designing experiments on farmers' fields there is in fact a lot to be said for including the supernumerary treatment of doing what the farmer says that he would do if left to himself. For one thing farmers do know their own business; for another, if it is alleged that the experiment has done harm, the farmer's plots will show whether or not it is the fault of the treatments that the scientists have applied. The author recalls a spraying experiment on apple trees in England, where the farmer insisted that sprays applied at that time would do damage, so a 'grower's control' was introduced in which everything was done as he said and a clause was put in the contract under which he would be compensated if yields fell below what it gave. In the event some treatments did cause damage though others proved decidedly advantageous. On balance the grower did slightly better and received a small amount of money. There are two interesting points here. First both the grower and the scientists learned something. Secondly, everything went off harmoniously, the incident leaving no hard feelings behind it, which cannot be said for all experiments conducted on growers' fields.

In fact, some treatments proposed by scientists can be quite effective but unpractical because no farmer would have labour available at that season of the year or be willing to spend money on equipment of limited usefulness. Again, ministries may be unwilling to permit imports of chemicals or equipment when something nearly as good can be made at home. All that points to the need to distinguish between agronomic experiments and botanical ones. If the conclusions are meant to apply to other people's way of life, the scientist must consider their wishes. If, on the other hand, the need is to understand how a plant reacts to some kind of stress or what makes it blossom when it does, it is all right then for the scientist to retreat to the laboratory and do things in his own way.

If scientists can form an uncomprehending sub-culture in their own society, matters are much worse when they go to a poorer country with different traditions of social, educational and religious values. A visiting expert may ask to be taken to the workshop of some craftsman to buy presents to take home. Only an idiot would complain of the lack of a sprinkler system to cope with fires or remark that in his country in such an industry everyone would be wearing hard hats. The reaction would rather be one of delight at superb objects made by a skilled craftsman despite little equipment. Nevertheless, in the field the same day the same expert may act the idiot in just that way. There is a story, probably apocryphal, of an expert who went to a country where wheat was harvested by cutting each stalk separately and carefully placing each head in a basket. Full of self-importance he introduced the sickle, which had two effects. First, he had caused an unemployment problem. Secondly, he had caused a famine, because some heads shattered and were lost that would have been preserved under the old system. The story is, however, scarcely credible, because small farmers have to be sensible to survive and they do not do silly things at the behest of silly foreigners.

In fact, they can be very resistant to suggestion. The author was once biometrical consultant to a team that was trying to improve banana production on a West Indian island, which at that time was a British colony and its scientific services staffed by Europeans. The team pointed out again and again the folly of letting the weeds grow under newly planted banana mats, but no one would take any notice. Recently he revisited the island and met the new director of research, who said with a twinkle, 'You remember all the trouble you people took to persuade smallholders to keep down weeds? You did not realize that their chief source of meat comes from rabbits, which have to be fed. Instead we have a research programme to find out what can safely be grown under new mats and we are getting an interested response from those same smallholders.' To take another example, there is little point in devising a better plough that will need only one ox to draw it if a farmer has to have two oxen anyway to take his cart to market. Further, what advantage would there be if local workmen cannot repair the new plough, whereas they know how to deal with the old sort? Such points are too easily overlooked.

It is not just that visiting experts do not know how the local people work. They do not know either what they want. For that matter, the local educated classes may be equally bad. There is a certain territory with a sound farming tradition accustomed to export food to neighbouring countries. Its farmers have a reputation for being hard working and receptive to new ideas. The government went to some pains to introduce better seed, etc. to increase yields and thus foreign earnings. Their efforts were greeted with great satisfaction by the smallholding community. As they pointed out, it became possible for a man to take a siesta during the heat of the day and still be able to support his family. Economists trained in developed countries assume too readily that small farmers think as they do and are trying to maximize income. Perhaps they are, but no one should assume as much. On the other side, it may be urged that some practices are so repugnant to local custom that there is no point in suggesting them, but there may be. There is the story of the tribe that indignantly rejected the suggestion that they should sow their crops earlier because that would mean working over the season that the gods had ordained for marriages. Without being cynical, it may be remarked that 'money talks' and they could perhaps be persuaded—or perhaps not. Foreigners do not understand such matters.

That is especially so when considering practices with a religious basis. Mention has already been made of astrology. The writer encountered the problem when local officials were lamenting the refusal of farmers to sow at the times advised, preferring to listen to others who claimed to be guided by the stars. His suggestion that it might be a good idea to try sowing when the local soothsayers advised, even if only to see what happened, caused great offence. The idea that gentlemen of education should take notice of ignorant charlatans was positively insulting. However, the suggestion was not meant in that way. Traditional wisdom is often sound. At least, it should be tried before it is rejected. If it is wrong, an experiment with a 'soothsayer's control' will establish the point. (That does sound rather like Elijah and the priests of Baal, but, after all, that confrontation did establish something.)

There is one point that must always be borne in mind when dealing with the agriculture of a poor country. The small farmer, untutored but resourceful within his limits, is just like his neighbour who makes beautiful objects for visitors to buy. His very lack of dependence upon mechanization gives him a flexibility of approach that needs to be valued. Similarly, where everything is done by hand crops can grow in close association to exploit limited rainfall or other scarce resources. The whole approach by intercropping (see Willey, 1979ab), for example, calls for an immense skill that outsiders do not understand very well, though after a time it commands their respect, so successful can it be. Above all, however, the need is for reliability. If a small farmer does have a large crop, so does everyone else and he can sell the excess, if at all, only for a poor price. Equally, if his crops fail, it is unlikely that his neighbours will have much to sell and, if they have, it will not be cheap. Of course, government buying schemes and good transport help the situation, especially if the crop is not perishable, but the problem remains. Too much thinking about agricultural improvement is directed to increasing mean crops over time, but to a poor man without financial resources it is the minimum that matters, not the mean. A new variety that fails in a dry season could lead to his death by starvation. That is why he diversifies. It also explains his conservatism. In conditions where those who practise poor methods die, those who survive have confidence in their practices and are not easily moved from them, nor should they be. It is a criticism of statistical methods that they pay too little attention to fluctuations in cropping, a result of their having been devised in other conditions. People trained in them should be aware of the deficiency.

This section may appear to deal with topics remote from the choice of treatments and the identification of contrasts of interest between them, but that is not so. Treatments should be realistic and relevant and the interest of a contrast derives likewise from the conditions that are under study.

1.8 THE SCHEME OF THIS BOOK

So far in this opening chapter a field experiment has been considered in various contexts. First there has been a historical survey to show how present practices have evolved—whether for good or bad is not the point. The experiment has then been considered in its physical context, defined plots in an actual field with its perplexing and often unknown pattern of fertilities. Also in Section 1.6 it has been shown how it is possible to specify what needs to be known, but before then in Section 1.2 a warning was given that information is needed not just for one field but for an area, a point that is too often neglected. How are all these threads to be woven together?

Everyone could have a different answer and the author can only describe the path he has himself chosen, while recognizing that others might have gone some other way. In the next chapter the simplest of all cases will be considered. The intention is to sample how a field would perform under a range of treatments. Clearly if the whole area is given over to one of them, it will not be possible to test the others, so the obvious solution is to divide it into plots and to allocate this treatment to a random selection of plots, that treatment to another random selection and so on. The resulting design is considered in some detail, partly because it is desired to exhibit the mathematical approach with a simple case and partly because the conclusions, though a little obvious, do show how contrasts are estimated and how it is possible to gauge the precision with which that has been done. (The reader is here reminded that the effective estimation of contrasts of interest is the keystone of the whole edifice here to be described.) It will also be shown that a simple randomization procedure validates the estimates. However, it may well be objected that such a design does nothing to allow for soil fertility, so various ways are presented of doing so. To the author it is a survey of lost opportunities. For forty years, largely on account of historical accident, all the emphasis has been on block experiments. Have they been so successful as to justify the continual neglect of all alternatives? The answer must surely be ‘No’.

However, it is block experiments that have been developed and there is a large body of knowledge about them. The general theory is set out in Chapter 3. Mathematically that is the crux of the whole argument. The reader who understands Sections 3.1, 3.2 and 3.3 will be able to understand the rest. One conclusion of great importance to what follows is the recognition that there can be contrasts that defy estimation. Also, some may be estimated more precisely than others. In a text that bases itself upon the estimation of contrasts of interest, that can be important also.

It also leads to Chapter 4, in which various results are presented that relate to precision of estimation. They are of various kinds and collectively have the appearance of suggestions for a research student in search of a project because the field is one that has been too little explored. There is one concept, however, that of general balance, presented in Section 4.8, that must be taken very seriously. Examination of design practice at agricultural research institutes shows that most trials do in fact exhibit the property, which shows how valuable the idea is. It would be a mistake to argue that only such designs should be used because some sets of contrasts of interest do not lend themselves to such an approach, though most do. It is not possible to make full use of the concept until something has been said about designs with more complicated blocking systems and that will be done in Chapter 6. However, the idea is introduced in Chapter 4 because that is where it belongs.

Chapter 5 sets out some suggestions for designs that could be used in practical situations. To be frank, it was a little tedious to write and the reader may find it so to read. If he does, nothing later depends very much upon it. This is the classical field of statistical endeavour in the subject of experimental design and the literature bristles with special cases studied in immense detail and presented without much reference to practical usefulness. Mostly they have been ignored, the purpose of this text being directed all the time to applications, even if some passages do look very theoretical. The broad classes presented are those that the author has found useful. Others do not have to agree.

In Chapter 6 the text breaks away from the simple block design. For example, there are designs in rows and columns, like Latin squares. Again, it is possible to use the plots of one experiment as the blocks of another, the result being said to be in 'split plots'. The possibilities are nearly endless and quite fascinating. It is here that general balance becomes prominent. There is no rule that all experimental designs should have the property and with simple block designs there are practical alternatives, but they are unusual in more complicated instances. Given general balance an alternative way exists for computing the analysis of variance, i.e. that of sweeping, which is the basis of GENSTAT, and that is explained in some detail.

However, the best plans go wrong from time to time and no practical text can ignore the spoiled experiment. One difficulty here is the large amount of work that has been done on missing data compared with other mishaps. Another is the general disregard of the reasons for the accident, though they must affect the assessment of data that can be salvaged. The approach in Chapter 7 is to relate the spoiled experiment as far as possible to the general case considered in Chapter 3. After all, a spoiled experiment is still an experiment and should be covered by a general method. (It can still exhibit some perverse features.) The chapter has been written to show the essential unity of the various approaches that have been suggested.

Chapter 8 very nearly did not get written. It concerns interactions, a topic about which so much has been said that there is little a modern author can do except give references. There is, however, an underlying unity in the literature that can usefully be emphasized.

Any approach leaves out something and Chapter 9 is really a collection of essays on topics that have been missed out in the present text so far. That does not imply that the subjects dealt with lack importance. With a different line of presentation they might have come first.

Chapter 10 is different because it deals with people—what they should do and what they can be expected to know. The subject is important because so many skills are required and no one person possesses them all. A book on theoretical statistics can confine itself to standard errors, least squares solutions and the rest, but the practical biometrician belongs to a larger community which comprises colleagues and friends who have to deal with the officials and committees that allocate funds, who have studied agronomy, botany, chemistry and other branches of science, who know how to apply differential treatments in the field and maintain good standards of husbandry in the difficult conditions of a field experiment, who understand surveying so that every plot goes where intended, who can record an experiment plot by plot accurately and with minimum disturbance both to the plants and to the other operations that have to go on and with others who understand computers and the ancillary equipment that goes with them. Unless all work together harmoniously and with agreed objectives the result will be pretentious chaos. Everyone will use apparatus and methods that may impress the visitor but the scientific results will impress no one, because each operation has been developed in isolation and not as part of the whole. There is another aspect of experimentation touched on in the chapter. Where so many people are involved it is easy to get into a rut. If a change in technique is required there are too many people who have to be consulted before anything is done so that everything goes on as it was. That is not good enough. Problems change and new forms of equipment become available. Not least, people have new ideas. Consequently experimental techniques at all stages have to be scrutinized continually in a spirit of constructive criticism. However, changes will not come about until everyone sees the point. That calls for open communications between all the various experts and a mutual respect between them. A serious and regular consideration of concluded experiments can be most salutary to prevent the same mistake occurring twice and to exploit new ideas to the full.

Finally, a word about mathematics. Anyone who wishes to enter into the argument of this book will have to cope with matrix algebra. Others may be content to accept the conclusions provided they can understand what they are. For them each chapter after the first concludes with worked examples, which show how the methods are used without showing why. Even here matrix algebra intrudes but only as a neat way of expressing calculations. For a reader who is content to miss the mathematical argument so long as he gets the practical point the examples should take him a long way, though he may find it useful to read the appendix first.

Chapter 2

Simple experiments and how they can be improved

2.1 COMPLETELY RANDOMIZED DESIGNS

The simplest way of designing an experiment is to decide the number of plots available, to apportion them among the treatments, either equally or in relation to some scheme of relative importance, and then to assign each treatment its number of plots at random. Indeed, there are many reasons for adopting the approach. It is simple both in concept and in the analysis of data. It may be objected that the design makes no attempt to control the environmental variation, but that is not necessarily a defect. For example, the experimenter may have reason to regard the area as virtually uniform. Even if he does suspect variation he may have no idea what form it will take so efforts to reduce its effect could be counter-productive if he were to make a bad guess. Also, he may not want to lose any of his error degrees of freedom. There are many instances, especially when the experiment is small, i.e., when environmental variation and error degrees of freedom are both minimal, in which he might well accept this simple form of trial.

Certainly there is no need to try to control environmental variation whatever the circumstances and there are risks in trying to do so. If there are e degrees of freedom for error, supposing that complete randomization is used, and if a scheme is proposed that will absorb c of them in efforts to isolate sources of variation, it might appear that no harm could result beyond having $(e-c)$ degrees of freedom for error instead of e. Even that could be serious, but a danger exists that the scheme will not only be unavailing but positively inappropriate. For example, the experimenter may assume that the field, after allowing for the effects of treatments, will show a certain fertility pattern, so he takes it into account. If the real pattern is quite different, his sum of squared deviations may be below expectation. Indeed, his control measures may remove scarcely any sum of squared deviations. At the worst his error variance will be increased by a multiplier of $e/(e-c)$, i.e. the more elaborate his scheme, the greater the harm it can do by failing. Further, by using parameters to control local variation he may be introducing instability. If his assumptions about fertility patterns are justified, he will reduce his error variance, perhaps considerably; if they are mistaken, he may increase it. Since he is probably going to determine the size of his experiment in the light of his worst experiences in the past, the habitual introduction of blocks, etc., only because 'good experiments always have them', can lead to an unhopeful view of possible errors and so prove wasteful. If there are features like slope, dampness, soil texture or soil depth, which must have an effect, it is good practice to allow for them, but the experimenter should avoid unbased assumptions. For that reason the completely randomized design has its advantages.

Mathematically such a design is very simple, but its properties will be shown with some rigour. If the full demonstration is a little tedious it is nonetheless important, because it will serve as a paradigm for some of the more complicated designs that follow. It will also introduce the notation, which will be followed consistently hereafter. Let $\mathbf{y}$ be a column vector of n data, the order of which is immaterial but, once chosen, may not be altered. Then $\mathbf{y}$ may be written

$$\mathbf{y} = \mathbf{\Delta}'\boldsymbol{\tau} + \boldsymbol{\eta}, \tag{2.1.1}$$

Here the prime after the $\mathbf{\Delta}$ indicates that the matrix has been transposed as described in Section 3 of the Appendix.

This expression is the first of many similar ones, so it will repay examination. First, the vector, $\boldsymbol{\tau}$, has v elements, one for each treatment. They are called 'parameters'. In statistical parlance the word has a meaning that some will find unfamiliar. It indicates a quantity that exists in nature and requires to be estimated from data. Here $\boldsymbol{\tau}$ represents the true yields of the various treatments averaged over the plots of the experimental area. Secondly, $\mathbf{\Delta}$ is the 'design matrix for treatments'. It has v rows, one for each treatment, and n columns, one for each plot. Its elements equal zero unless the plot indicated by the column receives the treatment indicated by the row; in that case the element equals one. It follows that the product, $\mathbf{\Delta}'\boldsymbol{\tau}$, is a vector in which each plot is assigned the treatment parameter appropriate to it. Finally, $\boldsymbol{\eta}$ is a vector of n random 'residuals', representing the various sources of error—lack of uniformity in the plant material, local variation in the soil, errors of measurement and perhaps much more—that make the experiment less than perfect. It will be assumed that the elements of $\boldsymbol{\eta}$ are all normally distributed with variance, σ^2. It will also be assumed that they are independent of one another. A cautious critic could well complain that elements for adjacent plots could be correlated. That is quite true but the reader is asked to suspend judgement for a while. It will appear later than the random allocation of treatments to plots does effectively meet the point. Finally, as a matter of definition, a set of values like the elements of $\mathbf{y}$ is called a 'variate' when, as here, it can be considered in two parts, one derived from parameters and the other representing the action of random error.

It will be seen that $\mathbf{\Delta 1}_n = \mathbf{r}$, where $\mathbf{1}_n$ represents a column vector with n elements, all equal to one, and $\mathbf{r}$ is a column vector of treatment replications. Also $\mathbf{\Delta}'\mathbf{1}_v = \mathbf{1}_n$ and $\mathbf{\Delta\Delta}' = \mathbf{r}^\delta$, i.e. the vector, $\mathbf{r}$, written as a diagonal matrix. (Those unfamiliar with the δ-notation will find an explanation in Section 24 of the Appendix.)

The problem now is to estimate $\boldsymbol{\tau}$, the solution being called $\hat{\boldsymbol{\tau}}$. A suitable statistical approach is to use the method of least squares. That is to say, the elements of $\boldsymbol{\eta}$ are squared and added to give $\boldsymbol{\eta}'\boldsymbol{\eta}$, the aim being to find the value of $\boldsymbol{\tau}$ that will minimize that quantity. Now,

$$\boldsymbol{\eta}'\boldsymbol{\eta} = (\mathbf{y}' - \boldsymbol{\tau}'\boldsymbol{\Delta})(\mathbf{y} - \boldsymbol{\Delta}\boldsymbol{\tau}) = \mathbf{y}'\mathbf{y} - 2\mathbf{T}'\boldsymbol{\tau} + \boldsymbol{\tau}'\mathbf{r}^{\delta}\boldsymbol{\tau}$$

where $\mathbf{T}(=\boldsymbol{\Delta}\mathbf{y})$ is a column vector of treatment totals. Minimizing for $\boldsymbol{\tau}$ gives

$$\mathbf{T} = \mathbf{r}^{\delta}\hat{\boldsymbol{\tau}}, \qquad \hat{\boldsymbol{\tau}} = \mathbf{r}^{-\delta}\mathbf{T}, \tag{2.1.2}$$

where the circumflex indicates an estimated value. In short, each treatment parameter is estimated by the corresponding treatment mean. Further,

$$\begin{aligned}\hat{\boldsymbol{\eta}} &= \mathbf{y} - \boldsymbol{\Delta}'\hat{\boldsymbol{\tau}} = \mathbf{y} - \boldsymbol{\Delta}'\mathbf{r}^{-\delta}\mathbf{T} \\ &= (\mathbf{I}_n - \boldsymbol{\Delta}'\mathbf{r}^{-\delta}\boldsymbol{\Delta})\mathbf{y} = \boldsymbol{\psi}\mathbf{y}, \text{ say.}\end{aligned} \tag{2.1.3}$$

Here $\boldsymbol{\psi}$ provides an example of a 'residuals matrix'. Such matrices are (1) symmetric, (2) singular, e.g.,

$$\boldsymbol{\psi}\mathbf{1}_n = \mathbf{1}_n - \boldsymbol{\Delta}'\mathbf{r}^{-\delta}\mathbf{r} = \mathbf{1}_n - \boldsymbol{\Delta}'\mathbf{1}_v = \mathbf{0}, \tag{2.1.4}$$

and (3) idempotent, e.g.

$$\begin{aligned}\boldsymbol{\psi}\boldsymbol{\psi} &= (\mathbf{I}_n - \boldsymbol{\Delta}'\mathbf{r}^{-\delta}\boldsymbol{\Delta})(\mathbf{I}_n - \boldsymbol{\Delta}'\mathbf{r}^{-\delta}\boldsymbol{\Delta}) \\ &= \mathbf{I}_n - 2\boldsymbol{\Delta}'\mathbf{r}^{-\delta}\boldsymbol{\Delta} + \boldsymbol{\Delta}'\mathbf{r}^{-\delta}\boldsymbol{\Delta}\boldsymbol{\Delta}'\mathbf{r}^{-\delta}\boldsymbol{\Delta} = \boldsymbol{\psi}.\end{aligned}$$

The minimized sum of squared deviations is therefore

$$\hat{\boldsymbol{\eta}}'\hat{\boldsymbol{\eta}} = \mathbf{y}'\boldsymbol{\psi}\mathbf{y} = \mathbf{y}'\mathbf{y} - \mathbf{T}'\mathbf{r}^{-\delta}\mathbf{T}. \tag{2.1.5}$$

For reasons that will appear later it has $(n-v)$ degrees of freedom.

To find the treatment sum of squared deviations it is necessary to minimize again under the null-hypothesis, which is that all the treatment parameters have the same value, i.e. $\boldsymbol{\Delta}'\boldsymbol{\tau} = \alpha\mathbf{1}_n$, where α is some constant. Writing

$$\mathbf{y} = \alpha\mathbf{1}_n + \boldsymbol{\eta}, \tag{2.1.6}$$

the new minimized value of $\hat{\boldsymbol{\eta}}'\hat{\boldsymbol{\eta}}$ is $\mathbf{y}'\boldsymbol{\phi}\mathbf{y}$ with $(n-1)$ degrees of freedom where $\boldsymbol{\phi}$, another residuals matrix, equals $\mathbf{I}_n - \mathbf{1}_n\mathbf{1}_n'/n$, i.e.

$$\mathbf{y}'\boldsymbol{\phi}\mathbf{y} = \mathbf{y}'\mathbf{y} - G^2/n, \tag{2.1.7}$$

where $G\,(=\mathbf{1}_n'\mathbf{y})$ is the grand total of data. Hence the treatment sum of squared deviations, which is the difference between these two minimizations, is

$$\mathbf{y}'(\boldsymbol{\phi} - \boldsymbol{\psi})\mathbf{y} = \mathbf{y}'(\boldsymbol{\Delta}'\mathbf{r}^{-\delta}\boldsymbol{\Delta} - \mathbf{1}_n\mathbf{1}_n'/n)\mathbf{y} = \mathbf{T}'\mathbf{r}^{-\delta}\mathbf{T} - G^2/n$$

with $(v-1)$ degrees of freedom. It should be noted that $\boldsymbol{\psi}\boldsymbol{\phi} = \boldsymbol{\psi}$. The three quantities, $\mathbf{y}'\mathbf{y}$, $\mathbf{T}'\mathbf{r}^{-\delta}\mathbf{T}$ and G^2/n, used in the calculations are examples of 'summation terms', which will be considered in more detail later (Section 4.1).

Since sums of squared deviations are available both for the error and treatments, an F-test is possible but its validity depends upon the randomization. Fisher's paradox on that matter was explained in Section 1.1.

It will be helpful to look at the plots in two ways. First, there are the 'design plots'. Suppose, for example, that there are to be four plots of Treatment A, called A1, A2, A3 and A4, then each will have its column in $\mathbf{\Delta}$. There are also the 'field plots', each with its own η_i and hence its own $\hat{\eta}_i$, since the latter is only the former diminished by the general mean of all η_i. (Note here that the whole argument is proceeding on the basis of the null-hypothesis that $\boldsymbol{\tau} = \alpha\mathbf{1}_v$, so the estimated residuals equal $\boldsymbol{\phi}\mathbf{y}$, not $\boldsymbol{\psi}\mathbf{y}$.)

Although what happens in practice is the allocation of a random permutation of design plots to the field plots, notionally it is easier to think of the field plots as being randomly assigned to the design plots, which comes to the same thing. As a result of all that, suppose that the pth design plot is allocated the jth field plot, then

$$\mathring{\eta}_p = \hat{\eta}_j$$

What is the expectation under randomization of $\mathring{\eta}_p^2$? Since all $\hat{\eta}_j$ are equally likely to be any particular $\mathring{\eta}_p$,

$$\text{ex}\,(\mathring{\eta}_p^2) = \frac{1}{n}\sum_j \hat{\eta}_j^2 = (n-1)\theta/n, \tag{2.1.8}$$

where $\mathbf{y}'\mathbf{y} - \mathbf{T}'\mathbf{r}^{-\delta}\mathbf{T}\ (=\sum_j \hat{\eta}_j^2)$ is set equal to $(n-1)\theta$. Now consider two different design plots, p and q. Since

$$\sum_q \mathring{\eta}_q = \sum_i \hat{\eta}_i = \mathbf{1}_n'\hat{\eta} = \mathbf{1}_n'\boldsymbol{\phi}\mathbf{y}$$
$$= \mathbf{0}'\mathbf{y} = 0,$$

then
$$\text{ex}\,(\mathring{\eta}_p \sum \mathring{\eta}_q) = 0$$

and
$$\text{ex}\,(\mathring{\eta}_p \sum_{q \neq p} \mathring{\eta}_q) = -\text{ex}\,(\mathring{\eta}_p)^2$$
$$= -(n-1)\theta/n.$$

But, on account of the symmetry imparted by the randomization, all $\mathring{\eta}_p\mathring{\eta}_q$ have the same expectation. Since for any $\mathring{\eta}_p$ there are $(n-1)$ possible $\mathring{\eta}_q$, it follows that

$$\text{ex}\,(\mathring{\eta}_p\mathring{\eta}_q) = -\theta/n. \tag{2.1.9}$$

In fact

$$\text{ex}\,(\mathring{\boldsymbol{\eta}}\mathring{\boldsymbol{\eta}}') = \theta\boldsymbol{\phi}$$

because all diagonal elements of $\boldsymbol{\phi}$ equal $(n-1)/n$ and all off-diagonal ones equal $-1/n$. Hence, on the null-hypothesis,

$$\text{ex}\,(\mathbf{y}'\boldsymbol{\psi}\mathbf{y}) = \text{ex}\,(\boldsymbol{\eta}'\boldsymbol{\psi}\boldsymbol{\eta}) = \text{ex}\,[\text{tr}\,(\boldsymbol{\psi}\boldsymbol{\eta}\boldsymbol{\eta}')]. \tag{2.1.10}$$

The last result depends upon two useful lemmas.

Lemma 2.1.A *If* **AB** *and* **BA** *both exist, then* tr (**AB**) = tr (**BA**)

Proof: $tr\,(\mathbf{AB}) = \sum_i \sum_j (A_{ij} B_{ji})$

$$= \sum_j \sum_i (B_{ji} A_{ij}) = \text{tr}\,(\mathbf{BA}). \quad \blacksquare$$

Lemma 2.1.B *If* **a** *is a vector and* **B** *is a matrix, if* **a′Ba** *exists it equals* tr (**Baa′**)
Proof: **a′Ba** = tr (**a′Ba**) because there is only one element. Hence the above lemma applies. ■

Continuing from (2.1.10)

$$\text{ex}\,[\text{tr}\,(\boldsymbol{\psi\eta\eta}')] = \text{ex}\,[\text{tr}\,(\boldsymbol{\psi}\mathring{\boldsymbol{\eta}}\mathring{\boldsymbol{\eta}}')] = \text{tr}\,(\boldsymbol{\psi\phi})\theta = \text{tr}\,(\boldsymbol{\psi})\theta.$$

Since a diagonal element of $\boldsymbol{\psi}$ equals $1 - 1/r_i$, where the plot comes from treatment i, so the sum of all such elements is $(n - v)$. Similarly, the expectation under randomization of $\mathbf{y}'\boldsymbol{\phi}\mathbf{y}$ is

$$\text{tr}\,(\boldsymbol{\phi\phi})\theta = \text{tr}\,(\boldsymbol{\phi})\theta = (n-1)\theta \tag{2.1.11}$$

because all diagonal elements of $\boldsymbol{\phi}$ equal $1 - 1/n$. In short, under the null-hypothesis the expectation of the partition of the sum of squared deviations in the analysis of variance is in proportion to the degrees of freedom given, so an F-test is unbiassed. Also, if σ^2 is estimated as $(\mathbf{y}'\mathbf{y} - \mathbf{T}'\mathbf{r}^{-\delta}\mathbf{T})/(n - v)$, then $\theta = \hat{\sigma}^2$. Those who would like a worked example will find one in Example 2A at the end of the chapter.

Of course, such an argument refers only to the mean values of the various sums of squared deviations under all possible permutations. What happens if the particular randomization leads to something apparently absurd, such as all plots of one treatment forming a cluster in one part of the field? The possibility serves as a warning. An experimenter should ask himself such questions before he begins. If he can see differences in his field he should ask himself if he is really wise to risk such happenings. If he genuinely does not see any differences, he ought not to care how the randomization goes, though fertility patterns can exist without being foreseen. However, that is to speak only of patterned variation, which may well be absent. Is he prepared to accept such randomizations in the knowledge that there will almost certainly be random variation but with adjacent residuals correlated? That is the question that has bedevilled the Fisherian field trial since it was first proposed. The truth is that most people, whatever their public stance on the matter, do have a limit of tolerance beyond which the randomization process starts again, but that limit should be set very high. Also, it ought to allow for randomizations in which treatments are too much dispersed as well as those in which they are agglomerated.

In any case, the above argument is concerned only with the lack of bias of the F-test. Preece *et al.* (1978) call that the 'weak criterion' as opposed to a 'strong criterion' in which each contrast is estimated with the correct variance, but that must be left till Section 4.9 after the discussion of natural and basic contrasts.

It may be noted that

$$\hat{\boldsymbol{\tau}} = \mathbf{r}^{-\delta}\boldsymbol{\Delta}\mathbf{y} = \mathbf{r}^{-\delta}\boldsymbol{\Delta}(\boldsymbol{\Delta}'\boldsymbol{\tau} + \boldsymbol{\eta}) = \boldsymbol{\tau} + \mathbf{r}^{-\delta}\boldsymbol{\Delta}\boldsymbol{\eta}.$$

Since Cov($\boldsymbol{\eta}$), the covariance matrix of the residuals, equals $\mathbf{I}_n\sigma^2$, it follows that

$$\text{Cov}(\hat{\boldsymbol{\tau}}) = (\mathbf{r}^{-\delta}\boldsymbol{\Delta})(\boldsymbol{\Delta}'\mathbf{r}^{-\delta})\sigma^2 = \mathbf{r}^{-\delta}\sigma^2. \tag{2.1.12}$$

In practice it is necessary to use

$$\text{Cov}(\hat{\boldsymbol{\tau}}) = \mathbf{r}^{-\delta}\hat{\sigma}^2 \tag{2.1.13}$$

and to adjust tests accordingly. That result, like those that have preceded it, will come as no surprise. Nevertheless the proofs just given will serve with little modification for much more complicated designs.

It is of interest to examine the assumptions that are implied in the mathematics just presented and to ask how far they are reasonable. First of all, it is implied that each treatment will have the same effect regardless of the plot to which it is applied. It is true that there are also random residuals, but they are assumed to be a characteristic of the field plots and unchanged by the treatments applied. Plainly that is not quite satisfactory. For example, a plot on which drainage was impeded might react unfavourably to irrigation, whereas another with poor moisture retention might benefit considerably. It is not suggested that the objection is usually serious in practice, but land should not be too variable especially with respect to characteristics likely to interact with the treatments. Also, it is assumed that all residuals have the same variance, but that is not necessarily so if some treatments involve more manipulation than others since the mere fact of applying them can be a source of error. Further, if one treatment puts the plants under stress in a way avoided by another, the two cannot lead to the same variability. Also, the variance of a plot may depend upon its response to the treatment, being perhaps higher for higher values, though that difficulty can commonly be met by using variance-stabilizing transformations. Not least, those who assume normality of distribution should remember the quip that normality is accepted by biologists because they understand that the mathematicians have proved it and by mathematicians because they understand that the biologists always find it. (In fact, it is an assumption of convenience on the part of the mathematicians, who would be in trouble without it.) The difficulties are mentioned for two reasons. One is to make the point that assumptions need to be watched. Usually they hold well enough but on occasion they can be absurd. The other is to emphasize the need for toleration if someone proposes an alternative approach. Too often a would-be innovator is accused of making unlikely assumptions, when really he is being no more rash than the proponents of conventional methods, who sometimes forget that a severe judge could find fault with them also. In fact, the precise and ordered world of the mathematician never does reflect very well the more exciting world of the biologist, who has to deal with living things with the power to adapt themselves. A skilful statistician is always gauging how far his assumptions are reasonable; only the foolish pretend that they are obviously justified and can be forgotten.

2.2 CHOICE OF REPLICATIONS

The chief design problem when treatments are completely randomized concerns the replications, which should depend upon the contrasts of interest. If there are four treatments and the relevant contrasts are those of (1.6.1), **r** should ideally take the form $(\rho_0, \rho, \rho, \rho)'$. (Although it is not essential that Treatments A, B and C should be equally replicated, they do have equivalent status in the experiment so it is reasonable to assign each to the same number of plots. Also, the experiment as a whole will be judged by its worst-estimated contrast of interest and that will depend upon the lowest replications among A, B and C.) Hence **V**, the covariance matrix of the treatment means, which (2.1.13) shows to equal $\mathbf{r}^{-\delta}\hat{\sigma}^2$, will be

$$\begin{bmatrix} 1/\rho_0 & \mathbf{0}' \\ \mathbf{0} & \mathbf{I}_3/\rho \end{bmatrix}\hat{\sigma}^2.$$

Calling the three contrasts of interest $\mathbf{c}_1$, $\mathbf{c}_2$ and $\mathbf{c}_3$ respectively, then

$$\mathbf{c}_1'\,\mathbf{V}\mathbf{c}_1 + \mathbf{c}_2'\mathbf{V}\mathbf{c}_2 + \mathbf{c}_3'\mathbf{V}\mathbf{c}_3 = \tfrac{3}{2}\left(\frac{1}{\rho_0} + \frac{1}{\rho}\right)\hat{\sigma}^2.$$

Suppose that 20 plots are available ($\rho_0 + 3\rho = 20$), then it is a question of minimizing $[1/(20-3\rho) + 1/\rho]$, which is achieved when $\rho = 4.23$ and $\rho_0 = 7.31$. Clearly some rounding is needed and, in order to retain symmetry between A, B and C, it would be best to assign each of them four plots, thus leaving eight for S. The variance of differences like (*S − *A) would then be $(\frac{1}{4} + \frac{1}{8})\sigma^2 = 0.375\sigma^2$. However, it could be argued that 19 plots would suffice since S appears to have a higher replication than necessary. The variance would then be $(\frac{1}{4} + \frac{1}{7})\sigma^2 = 0.393\sigma^2$. Admittedly that is a slight increase, but if it led to the removal of a plot that had been chosen with some reluctance in the first place, a small reduction in $\hat{\sigma}^2$ could be hoped for as partial compensation. Also, it might be thought that the trouble and expense of a 20th plot was not really justified if the reduction of variance was going to be so small.

To take the second set of contrasts, those in (1.6.2), it has already been remarked that they cannot completely sum up the desiderata or there would be no need of Treatment O. The position is perhaps better described by saying that both (1.6.1) and (1.6.2) are relevant, but the latter are more important. If, for example, it is to be expected that differences like (*A − *O) will be about twice as large as those like (*A − *B), it would be reasonable to plan for them to have four times the variance, i.e.

$$\left(\frac{1}{\rho_0} + \frac{1}{\rho}\right) = 4\left(\frac{2}{\rho}\right)$$

or ρ should be approximately $7\rho_0$. (It will be noticed that no minimization was here called for.) Hence if there are 20 plots, ρ should equal 6.36 and ρ_0 0.91. In practice most experimenters would want to have at least two plots of the control in case of accidents and they would also want to keep A, B and C equireplicate, so they would probably adopt $\rho = 6$, $\rho_0 = 2$.

In the last case, i.e. (1.6.3), it is clear that the four treatments should be equireplicate regardless of the relative importance of the three contrasts, so each should be assigned to five plots.

2.3 ADJUSTMENT BY CONCOMITANT VARIABLES

Although a completely randomized design makes no provision for removing environmental variation, it is often possible to allow for local characteristics provided that they can be given an objective value for each plot. For example:

(1) There could be measurable differences in soil depth or texture.
(2) When starting an experiment on established plants, such as fruit trees, there could be observable differences that could be taken into account. Indeed, in such circumstances, the 'calibration' of the plots is standard practice (Pearce, 1953, 1976a).

In such cases, it could well be that the good and bad plots form clearly defined areas and so suggest a blocking system. If, however, they are intermingled it would be most unwise to form blocks on any basis that was not territorial. If a further source of variation were to intervene, as is quite likely, such dispersed blocks would be unable to cope. For example, if a storm caused damage down one side of the experiment or if there was a drought and a dry patch appeared, the plots affected could well lie in several blocks, which might then do more harm than good. It would be better to obtain 'concomitant variables' at the start of the experiment by measuring the quantities likely to disturb it, i.e., the soil characteristics or present growth and past crop on the trees, and to make allowance for them when the data come to be analysed. Fortunately a method is available in the 'analysis of covariance'.

Sometimes a concomitant variable is derived from the location of the plots. It is then called a 'pseudo-variate'. For example, if plots lie in a row, it may be thought advisable to use the plot number as a concomitant variable to allow for a trend. (Since trends are rarely linear, there could be some advantage in using the square of the plot number also or perhaps its square root to allow for curvature.) Two relevant papers are those of Federer and Schottefeld (1954) and Outhwaite and Rutherford (1955), but the approach must have been used in many other instances. Also, experiments rarely consist of a single row of plots, so a second dimension should be considered and there is also the possibility of an interaction between the two directions. That implies regressing the data obtained from the experiment upon five variates, namely, x_1, x_2, the co-ordinates in two directions, x_1^2, x_2^2 and x_1x_2. The basic fertility pattern is thus assumed to be paraboloidal, which can scarcely be strictly true, though it may be a reasonable approximation. Such variables could allow fairly well for a large patch, whether of good or bad soil, or for a strip of different fertility not parallel to an axis of x_1 or x_2, both awkward features, but they could not be expected to allow for several such features, so their value is questionable when the experimental area is very large.

Another example of the use of concomitant variables is afforded by a pseudo-variate that is given the value one when the plot is exterior and zero when it is interior.

In Section 2.1 a warning was given against using over-elaborate methods for the control of environmental variation. If the attempt is unsuccessful, valuable degrees of freedom will have been lost from error. Worse, if the attempt later appears to have been perverse—as when blocks are introduced to allow for variation in one direction and the actual differences lie at right angles to it—the estimated error variance may well be increased. That warning applies here, but it is an advantage of concomitant variables that they usually require only few degrees of freedom in relation to what they set out to achieve.

Mathematically there is no great difficulty about the analysis of covariance. (Incidentally, the quantity under study is often called the 'dependent variate' and the concomitant variables the 'independent variates', the idea being that the dependent variate is derived from the others, which are free to take what values they choose.) If there is only one concomitant variable, the calculations are especially simple. The model (2.1.1) becomes

$$\mathbf{y} = \mathbf{\Delta}'\boldsymbol{\tau} + \theta\mathbf{x} + \boldsymbol{\eta},$$

where $\mathbf{x}$ holds the values of the concomitant variable and θ is a regression coefficient. (It would be more conventional to use β, but that symbol is being reserved for another purpose.) It will often be convenient to measure the values of $\mathbf{x}$ from some standard value, s, based on considerations that have nothing to do with the analysis, so $\mathbf{x}$ will be replaced by $\mathbf{z} = \mathbf{x} - s\mathbf{1}_n$. That makes the model

$$\mathbf{y} = \mathbf{\Delta}'\boldsymbol{\tau} + \theta\mathbf{z} + \boldsymbol{\eta}. \tag{2.3.1}$$

Minimizing $\boldsymbol{\eta}'\boldsymbol{\eta}$ for $\boldsymbol{\tau}$ and θ gives the equations

$$\mathbf{T} = \mathbf{r}^{\delta}\hat{\boldsymbol{\tau}} + \hat{\theta}\mathbf{\Delta z} \qquad (\mathbf{T} = \mathbf{\Delta y}), \tag{2.3.2a}$$

$$\mathbf{z}'\mathbf{y} = \mathbf{z}'\mathbf{\Delta}'\hat{\boldsymbol{\tau}} + \hat{\theta}\mathbf{z}'\mathbf{z}. \tag{2.3.2b}$$

Eliminating $\hat{\boldsymbol{\tau}}$ it appears that

$$\hat{\theta} = \frac{\mathbf{z}'\boldsymbol{\psi}\mathbf{y}}{\mathbf{z}'\boldsymbol{\psi}\mathbf{z}} = \frac{U}{V}, \text{ say,} \tag{2.3.3}$$

where $\boldsymbol{\psi}$ has the value assigned to it in (2.1.3). Further, from (2.3.2a)

$$\hat{\boldsymbol{\tau}} = \mathbf{r}^{-\delta}\mathbf{T} - \hat{\theta}(\mathbf{r}^{-\delta}\mathbf{\Delta z}). \tag{2.3.4}$$

The vector, $\mathbf{r}^{-\delta}\mathbf{\Delta z}$, of course represents the treatment means of the variable, $\mathbf{z}$. It will be seen that U and V and therefore $\hat{\theta}$ would have been the same if $\mathbf{x}$ had been used instead of $\mathbf{z}$ as originally proposed. The change has, however, led to all elements of $\hat{\boldsymbol{\tau}}$ being reduced by the constant, $\hat{\theta}s$.

With $\hat{\theta}$ and $\hat{\boldsymbol{\tau}}$ known, it appears that

$$\begin{aligned}\hat{\boldsymbol{\eta}} &= \mathbf{y} - \mathbf{\Delta}'\hat{\boldsymbol{\tau}} - \hat{\theta}\mathbf{z} \\ &= \boldsymbol{\psi}(\mathbf{y} - U\mathbf{z}/V),\end{aligned}$$

so the minimized value of $\boldsymbol{\eta}'\boldsymbol{\eta}$ is

$$\hat{\boldsymbol{\eta}}'\hat{\boldsymbol{\eta}} = (\mathbf{y}'\mathbf{y} - \mathbf{T}'\mathbf{r}^{-\delta}\mathbf{T}) - U^2/V \tag{2.3.5}$$

with one degree of freedom fewer than $(\mathbf{y}'\mathbf{y} - \mathbf{T}'\mathbf{r}^{-\delta}\mathbf{T})$, i.e. $(n - v - 1)$, because an additional parameter, θ, has been fitted. An estimate of σ^2 has accordingly been found.

Continuing from (2.3.4) it follows that

$$\hat{\boldsymbol{\tau}} = \mathbf{r}^{-\delta}\boldsymbol{\Delta}(\mathbf{y} - \hat{\theta}\mathbf{z}).$$

Here it may be noted that in $\hat{\theta} = U/V$ the value of V depends only on the concomitant variable. Although $\mathbf{x}$ and $\mathbf{z}$ are called independent variates they are not really variates at all because they lack any random component. They are in fact vectors of constants appended to the data in the hope that they will explain something. It is true that in origin they could be measured quantities, subject to error, but that does not affect their role in the analysis of covariance. Someone has said 'Adjust according to those values' and that is being done. On the other hand, U depends partly on $\mathbf{y}$ and is a variate because $\mathbf{y}$ is one. Hence

$$\begin{aligned}\hat{\boldsymbol{\tau}} &= \mathbf{r}^{-\delta}\boldsymbol{\Delta}(\mathbf{I}_n - \mathbf{z}\mathbf{z}'\boldsymbol{\psi}/V)\mathbf{y} \\ &= \mathbf{r}^{-\delta}\boldsymbol{\Delta}(\mathbf{I}_n - \mathbf{z}\mathbf{z}'\boldsymbol{\psi}/V)(\boldsymbol{\Delta}'\boldsymbol{\tau} + \theta\mathbf{z} + \boldsymbol{\eta}) \\ &= \boldsymbol{\tau} + \mathbf{r}^{-\delta}\boldsymbol{\Delta}(\mathbf{I} - \mathbf{z}\mathbf{z}'\boldsymbol{\psi}/V)\boldsymbol{\eta}\end{aligned} \tag{2.3.6}$$

becasue $\boldsymbol{\psi}\boldsymbol{\Delta}' = \mathbf{0}$. Hence $\hat{\boldsymbol{\tau}}$ is an unbiased estimator of $\boldsymbol{\tau}$ with variance

$$\begin{aligned}\text{Cov}(\hat{\boldsymbol{\tau}}) &= \mathbf{r}^{-\delta}\boldsymbol{\Delta}(\mathbf{I} - \mathbf{z}\mathbf{z}'\boldsymbol{\psi}/V)\text{Cov}(\boldsymbol{\eta})(I - \boldsymbol{\psi}\mathbf{z}\mathbf{z}'/V)\boldsymbol{\Delta}'\mathbf{r}^{-\delta} \\ &= [\mathbf{r}^{-\delta} + (\mathbf{r}^{-\delta}\boldsymbol{\Delta}\mathbf{z})(\mathbf{r}^{-\delta}\boldsymbol{\Delta}\mathbf{z})'/V]\sigma^2 = (\mathbf{r}^{-\delta} + \mathbf{h}\mathbf{h}'/V)\sigma^2,\end{aligned} \tag{2.3.7}$$

where $\mathbf{h} = \mathbf{r}^{-\delta}\boldsymbol{\Delta}\mathbf{z}$. (That is to say, the elements of $\mathbf{h}$ show the extent to which the treatment means of the concomitant variable deviate from the standard value, s.) It appears then that the covariance matrix of the estimated treatment parameters, which from (2.1.12) was $\mathbf{r}^{-\delta}\sigma^2$, has been inflated by the values of $\mathbf{h}$. If elements are small, i.e., the observed values of $\mathbf{x}$ are close to the standard value, s, the variances will be little affected. On the other hand, if some treatments have means that need a large adjustment, the variances may be much increased. Also, there could be appreciable correlations between means, which are no longer found independently. The last point has been examined in some detail by Preece (1980).

To form the analysis of variance it is necessary to revert to the null-hypothesis, i.e., to proceed as if all treatment parameters were the same, as in (2.1.6), and repeat. Writing

$$U_0 = \mathbf{z}'\boldsymbol{\phi}\mathbf{y}, \qquad V_0 = \mathbf{z}'\boldsymbol{\phi}\mathbf{z},$$

where $\boldsymbol{\phi}$ is the matrix defined above (2.1.7), the expression (2.3.5) becomes

$$(\mathbf{y}'\mathbf{y} - G^2/n) - U_0^2/V_0 \tag{2.3.8}$$

with $(n - 2)$ degrees of freedom, corresponding to (2.1.7), and the treatment sum of squared deviations is

$$(\mathbf{T}'\mathbf{r}^{-\delta}\mathbf{T} - G^2/n) - (U_0^2/V_0 - U^2/V) \tag{2.3.9}$$

with $(v-1)$ degrees of freedom. Expression (2.3.9) was obtained by subtracting (2.3.5) from (2.3.7).

Results are readily generalized to cover several concomitant variables. The model becomes

$$\mathbf{y} = \mathbf{\Delta}'\boldsymbol{\tau} + \mathbf{X}\boldsymbol{\theta} + \boldsymbol{\eta},$$

where $\mathbf{X}$ is a matrix that holds the jth concomitant variable in its jth column and $\boldsymbol{\theta}$ is a vector of regression coefficients. As before it will be used with $\mathbf{Z} = \mathbf{X} - \mathbf{1}_n\mathbf{s}'$ (where $\mathbf{s}$ is a vector of standard values, one for each concomitant variable) replacing $\mathbf{X}$. The analogues of (2.3.3), (2.3.4), (2.3.5) and (2.3.7) are respectively

$$\hat{\boldsymbol{\theta}} = \mathbf{V}^{-1}\mathbf{U}, \tag{2.3.10}$$

where $\mathbf{V} = \mathbf{Z}'\boldsymbol{\psi}\mathbf{Z}$ and $\mathbf{U} = \mathbf{Z}'\boldsymbol{\psi}\mathbf{y}$

$$\hat{\boldsymbol{\tau}} = \mathbf{r}^{-\delta}\mathbf{T} - \mathbf{r}^{-\delta}\mathbf{\Delta}\mathbf{Z}\hat{\boldsymbol{\theta}} = \mathbf{r}^{-\delta}\mathbf{T} - \mathbf{H}\hat{\boldsymbol{\theta}}, \tag{2.3.11}$$

where $\mathbf{H} = \mathbf{r}^{-\delta}\mathbf{\Delta}\mathbf{Z}$,

$$\hat{\boldsymbol{\eta}}'\boldsymbol{\eta} = \mathbf{y}'\mathbf{y} - \mathbf{T}'\mathbf{r}^{-\delta}\mathbf{T} - \mathbf{U}'\mathbf{V}^{-1}\mathbf{U}, \tag{2.3.12}$$

and

$$\text{Cov}\,(\hat{\boldsymbol{\tau}}) = [\mathbf{r}^{-\delta} + \mathbf{H}'\mathbf{V}^{-1}\mathbf{H}]\sigma^2. \tag{2.3.13}$$

The error variance is readily estimated from (2.3.12) as $\hat{\boldsymbol{\eta}}'\hat{\boldsymbol{\eta}}/(n-p-1)$, where p is the number of concomitant variables. Also, the treatment sum of squared deviations is

$$(\mathbf{T}'\mathbf{r}^{-\delta}\mathbf{T} - G^2/n) - (\mathbf{U}_0'\mathbf{V}_0^{-1}\mathbf{U}_0 - \mathbf{U}'\mathbf{V}^{-1}\mathbf{U}) \tag{2.3.14}$$

where $\mathbf{U}_0 = \mathbf{Z}'\boldsymbol{\phi}\mathbf{y}$ and $\mathbf{V}_0 = \mathbf{Z}'\boldsymbol{\phi}\mathbf{Z}$.

A worked example will be found in Example 2B at the end of the chapter. The simplified form appropriate to a single concomitant variable is used in Example 2C.

2.4 ADJUSTMENT BY NEIGHBOURING PLOTS

Although it is unwise to regard any model of fertility as completely reliable, few will question that there can be good and bad areas in a field. Consequently if a plot is surrounded by neighbours that are doing well, it can be expected to do well itself. If it does, the fact does not necessarily show its treatment to be a good one even though the yield in isolation would look encouraging. On the other hand, if all the neighbours are doing badly even a mediocre crop from a plot can be evidence that it has a good treatment. Obvious as these considerations are, there have been difficulties about building them into the structure of statistical theory. One suggestion, which no one has questioned but few have followed, is that of Yates (1963a), who proposed a grid of plots all with the same treatment. From their yields a response surface would be worked out for the area of the experiment and the performance of all other plots would be judged by reference to it using the analysis of covariance. Okigbo (1966) similarly used the crops from discard areas and improved his error variance considerably.

A more contentious suggestion is that of Papadakis (1937, 1940). First he took treatment means and worked out the extent to which the performance of each plot deviated from the appropriate mean. (For a completely randomized design those quantities are estimates of the residuals $\hat{\boldsymbol{\eta}}$. For that reason some call them by that name, but it is better to refer to them as 'the deviations' because there are more complicated designs for which the two are different.) Then for each plot a concomitant variable is worked out from the deviations of the neighbours. That quantity can be regarded as a measure of the inherent fertility of the plot. Finally, the data are analysed using a covariance adjustment on the concomitant variable. The whole method is so intuitively satisfying that one would expect to learn that it was adopted speedily and universally. In fact, more than forty years later it still has to make its way. One reason, no doubt, was the advent of the Second World War, when people had other things to think about, but a number of objections have been made and are still without answers.

The first systematic study was that of Bartlett (1938). Among other things he pointed out that the correlation between the performance of neighbouring plots was such that the covariance adjustment might remove too much variation. For that reason he recommended attributing two degrees of freedom instead of one for the regression coefficient and that has remained standard practice. It may go too far, but if it does, the error degrees of freedom will be too few in number and the error variance overestimated, both faults on the side of caution when estimating standard errors.

In essentials the method can be expressed thus: Treatment totals ($\boldsymbol{\Delta}\mathbf{y}$) are found and converted to means ($\mathbf{r}^{-\delta}\boldsymbol{\Delta}\mathbf{y}$), which are then allocated to the data ($\boldsymbol{\Delta}'\mathbf{r}^{-\delta}\boldsymbol{\Delta}\mathbf{y}$) and subtracted to give $(\mathbf{I}-\boldsymbol{\Delta}'\mathbf{r}^{-\delta}\boldsymbol{\Delta})\mathbf{y} = \mathbf{S}\mathbf{y}$, say. (It is true that $\mathbf{S}$ so defined is the same as $\boldsymbol{\psi}$ in Section 2.1, but a separate symbol is needed because $\boldsymbol{\psi}$ depends upon the design as later chapters will show, whereas the deviations are fixed.) A concomitant variable, $\mathbf{x}$, is then formed as the mean of deviations from the neighbouring plots; that operation also can be regarded as a matrix pre-multiplication, i.e. $\mathbf{x} = \mathbf{R}\mathbf{S}\mathbf{y}$, say. (The form of $\mathbf{R}$ will depend upon the neighbours chosen.) Hence the regression coefficient, θ, of $\mathbf{y}$ on $\mathbf{x}$ is $\mathbf{x}'\boldsymbol{\psi}\mathbf{y}/\mathbf{x}'\boldsymbol{\psi}\mathbf{x}$, where $\boldsymbol{\psi}$ is the residuals matrix of the design after allowing for treatments, i.e.

$$\theta = \frac{\mathbf{y}'\mathbf{S}\mathbf{R}'\boldsymbol{\psi}\mathbf{y}}{\mathbf{y}'\mathbf{S}\mathbf{R}'\boldsymbol{\psi}\mathbf{R}\mathbf{S}\mathbf{y}} \tag{2.4.1}$$

However, it is necessary to look at the consequences of the treatment means, from which the deviations are derived, being based on only a few plots. Bartlett (1978) has given the following result: A deviation equals

$$\begin{aligned} &y_i - (\text{a total of } r \text{ values including } y_i)/r \\ &= (r-1)y_i/r - [\text{a total of } (r-1) \text{ other values}]/r. \end{aligned}$$

Taking ρ to be the correlation coefficient between neighbouring plots, then from (1.5.4) θ is derived as $\rho/(1+\rho^2)$. Eliminating the effect of the plot on the relevant treatment mean, the regression is now estimated as

$$\theta' = \frac{2\rho}{1+\rho^2+\alpha} \tag{2.4.2}$$

where $\alpha = 1/(r-1)$ for the one-dimensional case. (The multiplier of two arises because the regression is on the mean of the neighbours, not their sum.) Also,

$$\theta' = \frac{2\rho}{1+\rho^2+\frac{1}{2}\alpha} \tag{2.4.3}$$

for the two-dimensional case.

At this point the objectors should be heard. The later paper by Bartlett (1978) was read at a meeting of the Royal Statistical Society and was published with the discussion that resulted. The comments of the other statisticians well repay study. First, what happens at the sides and corners? Papadakis's own solution was simply to use such neighbours as do exist. A plot would normally have four neighbours and its concomitant value would ordinarily be the mean of their four deviations; if it has only three neighbours, then the mean of those three deviations is used instead. The point in question is the form of **R**. Until someone looks into the matter and determines what characteristics are desirable, there seems to be no reason to demand one sort of **R** rather than another. (Some have sought to evade the difficulty by thinking of the experiment as being laid out on a cylinder, so that plots on opposite boundaries adjoin, and some have used a torus, so that the area, although limited, is without edges, but that is hardly practical. Given an outside row of plots, the least desirable substitute for their missing neighbours is the outside row on the other side.) One suggestion is that plots should be long and narrow and laid out side by side. It would then be necessary to consider only correlations over the long boundaries. Also, the problem of outside effects could be met by adding a few plots at the ends, which would serve to establish concomitant variables without themselves entering into the analysis of variance. Another occasion when the one-dimensional case could be useful is in the use of contour blocks. In hilly country a block may be formed by ploughing or terracing round a contour and it may twist its way round the hill in a confusing manner. Although its plots may well have much in common, some method of allowing for differences along their serpentine course is needed. An ability to adjust the performance of each by that of its neighbours could be useful in the extreme.

Other problems arise from the practice of allocating treatments at random. The concomitant value for a plot depends upon the deviations of its neighbours, which in turn depend upon the treatment means, and they are estimated quantities subject to error. Consequently it must matter what treatments are applied to the neighbours of any plot. Some have seen a need for systematic designs in preference to randomized ones, a subject that needs to be explored further (see Section 9.2).

Although the method itself appears to make no major assumptions about the pattern of field variation, the designer of an experiment has to do so. In particular, he has to choose between using all four neighbours of each plot or only two of them. Also, he could try double covariance, forming two variables from the neighbours, one from the sides and the other from the ends. Although such decisions may well be made on the basis of the dimensions of the experiment, so many plots one way and so many at right angles, they could also be influenced by ideas about fertility patterns. They must also depend upon the shape of plots. If they are square it may be reasonable to give all neighbours the same standing; if they are long and narrow, that may be unsound. Indeed, in that case the variable derived from the ends could perhaps better be ignored. It follows then that the method is not as independent of local conditions as might be supposed.

One objection can readily be countered. The deviations depend upon the estimated treatment means. After adjustment by covariance upon the concomitant variables, different treatment means are obtained, so there is a sense in which the method invalidates itself. That, however, is not really a problem. All that is needed is to enter the calculations with the revised treatment means and to go on until values are found that are confirmed by the next cycle. That is to say, on the first cycle $\mathbf{y}_1 = \mathbf{y}$ and $\mathbf{x}_1 = \mathbf{RSy}_1$ with $\boldsymbol{\theta}_1 = \mathbf{y}'_1\mathbf{SR}'\boldsymbol{\psi}\mathbf{y}_1/\mathbf{y}'_1\mathbf{SR}'\boldsymbol{\psi}\mathbf{RSy}_1$. As a consequence $\mathbf{y}_2$ is found, equal to $\mathbf{y}_1 - \theta_1(\mathbf{I}_n - \mathbf{1}_n\mathbf{1}'_n/n)\mathbf{x}_1$. From this $\mathbf{x}_2$ and $\boldsymbol{\theta}_2$ follow and so on until $\mathbf{x}_{j+1} = \mathbf{x}_j$ for some j. Example 2C at the end of the chapter provides a detailed worked example.

There is a certain amount of empirical experience of the method, some of which has been published (Pearce and Moore, 1976; Pearce, 1978, 1980; Lockwood, 1980; Kempton and Howes, 1981; see also Mendez, 1971). The method does appear to be promising as a means of reducing the error variance and it is improved in that respect by iteration. It appears that the first cycle sometimes gives erratic results, which is another reason for taking more. Little difficulty has been found with convergence, though there is a tendency for successive cycles to oscillate slowly to a solution. For that reason, it might be better to calculate cycles in pairs, and to take a mean before proceeding. A certain reservation arises from the values of the regression coefficient. Where the concomitant variable has been found as a mean of neighbouring deviations, the regression coefficient would be expected to lie between 0 and 1. In fact, that is not always so, especially after iteration.

2.5 BLOCK DESIGNS

The classical device for controlling local variation over the experimental area is to divide the land into 'blocks', such that plots are less variable within a block than within the area as a whole. Where the experimenter has reason to think that certain parts of his field are uniform the method has great potential and is commonly used. For example, if an experiment has to be conducted on a slope it will almost certainly be advantageous to form blocks in bands that keep to a limited range of contours. Many other examples can be suggested. Because parameters will be needed to allow for block effects, there will necessarily be a reduction in the degrees of freedom for error, but often that will not matter a lot.

It should be said right away and said with emphasis that block designs represent the normal and usual way of conducting field experiments. Whether they represent the only possible way is another matter. If events had taken a slightly different turn perhaps they would have rivals. However, events did not take that turn and it is necessary to know about block designs as the basis of most current practices.

They were developed in a context of English fields, which have mostly been farmed in one piece for years, if not for centuries. They were found to work very well. They were next used in the conditions of North America, mostly on land that had been wrested from the prairies and thereafter likewise farmed in one piece. Again they worked well. Thereafter it was supposed that they would work well anywhere, but that is not certain. In some places experimental farms have been formed by buying a lot of adjacent smallholdings and amalgamating them without even suveying the disappearing boundaries. It is not so clear that block designs are ideal for such conditions, nor is it clear that they should be used on the sides of volcanoes or on atolls in the Pacific or in other unusual sites, but they are.

There must then be reservations. Sometimes the land looks uniform and there is no reason to form blocks one way rather than another. The recommendation is often made to make them compact and nearly square, the argument being that plots close together will necessarily be similar, but that does not follow. The writer recalls one of the awkward moments of his career. In a certain country there are two research institutes engaged on similar work and both are on hilly land. At the larger a team of surveyors laid out all experiments, making careful adjustment to plot dimensions to keep areas constant despite varying slopes; at the smaller, where there were no surveyors and no one claimed any special skills, there were no refinements. Nevertheless, in comparable experiments the less favoured institute regularly obtained error variances about one-quarter of those at the main centre. It became obvious to the authorities that the workers at the subsidiary centre could not understand matters properly or they would not be making such absurd claims, so the writer was sent to put them right. He found that their low variances were quite authentic. When they wanted to lay down an experiment they took a stream coming down from the hills and disposed plots on either side of it at a uniform distance from it and regarded those plots as a block. Then they found another stream and formed another and so on till they had enough. At the main centre blocks were formed that spread over streams and even outcrops of rock. They looked magnificient on a plan but they did not control local variation. There is no need to point out the moral of the story.

Although it is unwise to introduce blocks to control unidentified sources of variation, in a large experiment they may be needed for another reason, namely, for purposes of administration. If the total area is too large to be harvested in one day, the designer of the experiment should provide blocks that correspond to a day's work. Similarly if the recording of stages of development, like germination or blossoming, is going to take a long time, division of the experimental area is called for. In some climatic conditions changes can take place quickly, so it is not good enough to record some plots today and some tomorrow without making allowance in the analysis of data for the lapse of time. Further, in places where the weather is changeable, a team of recorders or harvesters may start work on a fine morning and be driven off by a storm later in the day. It is then desirable to provide them with block boundaries where they can stop if the weather becomes threatening. In a large experiment blocks may be essential even though there is the danger of forming them wrongly and encapsulating environmental variation within them instead of between them. In a small one there may be little environmental variation to control, administratively there is no need of division and error degrees of freedom are precious. In such a case blocks are often better avoided.

Blocks have several other uses. For example, the best way of sampling a tract is to disperse the experiment over a number of sites. In that event, the plots at any site necessarily form a block. Such a dispersed experiment often provides a good demonstration to farmers, the plots at one site being easily comprehended.

The decision whether or not to rely upon blocks to control environmental variation must depend upon practicalities. If those with experience of the conditions aver that blocks are always advantageous, it is not for others to question their conclusions. Further, if they express a preference for compact blocks, no one else need object. Other people can have different experiences. The writer's early work was with fruit trees in England. He found that he had to contend with two quite different kinds of error. One came from the trees themselves and tended to eliminate itself over the years; the other came from the environment and tended to build up. His advice is therefore more complicated. For short-term experiments it is important to scrutinize carefully what comes from the nursery to ensure uniformity, but efforts to control local variation are commonly unavailing. For long-term experiments, on the other hand, there is less need to bother about the plant material, though features like twisted stems that will have a cumulative effect over the years are to be avoided. There is, however, every need to control depth of soil and exposure to wind which will be the same from year to year and can likewise build up large effects over a long period. Latterly he has been involved with tropical agriculture, which can be quite different again. Here the experimental area may have been formed by amalgamating a number of smallholdings. What is more, it may even include the site of a deserted village with former roads and pounds for animals. If it does, often no one knows exactly where they were. There are consequently great difficulties about deciding what fertility patterns have to be dealt with and a bad guess can do harm. There are therefore no rules of general application apart from the one that enjoins thought. It is not to be expected that practices evolved initially for English fields will necessarily apply without modification in other conditions. On the other hand, they could be ideal.

One point of nomenclature should be noted. It is misleading to speak of blocks as a 'factor' of the experiment. Strictly 'factorial' indicates a relationship between the treatments, i.e. that they are made up of all combinations of the levels of two or more factors. Where each treatment occurs once in each block, i.e. a randomized block design, there are admittedly resemblances to a two-factor experiment, but treatments are not necessarily disposed in that way. Usually it is better to speak of two 'classifications'. In any event, blocks have a characteristic role in the experiment, which needs to be recognized.

Some experimenters make a practice of separating blocks by an additional guard or especially wide discard areas. The need is not obvious. If sufficient provision has been made to overcome interference between plots of the same block, it is supposedly sufficient when the plots lie over a block boundary. However, where blocks are used for purposes of administration it is helpful to have some way of finding quickly where their boundaries come, whether by stakes or other markers, or by additional guards. For that reason and for that reason alone the practice can be commended, though it should be remembered that discard areas are always a nuisance. For that matter, stakes get in the way of farm machinery, so they also can be a nuisance. Altogether it is a choice of evils.

2.6 ROW-AND-COLUMN DESIGNS

Another well-established device is the use of two orthogonally disposed blocking systems, one called 'rows' and the other called 'columns', each intersection of a row and a column constituting a plot. (Of course, neither rows nor columns should be so long and straggling as to be ineffective as blocks.) The experimental area has to be rectangular, which can be something of a limitation. Obviously such a design, by providing for variation either along or across the area, does not call for so many awkward decisions about dominant sources of error. Also, it provides for cases where there are clearly two important sources, e.g., in an experiment that is to be laid down on a slope with a stream down one side. Not least of its advantages is its ability to cope with outside effects. Thus, bird damage often comes in over one boundary, wind damage affects one side and so on. Such disturbing influences are associated with the first or last row or the first and last column and their effect is removed by the design. Similarly any differences caused by an access road through the experiment become associated with the rows or columns on either side. Experience shows that designs of this kind are often extremely effective.

Nevertheless, they do have their lapses. Essentially the method assumes that the fertility of any plot can be derived from adding a parameter for rows to one for columns. In general that may work well enough, but a diagonal strip of high or low fertility or a large patch represents so different a pattern as almost to be disastrous. The estimate of error is then very high, often without the treatment sum of squared deviations following suit, and the F-value can be less than one, sometimes to a surprising degree. As was said about block designs, the designer of a future experiment has to bear in mind the misfortunes of the past, so a general level of success may have to be discounted if there are occasional serious lapses. With row-and-column designs it can truly be said that they do usually control local variation very well. Unfortunately, when they fail they fail badly (Pearce, 1978, 1980).

Other objections to row-and-column designs are a reputation for being mathematically difficult and fears that they often have low efficiency. Both matters will be dealt with in Section 6.6. Also, of course, they do not lend themselves to dispersed experiments: all plots have to be at one site.

Actually, row-and-column designs have a number of other uses besides controlling local variation in two directions. For example, a laboratory experiment may be set up to study the germination of spores or the emergence of larvae in a range of conditions. If each determination is going to take a long time, the experimenter is faced with a situation in which there could be differences between assistants, because several people will be needed, between the apparatus they will use and between occasions, because each assistant will have to make a succession of determinations. He can give each assistant a piece of apparatus for the whole study, thus associating two sources of variation, but determinations can only be made serially and the best course thereafter is to think of the assistant/apparatus complex as the rows and occasions (days, half-days or whatever) as the columns. Latin squares have been used in this way since the eighteenth century (Palluel, 1788). Another such situation arises in long-term experiments when a first set of treatments has been concluded, each treatment having occurred once in each block, and it is desired to apply a second set, which is expected not to interact with the first set. The topic has quite a large literature (Hoblyn *et al.* 1954; Freeman, 1957a, 1957b; Pearce, 1976a).

In cold stores and similar places, the problem can arise of how to remove variation in three dimensions. The problems have been discussed by Preece *et al.* (1973), while Preece (1975, 1979) has published a bibliography.

Where an experiment has to be designed on a single line of plots, some have seen merit in using blocks, which are then regarded as rows, the columns being the position within the block. Thus, the Latin square,

$$\begin{matrix} A & B & D & C \\ C & D & B & A \\ D & A & C & B \\ B & C & A & D, \end{matrix}$$

gives this linear design,

Block I	Block II	Block III	Block IV
A B D C	C D B A	D A C B	B C A D.

There is, however, an implied assumption that the fertility trend is going to be the same in all blocks. If the line of plots ran down a steady slope that might be plausible, but usually the assumption is far-fetched. Incidentally, if such a regular trend can reasonably be assumed there could be advantage in using an ordinary block design with a covariance adjustment on the number of the plot in the line. More often linear experiments arise when for practical reasons all plots are placed on a single contour, e.g. a terrace. In that case it would be most dangerous to make any assumption at all about a line of plots that wound its way across the land, but it might be worth considering adjustment by neighbouring plots.

Examples 2

A. Cochran and Cox (1950) present the following data from a completely randomized design. The purpose of the experiment was to study the effect of applications of sulphur in controlling scab on potatoes. Dressings of 0, 300, 600 or 1200 pounds per acre were disked into the soil either in the spring or in the fall (or autumn for those who prefer it that way). Since a nil application comes to the same thing whenever it is made, that gave eight replications of Treatment O and four each of the other treatments, which will be designated S3, F3, S6, F6, S12, F12. From each plot 100 potatoes were taken at random and recorded according to the percentage of surface scabbed. Data represent the mean for each plot of those 100 determinations.

F3	O	S6	F12	S6	S12	S3	F6
9	12	18	10	24	17	30	16
O	S3	F12	F6	S3	O	O	S6
10	7	4	10	21	24	29	12
F3	S12	F6	O	F6	S12	F3	F12
9	7	18	30	18	16	16	4
S3	O	S12	S6	O	F12	O	F3
9	18	17	19	32	5	26	4

Note: The treatments have a special structure, which will be examined in Section 8.7.

Answer: It is first necessary to work out the replications and data totals for the various treatments, i.e.

O	8, 181		
S3	4,67	F3	4,38
S6	4,73	F6	4,62
S12	4,57	S12	4,23

That makes the treatment means respectively O, 22.63; S3, 16.75; F3, 9.50; S6, 18.25; F6, 15.50; S12, 14.25 and F12, 5.75. The 32 plots in all have a grand total of 501.

Next, three quantities are needed:

(i) $\mathbf{y}'\mathbf{y} = 9^2 + 12^2 + 18^2 + \cdots + 4^2 = 9939.00$. It will be seen that two more decimal places are being used than have arisen from the squaring of the data, assuming that the precision of recording is about right. Of course, with a computer the number of places carried in the calculations will look after itself.

Usually it is best to have three significant figures in the data, so in this instance the first place of decimals would have helped, though it is not essential.

(ii) $\mathbf{T}'\mathbf{r}^{-\delta}\mathbf{T} = 181^2/8 + (67^2 + 73^2 + 57^2 + 38^2 + 62^2 + 23^2)/4$
$= 8816.13$

(iii) $G^2/n = 501^2/32 = 7843.78$

It follows that the sum of squared deviations for error, $\mathbf{y}'\mathbf{y} - \mathbf{T}'\mathbf{r}^{-\delta}\mathbf{T}$, with $(32-7)$ degrees of freedom equals 1122.87, while that for the treatments, $\mathbf{T}'\mathbf{r}^{-\delta}\mathbf{T} - G^2/n$, is 972.35 with $(7-1)$ degrees of freedom. The analysis of variance is therefore

Source	d.f.	s.s.	m.s.	F
Treatments	6	972.35	162.06	3.61*
Error	25	1122.87	44.91	
Total	31	2095.22		

The asterisk against the value of F shows that the figure could have arisen solely by chance on less than one occasion in 20 ($P = 0.05$). (Tables show that the critical value is 2.49 and that has been exceeded.) Two asterisks would indicate significance at one in 100 ($P = 0.01$) and three one in 1000 ($P = 0.001$). In fact, the critical value for $P = 0.01$ is 3.63 so the differences between treatments are nearly significant at that level.

Of course, prior knowledge also should be taken into account. No one should forget what he already knows. Some, for example, may never have doubted that applications of sulphur would reduce scab infection. Others may refuse to believe it even now, preferring to accept a result that could have arisen purely by chance on only about one occasion in a hundred. Such people are apt to make disconcerting remarks about using statistics as a substitute for thought and they can be quite right, even though their complaints are pointedly ignored by the indoctrinated. A significance level measures the weight of evidence from one particular experiment. If other evidence is available, and mostly it is, that also needs to be taken into account before reaching a definite conclusion.

In any case, analyses of variance are not carried out only to facilitate significance tests. In many instances no one needs them. For example, few people will need to be convinced that foliar sprays containing nitrogen lead to darker leaf colour and increased growth. A more extreme example is the introduction of two standard varieties chosen to be as different as possible, the purpose being to see if some cultural treatment is effective generally. There is no point in establishing that the varieties differ; if they did not, they would not have been included. On the other hand, there can be a legitimate difference of opinion whether a certain sort of treatment, e.g., the incorporation of native sulphur in the soil, could affect some quantity like the incidence of a fungal disease, and the above significance test shows that there is quite strong evidence that it can do so.

Even if a significance test is required, it does not follow that the F-test is the best. It asks a general question, i.e. whether there are differences between the treatments; sometimes more specific questions are called for. For example, it would be reasonable to enquire if at the highest dressing it made any difference when it was applied, i.e. whether S12 differed from F12. Using the asterisk notation to represent treatment means, *S12 − *F12 = 8.50. Its estimated variance is

$$44.91\left(\tfrac{1}{4}+\tfrac{1}{4}\right) = 22.455$$

and its standard error $4.74 = (22.455)^{\frac{1}{2}}$. (The multiplier, $\frac{1}{4}+\frac{1}{4}$, arises from the covariance matrix of estimated treatment means being $\mathbf{r}^{-\delta}\hat{\sigma}^2$.) Looking up the value of the criterion, t, for $P = 0.05$ and 25 degrees of freedom, a value of 2.06 is found. Since 8.50 is less than $4.74t$, the difference between S12 and F12 is not significant at the level chosen, i.e. $P = 0.05$.

At this point someone may suggest that a seasonal difference is perhaps more likely to show up when the effect is less definite. He therefore wants to compare *S6 and *F6 in the same way, but such an impulse must be restrained. If an investigator indulges in 'data snooping' i.e. testing one comparison after another till he finds something that can be declared significant, he will almost certainly succeed eventually, even with a table of random numbers. He has appealed unto Caesar and unto Caesar he must go. He has declared that (*S12 − *F12) will tell him what he wants to know and he has had his answer.

More complicated tests exist, e.g. Duncan (1955), Scheffé (1959), that will take the treatments and group them according to their means. Given a set of treatments in which one contrast is as important or as interesting as any other, they may have their uses, but in a good experiment the treatments are highly structured, as they are here, and other methods must be preferred.

Quite apart from significance tests, the analysis of variance has shown how precisely the means are determined. Thus, *O, which is based on eight data, has a standard error of $(44.91/8)^{\frac{1}{2}} = 2.37$, while for each of the others the figure is $(44.91/4)^{\frac{1}{2}} = 3.35$. Such values, however, relate to the precision with which the means are determined within one field. No one should suppose that they show the variability that will be found if the results are applied to other fields, or, for that matter, in other seasons, though greater constancy can be expected of treatment differences. Thus, (*S12 − *F12) has been found to equal 8.50 and it may well be that that value will apply better in other conditions than the means themselves.

Finally, the error variance, 44.91, should itself be examined. It may show that an experiment is more variable than usual and may suggest improvements in technique for future use if something has gone wrong. In the present instance no evidence is available to show what can be expected.

B. Analyse the data of Example 2A fitting the fertility pattern with a paraboloid.

Answer: The plan forms a rectangle, four plots by eight. Assign $x_1 = -3$ to all plots in the first row, $x_1 = -1$ to all in the second and $x_1 = +1$ and $+3$ to all in the third and fourth row respectively. In that way a concomitant variable is formed that should be closely related to a trend across the rows. Similarly assign x_2 the values $-7, -5, -3, -1, +1, +3, +5, +7$ to plots in the eight columns. Then for each plot assign $x_3 = x_1^2$, $x_4 = x_2^2$ and $x_5 = x_1 x_2$. Of these variables, x_3 and x_4 are intended to allow for curvature in the trends and x_5 for an interaction between them.

Now work out $\mathbf{V}$. Its element, V_{ij}, in the ith row and the jth column will equal $\mathbf{x}_i' \boldsymbol{\psi} \mathbf{x}_j = \mathbf{x}_i' \mathbf{x}_j - \mathbf{T}_i \mathbf{r}^{-\delta} \mathbf{T}_j$, where $\mathbf{T}_i = \boldsymbol{\Delta} \mathbf{x}_i$. (That is to say, $\mathbf{T}_i$ is the vector of treatment totals of the ith variate.) Treatment totals for x_1, x_2, etc. are:

	O	S3	F3	S6	F6	S12	F12
x_1	+4	−2	2	−4	−2	+2	0
x_2	−4	−6	−2	+4	+4	−2	+6
x_3	40	20	20	28	12	20	20
x_4	160	100	172	60	60	52	68
x_5	16	−32	40	−4	−22	−20	22

Hence, for example

$$V_{11} = 8[(-3)^2 + (-1)^2 + (+1)^2 + (+3)^2] - 4^2/8 - (2^2 + 2^2 + 4^2 + 2^2 + 2^2 + 0^2)/4 = 160 - 10 = 150.$$

$$V_{12} = 0 - (4)(-4)/8 - [(-2)(-6) + 2(-2) + (-4)4 + (-2)4 + 2(-2) + 0(6)]/4 = 7$$

and so on.

It will be noted that for a rectangular experiment, $\mathbf{x}_1' \mathbf{x}_2 = 0$. In all

$$\mathbf{V} = \begin{bmatrix} 150 & 7 & 4 & -52 & -49 \\ 7 & 642 & 0 & 120 & -37 \\ 4 & 0 & 480 & 0 & -36 \\ -52 & 120 & 0 & 8136 & -964 \\ -49 & -37 & -36 & -964 & 2326 \end{bmatrix}.$$

If a computer has not been used up to the present it will be a great convenience now to effect the inversion

$$\mathbf{V}^{-1} = \begin{bmatrix} 6748 & -76 & -44 & 64 & 167 \\ -76 & 1563 & 2 & -22 & 14 \\ -44 & 2 & 2086 & 4 & 33 \\ 64 & -22 & 4 & 130 & 55 \\ 167 & 14 & 33 & 55 & 457 \end{bmatrix} \times 10^{-6}.$$

The next step is the calculation of $\mathbf{U}$, where $U_i = \mathbf{x}_i' \boldsymbol{\psi} \mathbf{y}$. Thus,

$$U_1 = (-3 \times 9) + (-3 \times 12) + \cdots + (+3 \times 26) + (+3 \times 4) - (4 \times 181)/8 - [(-2)67 + (2)38 + \cdots + (0)23]/4 = 17 - 90.5 + 90 = 16.5$$

The calculation would have been simplified by first working out row totals. In all

$$\mathbf{U} = \begin{bmatrix} 16.5 \\ 388.0 \\ 102.0 \\ -1649.0 \\ -340.5 \end{bmatrix} \quad \text{and} \quad \hat{\boldsymbol{\theta}} = \mathbf{V}^{-1}\mathbf{U} = \begin{bmatrix} -0.0850 \\ +0.6369 \\ +0.1950 \\ -0.2402 \\ -0.2348 \end{bmatrix}.$$

The error sum of squared deviations is shown by (2.3.11) to be

$$9939.00 - 8816.13 - 741.64 = 381.23$$

That gives $381.23/(25-5) = 19.06$ as the estimate of the error variance, i.e. $\hat{\sigma}^2$.

The time has now come to find the adjusted treatment means, but that requires fixing the standard values, **s**, to which the concomitant variates are to be adjusted. Since there should be no change in the general level of values, anything added to one treatment mean being subtracted from the others, the best procedure will be to adjust to the general means of the concomitant variables, i.e.

$$\mathbf{s} = (0,\ 0,\ 5,\ 21,\ 0)'$$

Starting with the treatment totals for x_1, x_2, etc. already given, changing them to means, and subtracting the appropriate value of **s**, gives the following for $\mathbf{H} = \mathbf{r}^{-\delta}\boldsymbol{\Delta}\mathbf{Z}$:

$$\mathbf{H} = \begin{bmatrix} 0.5 & -0.5 & 0.0 & -1.0 & 2.0 \\ -0.5 & -1.5 & 0.0 & 4.0 & -8.0 \\ 0.5 & -0.5 & 0.0 & 22.0 & 10.0 \\ -1.0 & 1.0 & 2.0 & -6.0 & -1.0 \\ -0.5 & 1.0 & -2.0 & -6.0 & -5.5 \\ 0.5 & -0.5 & 0.0 & -8.0 & -5.0 \\ 0.0 & 1.5 & 0.0 & -4.0 & 5.5 \end{bmatrix} \begin{matrix} \text{O} \\ \text{S3} \\ \text{F3} \\ \text{S6} \\ \text{F6} \\ \text{S12} \\ \text{F12} \end{matrix}$$

Then the adjusted treatment means are $\mathbf{r}^{-\delta}\mathbf{T} - \mathbf{H}\hat{\boldsymbol{\theta}}$. (The unadjusted values, i.e. $\mathbf{r}^{-\delta}\mathbf{T}$, are those given in Example 2A.) They are therefore evaluated as:

O,	23.15;		
S3,	16.98;	F3,	16.93;
S6,	15.57;	F6,	12.74;
S12,	11.75;	F12,	4.98.

Such means certainly make better sense than those before adjustment. In particular Treatments F3 and F6 have now fallen into place. Of course, no one should adopt a method of analysis just because it leads to confirmation of his prejudices, but it is hard to see how applying 300 pounds can be more effective than applying 600 pounds and that difficulty is removed.

The adjusted treatment means have the covariance matrix given by (2.3.12), namely $(\mathbf{r}^{-\delta} + \mathbf{H}\mathbf{V}^{-1}\mathbf{H}')\hat{\sigma}^2 =$

$$19.06 \begin{bmatrix} 0.1291 & -0.0081 & 0.0112 & -0.0052 & -0.0077 & -0.0023 & 0.0040 \\ -0.0081 & 0.2849 & -0.0314 & 0.0036 & 0.0181 & 0.0150 & -0.0239 \\ 0.0112 & -0.0311 & 0.3779 & -0.0313 & -0.0539 & -0.0488 & 0.0149 \\ -0.0052 & 0.0036 & -0.0313 & 0.2735 & 0.0065 & 0.0062 & 0.0020 \\ -0.0077 & 0.0181 & -0.0537 & 0.0065 & 0.2841 & 0.0191 & -0.0089 \\ -0.0023 & 0.0150 & -0.0488 & 0.0062 & 0.0191 & 0.2731 & -0.0100 \\ 0.0040 & -0.0239 & 0.0149 & 0.0020 & -0.0089 & -0.0100 & 0.2679 \end{bmatrix}.$$

To find the corresponding figures when treatments are merged with error, it is necessary to repeat the calculations using $\boldsymbol{\phi}$ instead of $\boldsymbol{\psi}$. However, since $\mathbf{y}'\boldsymbol{\phi}\mathbf{y}$, to take an example, equals $\mathbf{y}'\mathbf{y} - G^2/n$, where G is the grand total of the data, and since most of the concomitant variates sum to zero, the effort is much less. It appears that

$$\mathbf{V}_0 = \begin{bmatrix} 160 & 0 & 0 & 0 & 0 \\ 0 & 672 & 0 & 0 & 0 \\ 0 & 0 & 512 & 0 & 0 \\ 0 & 0 & 0 & 10752 & 0 \\ 0 & 0 & 0 & 0 & 3360 \end{bmatrix} \quad \text{and} \quad \mathbf{U}_0 = \begin{bmatrix} 17 \\ 319 \\ 124 \\ -2084 \\ -707 \end{bmatrix},$$

so $\mathbf{U}_0'\mathbf{V}_0^{-1}\mathbf{U}_0$ equals

$$\frac{17^2}{160} + \frac{319^2}{672} + \frac{124^2}{512} + \frac{2084^2}{10752} + \frac{707^2}{3360} = 735.96$$

and $\mathbf{y}'\mathbf{y} - G^2/n - \mathbf{U}_0'\mathbf{V}_0^{-1}\mathbf{U}_0$ is $2095.22 - 735.96 = 1359.26$ with 26 ($= 31 - 5$) degrees of freedom. The analysis of variance is therefore

Source	d.f.	s.s.	m.s.	F
Treatments	6	978.03	163.01	8.55***
Error	20	381.23	19.26	
Treatments + Error	26	1359.26		

The conclusions are now much clearer.

C. Analyse the data of Example 2A using adjustment by neighbouring plots. *Answer:* It is first necessary to find the extent to which each datum differs from the appropriate treatment mean, thus:

F3	O	S6	F12	S6	S12	S3	F6
−0.50	−10.63	−0.25	4.25	5.75	2.75	13.25	0.50
O	S3	F12	F6	S3	O	O	S6
−12.63	−9.75	−1.75	−5.50	4.25	1.37	6.37	−6.25
F3	S12	F6	O	F6	S12	F3	F12
−0.50	−7.25	2.50	7.37	2.50	1.75	6.50	−1.75
S3	O	S12	S6	O	F12	O	F3
−7.75	−4.63	2.75	0.75	9.37	−0.75	3.37	−5.50

Some might object that the deviations are not all of equal precision because those for Treatment O are derived from a mean of eight values and the rest from four. The point is valid but it is not yet known how seriously it is to be regarded. The deviations once obtained show that the area is far from uniform, since the negative deviations, which show high infectivity, form clusters, as do the positive ones.

The next step is to form a concomitant value for each plot, derived as a mean of deviations for neighbouring plots. There could be questions about the definition of neighbourliness. Here convention will be followed and two plots will be said to be neighbours if they adjoin along either an end or a side. Thus, an inside plot has four neighbours, an outside one has three unless it is on a corner, when it will have only two. The concomitant variable is therefore

F3	O	S6	F12	S6	S12	S3	F6
−11.63	−3.50	−2.71	0.00	3.75	6.79	3.21	3.50
O	S3	F12	F6	S3	O	O	S6
−3.58	−8.07	−3.25	3.53	1.03	3.78	3.72	1.71
F3	S12	F6	O	F6	S12	F3	F12
−9.21	−3.10	0.28	0.06	5.69	2.41	2.44	−1.75
S3	O	S12	S6	O	F12	O	F3
−2.57	−4.08	0.46	6.50	0.83	4.83	0.08	0.81

The figures so found, which will be written **x**, provide an even better indication of fertility patterns. They are used to adjust **y**, the actual data, by way of an analysis of covariance. Here V and U have each only one element, since there is only one concomitant variable, and it will be convenient in that case to write analyses of x^2 and xy to supplement that for y^2 already obtained in Example 2A. That for x^2 is obtained in exactly the same way as that for y^2. In the case of the figures for xy they are obtained by multiplying corresponding values for x and y instead of squaring the value of x or y as the case may be. Thus:

(i) $\mathbf{x'y} = (9 \times -11.63) + (12 \times -3.50) + \cdots + (4 \times 0.81)$
$= 499.09$

(ii) $\mathbf{x'x} = (-11.63)^2 + (-3.50)^2 + \cdots + (0.81)^2 = 602.76.$

Before going further it will be necessary to find the treatment totals and means for x. They are

O	−2.69	−0.336
S3	−6.40	−1.600
F3	−17.59	−4.398
S6	9.25	2.313
F6	13.00	3.250
S12	6.56	1.640
F12	−0.17	−0.043
	1.96	0.061 = General mean.

The vector of treatment totals will be written $\mathbf{T}_x$. Clearly, by ill chance the treatments have been allocated rather unevenly to the good and bad patches of the field. Treatment F6 in particular is on highly infested soil, whereas F3 has been fortunate. It would be a mistake to blame the designer of the experiment; if he did not know where areas of high and low infectivity were going to occur, he was right not to make any guesses but to randomize completely.

With totals known, the analysis proceeds thus:

(iii) $\mathbf{T}'_x\mathbf{r}^{-\delta}\mathbf{T} = (-2.69 \times 118)/8 + (-6.40 \times 7 - 17.59 \times 73 + \cdots$
$-0.17 \times 23)/4$
$= 127.65$

(iv) $\mathbf{T}'_x\mathbf{r}^{-\delta}\mathbf{T}_x = (-2.69)^2/8 + (6.40^2 + 17.59^2 + \cdots + 0.17^2)/4 = 162.90$

From these results it appears that

$$U = \mathbf{x'y} - \mathbf{T}'_x\mathbf{r}^{-\delta}\mathbf{T} = 371.44$$
$$V = \mathbf{x'x} - \mathbf{T}'_x\mathbf{r}^{-\delta}\mathbf{T}_x = 439.86$$

whence it appears that the regression coefficient of y on x is 371.44/439.86 = 0.845. That is to say, if the infectivity of a plot as shown by the performance of its neighbours exceeds expectation by one, it can itself be expected to be raised by 0.845.

By use of the regression coefficient these means can be adjusted. For example, Treatment O gave a mean x of −0.336, which differs by −0.336 − (0.061) = −0.397 from the general mean. Thinking of the infectivity as evened out over the area, that implies an increase of 0.397 × 0.845 = 0.34 in the treatment mean of y, which should therefore be 22.97 rather than 22.63.

Deviations for other treatments are S3, -1.661; F3, -4.459; S6, 2.252; F6, 3.189; S12, 1.579 and F12, -0.104. Those seven quantities will be called the vector $\mathbf{h}$. Similarly adjusted means for the other treatments are:

S3, 18.15; F3, 13.27;
S6, 16.35; F6, 12.81;
S12, 12.92; F12, 5.84.

Certainly the new figures are again more intelligible than the original, even if only because the rather puzzling results for F3 and F6 have again fallen into place.

It is now possible to construct the analysis of variance. The calculation begins by finding the sum of squared deviations assuming the null-hypothesis that there are no treatment effects. That requires G^2/n and its analogues. However, $(501)^2/32 = 7843.78$ has already been found. The others are $(501)(1.96)/32 = 30.69$ and $(1.96)^2/32 = 0.12$. From (2.3.7) $(\mathbf{r}^{-\delta} + \mathbf{hh}'/V)\hat{\sigma}^2$ is needed, i.e.

$$30.28\begin{bmatrix} 0.1254 & 0.0015 & 0.0040 & -0.0020 & -0.0029 & -0.0014 & 0.0001 \\ 0.0015 & 0.2563 & 0.0168 & -0.0085 & -0.0120 & -0.0060 & 0.0004 \\ 0.0040 & 0.0168 & 0.2952 & -0.0228 & -0.0323 & -0.0160 & 0.0011 \\ -0.0020 & -0.0085 & -0.0228 & 0.2615 & 0.0163 & -0.0081 & -0.0005 \\ -0.0029 & -0.0120 & -0.0323 & 0.0164 & 0.2731 & 0.0114 & -0.0008 \\ -0.0014 & -0.0060 & -0.0160 & 0.0081 & 0.0114 & 0.2557 & -0.0004 \\ 0.0001 & 0.0004 & 0.0011 & -0.0005 & -0.0008 & -0.0004 & 0.2500 \end{bmatrix}.$$

The value of $\hat{\sigma}^2$ was obtained from $\mathbf{y}'\mathbf{y} - \mathbf{T}'\mathbf{r}^{-\delta}\mathbf{T} - U^2/V$ with $(25-2)$ degrees of freedom. Although the off-diagonal elements are not very large and the others are not much inflated, the example gives warning of what could have happened if the elements of $\mathbf{h}$ had been larger relative to $\mathbf{V}$.

However, the analysis has in a sense contradicted itself. First, deviations were worked out using treatment means, 22.63, 16.75, etc. and they led to the conclusion that the values should have been taken rather as 22.97, 18.15, etc. The only remedy is to iterate, i.e. to repeat the calculation with the new values and to go on doing so until the treatment parameters used to find the deviations return themselves. To do that a computer is needed. In the present example everything settled down after four cycles to give a regression coefficient of 0.91 and adjusted treatment means, as follows:

O, 23.32
S3, 18.10; F3, 13.80;
S6, 15.95; F6, 12.49;
S12, 12.92; F12, 5.35.

D. An experimenter can use up to 24 plots. His object is to test if the response to increasing dosage is indeed linear over the range a to b. He seeks advice about the doses to study and the best replication for each.

Answer: The experimenter may think that he is asking a clear question, but he is not. Linearity is a definable property; departure from it may mean anything. If he is implying that the response may have a marked parabolic component and he is not interested in anything else, then, for reasons of symmetry it would be sensible to place r plots at doses of a and b, leaving $(24 - 2r)$ for $\frac{1}{2}(a+b)$. In that case, the variance of the contrast $(\frac{1}{2}, -1, \frac{1}{2})'$, which measures the deviation of the middle response from the mean of the other two, will be

$$\left(\frac{1}{2r} + \frac{1}{24-2r}\right)\sigma^2,$$

which has a minimum when $r = 6$, $24 - 2r = 12$.

E. The experimenter replies that such a design would not suit his purposes at all. If there is a departure from linearity it will be sigmoidal, the response falling initially below that given by a straight line and rising above it later. In the middle of the range, where the suggested design would be most effective, there may be little difference. What should be done in those circumstances?
Answer: He should have said all that before. One possibility is to think in terms of five points, a, $\frac{1}{4}(3a+b)$, $\frac{1}{2}(a+b)$, $\frac{1}{4}(a+3b)$, b. Preserving symmetry, the first lobe will be indicated by the deviation of the second point from the mean of the first and third, the second by the deviation of the fourth point from the mean of the third and fifth. Since these deviations will be of opposite sign they will be best aggregated by taking a difference, i.e. by studying $(\frac{1}{2}, -1, 0, 1, -\frac{1}{2})'$. It appears that the middle point, where he thinks he knows what will happen, is not going to be needed. Again for reasons of symmetry, r plots should be assigned to each outside point, leaving $(12 - r)$ for each of the two inside. That gives a variance of

$$\left(\frac{1}{2r} + \frac{2}{12-r}\right)\sigma^2,$$

which has a minimum when $r = 4$, $12 - r = 8$.

Comment: The two examples above have been included primarily as an indication of method. They serve also to show the impossibility of designing a good experiment when the purpose is vague. (It is true that one could assign six plots to each of the points, a, $\frac{1}{3}(2a+b)$, $\frac{1}{3}(a+2b)$ and b, and hope for the best, but that would be pusillanimous.)

F. Assuming Fairfield Smith's Law $(0 < b < 1)$, show that maximum information from a given area is obtained by using the smallest plot possible, whereas the greatest information from a given number of plots is obtained by using the largest possible.
Answer: The second part is trivial. For a given number of plots, the larger they are the less will σ^2 be. The first part is more difficult. Let the total area be R and let replications be arranged so that Treatment X is to be assigned an area, xR, and Treatment Y is to have yR. If plots are to have an area of α, that

means a replication of xR/α for X and yR/α for Y. The variance of their contrast will therefore be

$$\left(\frac{1}{x}+\frac{1}{y}\right)\frac{\alpha\sigma_\alpha^2}{R}$$

But $\sigma_\alpha^2 = \sigma^2/\alpha^b$, so the variance is

$$\left(\frac{1}{x}+\frac{1}{y}\right)\frac{\alpha^{1-b}\sigma^2}{R}$$

which is proportionate to α^{1-b}, where $1-b>0$.

Note: At a meeting of the Royal Statistical Society held on 8th December 1982 a paper was presented on the subject covered by Section 2.4. It was entitled 'Nearest neighbour (NN) analysis of field experiments' and was written by G. N. Wilkinson, S. R. Eckert, T. W. Hancock and O. Mayo. Both it and the extensive discussion that followed are to be published in the Society's journal, Volume B45. It appears that an enormous amount of work has been done on the subject since the publication of Bartlett (1978). Also, the accepted wisdom of (2.4.2) and (2.4.3) is slightly at fault on account of an algebraic slip.

Chapter 3

The general case of block designs

3.1 THE LEAST SQUARES SOLUTION FOR A BLOCK DESIGN

As was said in Section 2.5 quite the most usual way of designing an experiment on variable land is to use blocks. If there are b of them (2.1.1) is augmented to become

$$\mathbf{y} = \mathbf{D}'\boldsymbol{\beta} + \boldsymbol{\Delta}'\boldsymbol{\tau} + \boldsymbol{\eta}, \tag{3.1.1}$$

where $\mathbf{D}$ is the 'design matrix for blocks' analogous to $\boldsymbol{\Delta}$ for treatments, while $\boldsymbol{\beta}$ is a vector of block parameters. That is to say, $\mathbf{D}'\mathbf{1}_b = \mathbf{1}_n$, $\mathbf{D}\mathbf{1}_n = \mathbf{k}$, the vector of block sizes and $\mathbf{DD}' = \mathbf{k}^{\delta}$. In what follows no assumptions will be made as to the manner in which treatments have been assigned to blocks.

Simple as this augmentation appears, it does in fact give rise to a number of difficulties and it will be instructive to look at them in some detail. First, the expression (3.1.1) has too many parameters. If a constant is added to all elements of $\boldsymbol{\beta}$ and subtracted from all those of $\boldsymbol{\tau}$, the model is unchanged in essentials. Several solutions present themselves. One is to use fewer parameters. Those for treatments are better left alone because it is contrasts between treatments that are under study, but those for blocks can easily be amended to reduce their number. For example, they could be transformed to

$$\gamma_1 = \beta_1 - \beta_2$$
$$\gamma_2 = \beta_1 + \beta_2 - 2\beta_3, \text{ etc.}$$

Alternatively, constraints could be imposed. Because the sum of the data, $\mathbf{1}_n'\mathbf{y}$, equals G, an obvious possibility is to write $\mathbf{k}'\boldsymbol{\beta} = G$. Since

$$\begin{aligned} G = \mathbf{1}_n'\mathbf{y} &= \mathbf{1}_n'\mathbf{D}'\boldsymbol{\beta} + \mathbf{1}_n'\boldsymbol{\Delta}'\boldsymbol{\tau} + \mathbf{1}_n'\boldsymbol{\eta} \\ &= \mathbf{k}'\boldsymbol{\beta} + \mathbf{r}'\boldsymbol{\tau} + \mathbf{1}_n'\boldsymbol{\eta} \end{aligned}$$

the expectation of $\mathbf{r}'\boldsymbol{\tau} = 0$, i.e. the treatment parameters are to be such that the overall effect of treatments, weighting by the frequency with which each one occurs, is to be zero. Although these possibilities are quite reasonable the course chosen here is to disregard the over-parameterization and to proceed until difficulties actually arise. Although that may seem rather dangerous, it is justified by the insight it gives as to what can be estimated and what cannot.

Apart from the over-parameterization, the model (3.1.1) is open to criticism on another ground. It assumes that the block and treatment parameters can just be added, as if two treatments had the same relative effect whatever the environmental conditions represented by the block parameter. If blocks really have been chosen to be different, it is scarcely conceivable that that should be so. The best fertilizer treatment where soil is rich and deep may well be inferior where it is poor and shallow, while a variety that does well in exposed conditions at the top of a slope may be relatively less good at the bottom where there is more shelter and its rivals thrive. It is sometimes asserted that the assumption of no interaction between treatments and blocks is well confirmed by experience but convincing evidence is rarely presented.

Returning to the mathematics, it may be noted that special importance should be attached to the 'incidence matrix', $\mathbf{N}$, i.e.

$$\mathbf{N} = \mathbf{\Delta D}'. \tag{3.1.2}$$

It has v rows and b columns and its elements show the number of times each treatment occurs in each block. Accordingly it provides a succinct description of the design. Also, $\mathbf{N1}_b = \mathbf{r}$ and $\mathbf{N}'\mathbf{1}_v = \mathbf{k}$.

To proceed with (3.1.1) unmodified, it follows that

$$\boldsymbol{\eta} = \mathbf{y} - \mathbf{D}'\boldsymbol{\beta} - \mathbf{\Delta}'\boldsymbol{\tau}.$$

Hence, the sum of squares of the residuals, i.e. $\boldsymbol{\eta}'\boldsymbol{\eta}$, may be written

$$\boldsymbol{\eta}'\boldsymbol{\eta} = \mathbf{y}'\mathbf{y} + \boldsymbol{\beta}'\mathbf{k}^{\delta}\boldsymbol{\beta} + \boldsymbol{\tau}'\mathbf{r}^{\delta}\boldsymbol{\tau} - 2\mathbf{B}'\boldsymbol{\beta} - 2\mathbf{T}'\boldsymbol{\tau} + 2\boldsymbol{\beta}'\mathbf{N}'\boldsymbol{\tau}.$$

That was obtained by using the values of $\mathbf{DD}'$, $\mathbf{\Delta D}'$ and $\mathbf{\Delta\Delta}'$ already given and writing the block totals, i.e. $\mathbf{Dy}$ as $\mathbf{B}$ and the treatment totals $\mathbf{\Delta y}$, as $\mathbf{T}$. Now differentiating $\boldsymbol{\eta}'\boldsymbol{\eta}$ first by $\boldsymbol{\beta}$ and then by $\boldsymbol{\tau}$ gives the so-called 'normal equations'

$$\mathbf{B} = \mathbf{k}^{\delta}\hat{\boldsymbol{\beta}} + \mathbf{N}'\hat{\boldsymbol{\tau}} \tag{3.1.3a),}$$

$$\mathbf{T} = \mathbf{N}\hat{\boldsymbol{\beta}} + \mathbf{r}^{\delta}\hat{\boldsymbol{\tau}} \tag{3.1.3b}$$

A simple solution is afforded by noting that

$$\begin{aligned} \mathbf{Q} &= \mathbf{T} - \mathbf{Nk}^{-\delta}\mathbf{B} \\ &= (\mathbf{N}\hat{\boldsymbol{\beta}} + \mathbf{r}^{\delta}\hat{\boldsymbol{\tau}}) - \mathbf{Nk}^{-\delta}(\mathbf{k}^{\delta}\hat{\boldsymbol{\beta}} + \mathbf{N}'\hat{\boldsymbol{\tau}}) \\ &= (\mathbf{r}^{\delta} - \mathbf{Nk}^{-\delta}\mathbf{N}')\hat{\boldsymbol{\tau}} = \mathbf{C}\boldsymbol{\tau}, \text{ say,} \end{aligned} \tag{3.1.4}$$

where $\mathbf{C} = \mathbf{r}^{\delta} - \mathbf{Nk}^{-\delta}\mathbf{N}'$. In a sense (3.1.4) is the crux of the whole argument, both in its usefulness and in the difficulties it raises.

At first sight (3.1.4) gives a ready solution for $\hat{\boldsymbol{\tau}}$, namely $\hat{\boldsymbol{\tau}} = \mathbf{C}^{-1}\mathbf{Q}$. Examination shows that solution to be spurious because $\mathbf{C}$ is singular, i.e.

$$\begin{aligned} \mathbf{Cl}_v &= \mathbf{r}^{\delta}\mathbf{1}_v - \mathbf{Nk}^{-\delta}\mathbf{N}'\mathbf{1}_v \\ &= \mathbf{r} - \mathbf{Nk}^{-\delta}\mathbf{k} = \mathbf{r} - \mathbf{N1}_b = \mathbf{0} \end{aligned}$$

Suppose, however, that some solution exists, i.e. $\hat{\boldsymbol{\tau}} = \mathbf{XQ}$, where $\mathbf{X}$ is some matrix, then

$$\mathbf{Q} = \mathbf{C}\hat{\boldsymbol{\tau}} = \mathbf{CXQ} = \mathbf{CXC}\hat{\boldsymbol{\tau}}$$

It follows that $\mathbf{X}$ must be a generalized inverse of $\mathbf{C}$, i.e. $\mathbf{CXC} = \mathbf{C}$. The argument will be continued in Section 3.3, but first it will be shown in Section 3.2 that for any $\mathbf{C}$ an infinitude of solutions exists for $\mathbf{X}$ but that they have certain characteristics in common. By basing further work on those invariant properties it will be possible to find a truly general solution.

Before doing that, however, it will be useful to look further at the quantities involved in (3.1.4). To start with $\mathbf{Q}$, it is called the vector of 'adjusted treatment totals'. A glance at its structure will show why. The first term, $\mathbf{T}$, represents unadjusted treatment totals. The block means are given by $\mathbf{k}^{-\delta}\mathbf{B}$ so the effect of subtracting $\mathbf{Nk}^{-\delta}\mathbf{B}$ is to remove from an element of $\mathbf{T}$ the block means as often as the treatment occurs in each. Further, the elements of $\mathbf{Q}$ sum to zero, because

$$\begin{aligned} \mathbf{1}_v'\mathbf{Q} &= \mathbf{1}_v'(\mathbf{T} - \mathbf{Nk}^{-\delta}\mathbf{B}) = \mathbf{1}_v'\mathbf{T} - \mathbf{k}'\mathbf{k}^{-\delta}\mathbf{B} \\ &= \mathbf{1}_v'\mathbf{T} - \mathbf{1}_b'\mathbf{B} = G - G = 0, \end{aligned} \qquad (3.1.5)$$

The matrix, $\mathbf{C}$, introduced by Chakrabati (1962), is here called the 'coefficient matrix'. Just as $\mathbf{Q}$ sums up the data, so $\mathbf{C}$ sums up the design. Actually $\mathbf{Q}$ and $\mathbf{C}$ are closely related. Writing

$$\boldsymbol{\phi} = \mathbf{I} - \mathbf{D}'\mathbf{k}^{-\delta}\mathbf{D}, \qquad (3.1.6)$$

then

$$\mathbf{C} = \boldsymbol{\Delta}\boldsymbol{\phi}\boldsymbol{\Delta}' \qquad (3.1.7)$$

$$\mathbf{Q} = \boldsymbol{\Delta}\boldsymbol{\phi}\mathbf{y} \qquad (3.1.8)$$

The $\boldsymbol{\phi}$ so defined bears the same relationship to a block design that the matrix in (2.1.7) bears to one that is completely randomized. That is to say, if $\boldsymbol{\eta}'\boldsymbol{\eta}$ is estimated under the null-hypothesis, the result is $\mathbf{y}'\boldsymbol{\phi}\mathbf{y}$. It will be seen that

$$\boldsymbol{\phi} = \boldsymbol{\phi}', \qquad \boldsymbol{\phi}\mathbf{1}_n = \mathbf{0}, \qquad \boldsymbol{\phi}\boldsymbol{\phi} = \boldsymbol{\phi} \qquad (3.1.9)$$

as before. Also

$$\boldsymbol{\phi}\mathbf{D}' = \mathbf{D}' - \mathbf{D}'\mathbf{k}^{-\delta}\mathbf{k}^{\delta} = \mathbf{0} \qquad (3.1.10)$$

Before going further, it may be helpful to give an example of the calculation of $\mathbf{C}$ and to do so for a design that may fairly be described as 'general', the word in this instance implying that it is without pattern. (It is therefore unlikely to arise in practice.) Suppose that an experimenter had three treatments A, B and C, disposed in five blocks of unequal size, thus:

Block	I	A	B	B		
	II	A	A	B	C	C
	III	A	B	B	C	
	IV	A	B	C	C	
	V	A	A	B	B	C

Then the incidence matrix, **N**, shows how treatments are applied to blocks, namely

$$\mathbf{N} = \begin{bmatrix} 1 & 2 & 1 & 1 & 2 \\ 2 & 1 & 2 & 1 & 2 \\ 0 & 2 & 1 & 2 & 1 \end{bmatrix}.$$

The row-totals of **N** give the replications ($\mathbf{r} = \mathbf{N1}_n$), i.e., 7, 8, 6. To find **C** it is required first to write **r** as a diagonal matrix ($\mathbf{r}^\delta$) and then to subtract

$$\tfrac{1}{3}\mathbf{U}_3\mathbf{U}_3' + \tfrac{1}{4}\mathbf{U}_4\mathbf{U}_4' + \tfrac{1}{5}\mathbf{U}_5\mathbf{U}_5',$$

where $\mathbf{U}_j$ holds those columns of **N** that sum to j, i.e.

$$\mathbf{U}_3 = \begin{bmatrix} 1 \\ 2 \\ 0 \end{bmatrix}, \quad \mathbf{U}_4 = \begin{bmatrix} 1 & 1 \\ 2 & 1 \\ 1 & 2 \end{bmatrix}, \quad \mathbf{U}_5 = \begin{bmatrix} 2 & 2 \\ 1 & 2 \\ 2 & 1 \end{bmatrix}.$$

Hence

$$\mathbf{C} = \begin{bmatrix} 7 & 0 & 0 \\ 0 & 8 & 0 \\ 0 & 0 & 6 \end{bmatrix} - \tfrac{1}{3}\begin{bmatrix} 1 & 2 & 0 \\ 2 & 4 & 0 \\ 0 & 0 & 0 \end{bmatrix} - \tfrac{1}{4}\begin{bmatrix} 2 & 3 & 3 \\ 3 & 5 & 4 \\ 3 & 4 & 5 \end{bmatrix} - \tfrac{1}{5}\begin{bmatrix} 8 & 6 & 6 \\ 6 & 5 & 4 \\ 6 & 4 & 5 \end{bmatrix}$$

$$= \tfrac{1}{60}\begin{bmatrix} 274 & -157 & -117 \\ -157 & 265 & -108 \\ -117 & -108 & 225 \end{bmatrix}.$$

3.2 A DIGRESSION ABOUT GENERALIZED INVERSES

Although there are good books on generalized inverses it will be convenient to summarize here the main results to be used later, especially as only symmetric matrices will be used, thus allowing for simplification. A matrix, $\mathbf{S}^-$, will be said to be a generalized inverse of **S** if

$$\mathbf{SS}^-\mathbf{S} = \mathbf{S} \tag{3.2.1}$$

If **S** is non-singular, the only solution for $\mathbf{S}^-$ is $\mathbf{S}^{-1}$; interest attaches rather to the case when **S** is singular. It will be assumed that $\mathbf{S} = \mathbf{S}'$.

Lemma 3.2.A *If* **S** *is singular, it has infinitely many generalized inverses.*

Method: Let h be the rank of **S** and let v be its order, then there are $(v-h)$ orthogonal eigenvectors, $\mathbf{c}_i$ $(1 < h < i \leqslant v)$, such that $\mathbf{Sc}_i = \mathbf{0}$, $\mathbf{c}_i'\mathbf{c}_i = 1$ and $\mathbf{c}_i'\mathbf{c}_j = 0$ $(j \neq i)$. Let $\mathbf{p}_i$ be any vector with v elements. Then it can be shown that

$$\mathbf{X} = \left(\mathbf{S} + \sum_{i=h+1}^{v} \mathbf{p}_i\mathbf{c}_i'\right)^{-1}$$

is a generalized inverse of **S** regardless of the choice of the $\mathbf{p}_i$.

Proof: $\mathbf{S} = \mathbf{X}^{-1} - \sum_i \mathbf{p}_i \mathbf{c}_i'$. Hence, if $h < j \leqslant v$,

$$\mathbf{0} = \mathbf{S}\mathbf{c}_j = \mathbf{X}^{-1}\mathbf{c}_j - \mathbf{p}_j,$$

so
$$\mathbf{p}_j = \mathbf{X}^{-1}\mathbf{c}_j$$

and
$$\mathbf{S} = \mathbf{X}^{-1}\left(\mathbf{I} - \sum_i \mathbf{c}_i \mathbf{c}_i'\right).$$

Hence
$$\mathbf{S}(\mathbf{XS}) = \mathbf{S}\left(\mathbf{I} - \sum_i \mathbf{c}_i \mathbf{c}_i'\right) = \mathbf{S}. \quad \blacksquare$$

It will be noticed that the generalized inverse of a symmetric matrix is not necessarily itself symmetric. Also, there could be other solutions beside those just found.

Lemma 3.2.B *If* $\mathbf{X}$ *is a generalized inverse of* $\mathbf{S}$, *so also is* $\mathbf{X}'$.

Proof: $\mathbf{S} = \mathbf{SXS}$. Transposing both sides gives $\mathbf{S} = \mathbf{SX'S}$. $\blacksquare$

Lemma 3.2.C *If* $\mathbf{Sc} = \lambda\mathbf{c}\,(\lambda \neq 0)$, *then* $\mathbf{SS}^-\mathbf{c} = \mathbf{c}$.

Proof: $\mathbf{c} = \mathbf{Sc}/\lambda$. Hence $\mathbf{SS}^-\mathbf{c} = \mathbf{SS}^-\mathbf{Sc}/\lambda = \mathbf{Sc}/\lambda = \mathbf{c}$. $\blacksquare$

Lemma 3.2.D *Regardless of the choice of* $\mathbf{C}^-$, $\mathbf{CC}^-\boldsymbol{\Delta\phi} = \boldsymbol{\Delta\phi}$.

Proof: Let $\mathbf{E} = \mathbf{CC}^-\boldsymbol{\Delta\phi} - \boldsymbol{\Delta\phi}$, then from (3.1.9) and (3.1.7)

$$\mathbf{EE}' = \mathbf{CC}^-\mathbf{CC}^-\mathbf{C} - \mathbf{CC}^-\mathbf{C} - \mathbf{C}(\mathbf{C}^-)'\mathbf{C} + \mathbf{C} = \mathbf{0},$$

so $\mathbf{E}$ must be a null-matrix. $\blacksquare$

Lemma 3.2.E *Regardless of the choice of* $\mathbf{C}^-$, $\boldsymbol{\phi\Delta}'\mathbf{C}^-\boldsymbol{\Delta\phi}$ *is invariant.*

Proof: Let $\mathbf{X}_1$ and $\mathbf{X}_2'$ be any two generalized inverses of $\mathbf{C}$; then from the last lemma

$$\boldsymbol{\phi\Delta}'\mathbf{X}_1\boldsymbol{\Delta\phi} = (\boldsymbol{\phi\Delta}'\mathbf{X}_2'\mathbf{C})\mathbf{X}_1\boldsymbol{\Delta\phi} = \boldsymbol{\phi\Delta}'\mathbf{X}_2'(\mathbf{CX}_1\boldsymbol{\Delta\phi}) = \boldsymbol{\phi\Delta}'\mathbf{X}_2'\boldsymbol{\Delta\phi}. \quad \blacksquare$$

Lemma 3.2.F *If* $\mathbf{Sc}_1 = \lambda_1\mathbf{c}_1$ *and* $\mathbf{Sc}_2 = \lambda_2\mathbf{c}_2$, *neither* λ_1 *nor* λ_2 *being zero, then* $\mathbf{c}_1'\mathbf{S}^-\mathbf{c}_2$ *is invariant, regardless of the choice of* $\mathbf{S}^-$.

Proof: From Lemma 3.2.C if $\mathbf{X}_1$ and $\mathbf{X}_2'$ are two generalized inverses of $\mathbf{S}$, then

$$\mathbf{c}_1'\mathbf{X}_1\mathbf{c}_2 = (\mathbf{c}_1'\mathbf{X}_2'\mathbf{S})\mathbf{X}_1\mathbf{c}_2 = \mathbf{c}_1'\mathbf{X}_2'(\mathbf{SX}_1\mathbf{c}_2) = \mathbf{c}_1'\mathbf{X}_2'\mathbf{c}_2. \quad \blacksquare$$

Lemma 3.2.G *The trace of* $\mathbf{CC}^-$ *equals the number of non-zero eigenvalues of* $\mathbf{C}$, *i.e.* h.

Proof: The rank of a product of matrices cannot exceed the rank of either of the matrices multiplied, so $\mathbf{CC}^-$ cannot have more than h non-zero eigenvalues. From Lemma 3.2.C it does have h, all equal to unity. Their sum equals its trace. $\blacksquare$

In what follows it will be helpful to bear in mind also that

$$\operatorname{tr}(\boldsymbol{\phi\Delta}'\mathbf{C}^-\boldsymbol{\Delta\phi}) = \operatorname{tr}(\mathbf{CC}^-) = h, \tag{3.2.2}$$

from Lemma 2.1.A and (3.1.6).

3.3 CONTINUATION OF THE LEAST SQUARES SOLUTION

Now that generalized inverses have been considered, it is possible to examine further the equation

$$\hat{\boldsymbol{\tau}} = \mathbf{C}^{-}\mathbf{Q}. \tag{3.3.1}$$

It is known that $\mathbf{C}$ has at least one zero eigenvalue because $\mathbf{C}\mathbf{1}_v = \mathbf{0}$, but it may have several. Using (3.1.8), (3.1.1), (3.1.10) and (3.1.7)

$$\begin{aligned}\hat{\boldsymbol{\tau}} &= \mathbf{C}^{-}\boldsymbol{\Delta\phi}\mathbf{y} = \mathbf{C}^{-}\boldsymbol{\Delta\phi}(\mathbf{D}'\boldsymbol{\beta}+\boldsymbol{\Delta}'\boldsymbol{\tau}+\boldsymbol{\eta})\\ &= \mathbf{C}^{-}\mathbf{C}\boldsymbol{\tau}+\mathbf{C}^{-}\boldsymbol{\Delta\phi\eta}.\end{aligned} \tag{3.3.2}$$

Hence $\hat{\boldsymbol{\tau}}$ is not necessarily an estimator of $\boldsymbol{\tau}$, though in view of the over-parameterization that comes as no surprise. Since, however, only contrasts are to be estimated, as was explained in Section 1.6, that is not necessarily disastrous. It is important rather to know about $\mathbf{c}'\hat{\boldsymbol{\tau}}$, where $\mathbf{c}$ represents some contrast of interest. First, let it be written in terms of the eigenvectors, $\mathbf{u}_i$, of $\mathbf{C}$; then if

$$\mathbf{c} = \sum_i (\alpha_i \mathbf{u}_i), \tag{3.3.3}$$

from (3.3.2)

$$\begin{aligned}\mathbf{c}'\hat{\boldsymbol{\tau}} &= \mathbf{c}'\mathbf{C}^{-}\mathbf{C}\boldsymbol{\tau}+\mathbf{c}'\mathbf{C}^{-}\boldsymbol{\Delta\phi\eta}\\ &= \mathbf{c}'\boldsymbol{\tau}+\mathbf{c}'\mathbf{C}^{-}\boldsymbol{\Delta\phi\eta}\end{aligned} \tag{3.3.4}$$

provided all the $\mathbf{u}_i$ in (3.3.3) have non-zero eigenvalues (Lemma 3.2.C). In short, contrasts can be estimated without bias if they can be derived from eigenvectors not involved in the singularity of $\mathbf{C}$. Further, the variance of such a contrast is given by

$$\begin{aligned}&(\mathbf{c}'\mathbf{C}^{-}\boldsymbol{\Delta\phi})\,\mathrm{Cov}\,(\boldsymbol{\eta})\,(\mathbf{c}'\mathbf{C}^{-}\boldsymbol{\Delta\phi})'\\ &= \mathbf{c}'\mathbf{C}^{-}\boldsymbol{\Delta\phi\phi\Delta}'\,\mathbf{C}^{-}\mathbf{c}\sigma^2\end{aligned}$$

because Cov $(\eta) = \mathbf{I}_n\sigma^2$. Hence, from (3.1.9), (3.1.7) and Lemma 3.2.C the variance is

$$\mathbf{c}'\mathbf{C}^{-}\mathbf{c}\sigma^2. \tag{3.3.5}$$

Further, from (3.3.3) and Lemma 3.2.F that quantity does not depend upon the choice of $\mathbf{C}^{-}$ but is invariant. In short, both the contrast and its variance have been found without ambiguity.

Such contrasts, i.e. those that can be obtained as a linear combination of the eigenvectors of $\mathbf{C}$, avoiding those that have a zero eigenvalue, are said to be 'estimable'. The term is unfortunate because in ordinary English the word implies that such contrasts are worthy of esteem, which is not necessarily so. For that reason, the term will be avoided here. Other contrasts are said to be 'confounded', but more of that in Section 3.7.

Now that $\hat{\boldsymbol{\tau}}$ can be found, albeit by an arbitrary choice of $\mathbf{C}^{-}$ in (3.3.1), it becomes possible to estimate $\boldsymbol{\beta}$ also, because from (3.1.3a)

$$\begin{aligned}\hat{\boldsymbol{\beta}} &= \mathbf{k}^{-\delta}\mathbf{B}-\mathbf{k}^{-\delta}\mathbf{N}'\mathbf{C}^{-}\mathbf{Q}\\ &= (\mathbf{k}^{-\delta}\mathbf{D}-\mathbf{k}^{-\delta}\mathbf{N}'\mathbf{C}^{-}\boldsymbol{\Delta\phi})\mathbf{y}.\end{aligned} \tag{3.3.6}$$

From (3.1.1)

$$\begin{aligned}\hat{\boldsymbol{\eta}} &= \mathbf{y} - \mathbf{D}'\hat{\boldsymbol{\beta}} - \boldsymbol{\Delta}'\hat{\boldsymbol{\tau}} \\ &= (\mathbf{I} - \mathbf{D}'\mathbf{k}^{-\delta}\mathbf{D} + \mathbf{D}'\mathbf{k}^{-\delta}\mathbf{N}'\mathbf{C}^{-}\boldsymbol{\Delta}\boldsymbol{\phi} - \boldsymbol{\Delta}'\mathbf{C}^{-}\boldsymbol{\Delta}\boldsymbol{\phi})\mathbf{y} \\ &= (\boldsymbol{\phi} - \boldsymbol{\phi}\boldsymbol{\Delta}'\mathbf{C}^{-}\boldsymbol{\Delta}\boldsymbol{\phi})\mathbf{y} \\ &= \boldsymbol{\psi}\mathbf{y}, \text{ say,} \end{aligned} \tag{3.3.7}$$

where $\boldsymbol{\phi}$ is the matrix defined in (3.1.6), i.e. $\mathbf{I}_n - \mathbf{D}'\mathbf{k}^{-\delta}\mathbf{D}$. Here then is another residual matrix, $\boldsymbol{\psi}$. It has been written with that symbol to point the analogy with (2.1.3). Note that $\boldsymbol{\psi}\boldsymbol{\phi} = \boldsymbol{\psi}$, $\boldsymbol{\psi}\mathbf{D}' = \mathbf{0} = \boldsymbol{\psi}\boldsymbol{\Delta}'$ as well as the usual relationships, $\boldsymbol{\psi} = \boldsymbol{\psi}'$, $\boldsymbol{\psi}\mathbf{1}_n = \mathbf{0}$ and $\boldsymbol{\psi}\boldsymbol{\psi} = \boldsymbol{\psi}$. Hence, from (3.3.7)

$$\mathbf{D}'\hat{\boldsymbol{\eta}} = \mathbf{D}'\boldsymbol{\psi}\mathbf{y} = \mathbf{0}, \qquad \boldsymbol{\Delta}'\hat{\boldsymbol{\eta}} = \boldsymbol{\Delta}'\boldsymbol{\psi}\boldsymbol{\eta} = \mathbf{0}. \tag{3.3.8}$$

That is to say, the residuals found in (3.3.7) sum to zero over any block or over any treatment. Also, from Lemma 3.2.E, $\boldsymbol{\psi}$ does not depend upon the choice of $\mathbf{C}^{-}$.

The most important conclusion to be derived from (3.3.7) is, however, the knowledge of the error sum of squared deviations, because

$$\hat{\boldsymbol{\eta}}'\hat{\boldsymbol{\eta}} = \mathbf{y}'\boldsymbol{\psi}\mathbf{y} = \mathbf{y}'\boldsymbol{\phi}\mathbf{y} - \mathbf{Q}'\mathbf{C}^{-}\mathbf{Q}. \tag{3.3.9}$$

On the null-hypothesis

$$\mathbf{y} = \mathbf{D}'\boldsymbol{\beta} + \boldsymbol{\eta} \tag{3.3.10}$$

and the corresponding value is $\mathbf{y}'\boldsymbol{\phi}\mathbf{y}$. (That is why the new $\boldsymbol{\phi}$ at its introduction was referred to as a residuals matrix.) The difference between $\mathbf{y}'\boldsymbol{\phi}\mathbf{y}$ and $\mathbf{y}'\boldsymbol{\psi}\mathbf{y}$ must be the sum of squared deviations for treatments, i.e.

$$\mathbf{Q}'\mathbf{C}^{-}\mathbf{Q}. \tag{3.3.11}$$

On account of (3.1.8) and Lemma 3.2.E, that quantity is invariant.

So far the question of degrees of freedom has been avoided because it is best resolved by considering the randomization process. Ordinarily in a block design a set of treatments is allotted to a block and then its members are assigned at random to the plots within that block. By analogy with the argument presented between (2.1.8) and (2.1.11), the expectation under randomization of any $\hat{\eta}^2$ is $(k_i - 1)\sigma^2/k_i$, where k_i is the number of plots in the block. The expectation of the product of two estimated residuals is similarly $-\sigma^2/k_i$ if the two come from the same block. If they do not, it is zero. In fact, as before ex $(\hat{\boldsymbol{\eta}}\hat{\boldsymbol{\eta}}') = \boldsymbol{\phi}\theta$, adopting the redefinition of $\boldsymbol{\phi}$ made at (3.1.6). Hence, as before, the expectation of the sum of squared deviations for error under randomization will be tr $(\boldsymbol{\psi})\sigma^2$, while that for treatments will be

$$\begin{aligned}[\text{tr}\,(\boldsymbol{\phi}) - \text{tr}\,(\boldsymbol{\psi})]\sigma^2 &= \text{tr}\,(\boldsymbol{\phi}\boldsymbol{\Delta}\mathbf{C}^{-}\boldsymbol{\Delta}'\boldsymbol{\phi})\sigma^2 \\ &= \text{tr}\,(\mathbf{C}^{-}\boldsymbol{\Delta}\boldsymbol{\phi}\boldsymbol{\Delta}')\sigma^2 = \text{tr}\,(\mathbf{C}^{-}\mathbf{C})\sigma^2 = h\sigma^2, \end{aligned} \tag{3.3.12}$$

where h is the number of non-zero eigenvalues of $\mathbf{C}$. (That result follows from Lemma 3.2.G.) In other words, h is the number of independent unconfounded contrasts that can be estimated. To continue, the trace of $\boldsymbol{\phi}$ causes little trouble, because each of the k_i diagonal elements of $\boldsymbol{\phi}$ relating to block i has the value $(k_i - 1)/k_i$, making $(k_i - 1)$ in all from that sub-matrix. Hence

$$\mathrm{tr}\,(\boldsymbol{\phi}) = n - b.$$

Although the line for blocks is not really needed, it is usually included for the sake of completeness. Actually it could be misleading because its sum of squared deviations has not been obtained by any process of minimization and it could well be too large, so no one should use it for an F-test or anything similar.

At the end of this chapter Example 3A shows the complete calculations for data from an actual experiment.

3.4 CALCULATING A GENERALIZED INVERSE

It appears then that a general solution is always possible, whatever the design. First, it is necessary to calculate **Q** to sum up the data and **C** to sum up the design. Then, it is required to find one of the generalized inverses of **C**, namely $\mathbf{C}^-$, and everything else follows. At this point the reader, though possibly encouraged by the knowledge that an infinitude of such inverses exists, could reasonably ask how he is to find one of them. In this section three solutions will be proffered and illustrated using the following design, which has six blocks (I–VI) each with four plots, and six treatments (A–F) each with four-fold replication, namely

Block	I	A	B	C	D
	II	A	B	C	E
	III	A	B	C	F
	IV	A	D	E	F
	V	B	D	E	F
	VI	C	D	E	F

The design has two properties that are usual without being essential. For one thing, it is 'proper', i.e. all its blocks are of the same size. (It is not clear why adapting block size to the nature of the materials should be regarded as 'improper'.) For another, it is 'equireplicate', i.e. all the treatments are applied to the same number of plots. For reasons already given, that is not necessarily a good thing, though it can be.

The incidence matrix, **N**, has six rows, one for each treatment and six columns, one for each block:

$$\mathbf{N} = \begin{bmatrix} 1 & 1 & 1 & 1 & 0 & 0 \\ 1 & 1 & 1 & 0 & 1 & 0 \\ 1 & 1 & 1 & 0 & 0 & 1 \\ 1 & 0 & 0 & 1 & 1 & 1 \\ 0 & 1 & 0 & 1 & 1 & 1 \\ 0 & 0 & 1 & 1 & 1 & 1 \end{bmatrix}.$$

If T_i is the sum of the four data from Treatment i and b_j is the mean of the four data from Block j, then, since $\mathbf{Q} = \mathbf{T} - \mathbf{N}\mathbf{k}^{-\delta}\mathbf{B} = \mathbf{T} - \mathbf{N}\mathbf{b}$,

$$\begin{aligned}
Q_1 &= T_1 - b_1 - b_2 - b_3 - b_4 \\
Q_2 &= T_2 - b_1 - b_2 - b_3 - b_5 \\
Q_3 &= T_3 - b_1 - b_2 - b_3 - b_6 \\
Q_4 &= T_4 - b_1 - b_4 - b_5 - b_6 \\
Q_5 &= T_5 - b_2 - b_4 - b_5 - b_6 \\
Q_6 &= T_6 - b_3 - b_4 - b_5 - b_6.
\end{aligned}$$

The elements of $\mathbf{Q}$ should sum to zero, which provides a check.

It is now necessary to find $\mathbf{C} = \mathbf{r}^{\delta} - \mathbf{N}\mathbf{k}^{-\delta}\mathbf{N}' = 4\mathbf{I}_6 - \frac{1}{4}\mathbf{N}\mathbf{N}'$, i.e.,

$$\mathbf{C} = \begin{bmatrix}
4-1 & -\frac{3}{4} & -\frac{3}{4} & -\frac{1}{2} & -\frac{1}{2} & -\frac{1}{2} \\
-\frac{3}{4} & 4-1 & -\frac{3}{4} & -\frac{1}{2} & -\frac{1}{2} & -\frac{1}{2} \\
-\frac{3}{4} & -\frac{3}{4} & 4-1 & -\frac{1}{2} & -\frac{1}{2} & -\frac{1}{2} \\
-\frac{1}{2} & -\frac{1}{2} & -\frac{1}{2} & 4-1 & -\frac{3}{4} & -\frac{3}{4} \\
-\frac{1}{2} & -\frac{1}{2} & -\frac{1}{2} & -\frac{3}{4} & 4-1 & -\frac{3}{4} \\
-\frac{1}{2} & -\frac{1}{2} & -\frac{1}{2} & -\frac{3}{4} & -\frac{3}{4} & 4-1
\end{bmatrix}.$$

A check is afforded by each row and column summing to zero.

The oldest solution of (3.3.1) is that of Tocher (1952), who suggested using

$$\mathbf{C}^{-} = (\mathbf{C} + \mathbf{r}\mathbf{r}'/n)^{-1} \tag{3.4.1}$$

which is usually written $\boldsymbol{\Omega}$. Clearly

$$\mathbf{C}\boldsymbol{\Omega}\mathbf{C} = (\boldsymbol{\Omega}^{-1} - \mathbf{r}\mathbf{r}'/n)\boldsymbol{\Omega}(\boldsymbol{\Omega}^{-1} - \mathbf{r}\mathbf{r}'/n) = \boldsymbol{\Omega}^{-1} - \mathbf{r}\mathbf{r}'/n = \mathbf{C}$$

because $\quad \boldsymbol{\Omega}^{-1}\mathbf{1}_v = \mathbf{r}$, so $\mathbf{1}_v = \boldsymbol{\Omega}\mathbf{r}$. Also $\mathbf{1}_v'\mathbf{r} = n$. (3.4.2)

However, Tocher did not find his matrix as a generalized inverse, but by applying the constraint, $\mathbf{r}'\boldsymbol{\tau} = 0$. That suggests that there is a relationship between choosing a constraint, which was rejected in Section 3.1, and choosing a generalized inverse. That is indeed so. From (3.3.2) it follows that $\mathbf{u}'\mathbf{C}^{-}\mathbf{C}\hat{\boldsymbol{\tau}}$ is an unbiassed estimate of $\mathbf{u}'\hat{\boldsymbol{\tau}}$ provided $\mathbf{u}$ is not confounded. If it is, however, some decision has to be made and the simplest is to constrain $\mathbf{u}'\hat{\boldsymbol{\tau}}$ to equal zero, since there can be no information about it and one value is as good as another. Thus, here, on account of (3.3.1) and (3.4.2),

$$\mathbf{r}'\hat{\boldsymbol{\tau}} = \mathbf{r}'\boldsymbol{\Omega}\mathbf{Q} = \mathbf{1}_v'\mathbf{Q} = 0$$

which is effectively a constraint upon the value of $\hat{\boldsymbol{\tau}}$. However, that does not evaluate $\boldsymbol{\Omega}$. All that is needed is to add $r^2/n = \frac{16}{24} = \frac{2}{3}$ to each element of $\mathbf{C}$ and to invert. Hence,

$$\mathbf{\Omega}^{-1} = \tfrac{1}{12}\begin{bmatrix} 44 & -1 & -1 & 2 & 2 & 2 \\ -1 & 44 & -1 & 2 & 2 & 2 \\ -1 & -1 & 44 & 2 & 2 & 2 \\ 2 & 2 & 2 & 44 & -1 & -1 \\ 2 & 2 & 2 & -1 & 44 & -1 \\ 2 & 2 & 2 & -1 & -1 & 44 \end{bmatrix},$$

$$\mathbf{\Omega} = \tfrac{1}{360}\begin{bmatrix} 99 & 3 & 3 & -5 & -5 & -5 \\ 3 & 99 & 3 & -5 & -5 & -5 \\ 3 & 3 & 99 & -5 & -5 & -5 \\ -5 & -5 & -5 & 99 & 3 & 3 \\ -5 & -5 & -5 & 3 & 99 & 3 \\ -5 & -5 & -5 & 3 & 3 & 99 \end{bmatrix}.$$

Checks are provided by (3.4.2). For most purposes $\mathbf{\Omega}$ remains the most convenient solution for $\mathbf{C}^{-}$. The approach is illustrated with a worked example in Example 3A.

An alternative (Pearce, 1976d) is of some theoretical importance, because it shows that precision depends solely upon the off-diagonal elements of $\mathbf{N}\mathbf{k}^{-\delta}\mathbf{N}'$, known as the 'weighted concurrences'. In this instance,

$$\mathbf{W} = \tfrac{1}{4}\begin{bmatrix} 0 & 3 & 3 & 2 & 2 & 2 \\ 3 & 0 & 3 & 2 & 2 & 2 \\ 3 & 3 & 0 & 2 & 2 & 2 \\ 2 & 2 & 2 & 0 & 3 & 3 \\ 2 & 2 & 2 & 3 & 0 & 3 \\ 2 & 2 & 2 & 3 & 3 & 0 \end{bmatrix},$$

i.e. $\mathbf{W} = \mathbf{N}\mathbf{k}^{-\delta}\mathbf{N}'$ with diagonal elements set equal to zero. The elements of $\mathbf{W}\mathbf{1} = \mathbf{q}$ form the diagonal elements of $\mathbf{C}$ and are called the 'quasi-replications'. (Here $\mathbf{q} = \mathbf{3} \times \mathbf{1}_6$.) They measure the extent to which each treatment has concurred with other treatments in the same block and has thus contributed to building up information about contrasts. Also, $\mathbf{q}'\mathbf{1}_v$ ($= u$) is called the 'size' of the experiment, indicating as it does how far the design has succeeded in bringing treatments together for purposes of comparison. In this notation

$$\mathbf{C} = \mathbf{q}^{\delta} - \mathbf{W} \tag{3.4.3}$$

and the suggested solution for $\mathbf{C}^{-}$ is

$$\mathbf{\Xi} = (\mathbf{C} + \mathbf{q}\mathbf{q}'/u)^{-1}. \tag{3.4.4}$$

It may be justified as for $\mathbf{\Omega}$. In this instance, where $u = 18$, $q^2/u = \frac{1}{2}$ is added to each element of $\mathbf{C}$ to give

$$\Xi^{-1} = \tfrac{1}{4}\begin{bmatrix} 14 & -1 & -1 & 0 & 0 & 0 \\ -1 & 14 & -1 & 0 & 0 & 0 \\ -1 & -1 & 14 & 0 & 0 & 0 \\ 0 & 0 & 0 & 14 & -1 & -1 \\ 0 & 0 & 0 & -1 & 14 & -1 \\ 0 & 0 & 0 & -1 & -1 & 14 \end{bmatrix},$$

$$\Xi = \tfrac{1}{45}\begin{bmatrix} 13 & 1 & 1 & 0 & 0 & 0 \\ 1 & 13 & 1 & 0 & 0 & 0 \\ 1 & 1 & 13 & 0 & 0 & 0 \\ 0 & 0 & 0 & 13 & 1 & 1 \\ 0 & 0 & 0 & 1 & 13 & 1 \\ 0 & 0 & 0 & 1 & 1 & 13 \end{bmatrix}.$$

It will be seen that corresponding elements of $\mathbf{\Omega}$ and Ξ differ by $\frac{1}{72}$, so for purposes of working out the variance of a contrast, one matrix is as good as the other. The constraint implied is $\mathbf{q}'\boldsymbol{\tau} = 0$. Typically Ξ gives rise to more complicated arithmetic than $\mathbf{\Omega}$; the above example provides an exception. Checks are given by the relationships $\Xi^{-1}\mathbf{1}_v = \mathbf{q}$ and $\Xi\mathbf{q} = \mathbf{1}_v$ analogous to (3.4.2). The whole approach is exemplified in Example 3B.

A third possibility is of use when the eigenvectors of $\mathbf{C}$ are known or can easily be derived. Thus here there is clearly a contrast between two groups (A, B, C *versus* D, E, F), treatments within a group all being of the same status. It can readily be guessed that the eigenvectors of $\mathbf{C}$ are

$$\mathbf{u}_1 = \frac{1}{\sqrt{6}}\begin{bmatrix} 1 \\ 1 \\ 1 \\ -1 \\ -1 \\ -1 \end{bmatrix}, \quad \mathbf{u}_2 = \frac{1}{\sqrt{2}}\begin{bmatrix} 1 \\ -1 \\ 0 \\ 0 \\ 0 \\ 0 \end{bmatrix}, \quad \mathbf{u}_3 = \frac{1}{\sqrt{6}}\begin{bmatrix} 1 \\ 1 \\ -2 \\ 0 \\ 0 \\ 0 \end{bmatrix},$$

$$\mathbf{u}_4 = \frac{1}{\sqrt{2}}\begin{bmatrix} 0 \\ 0 \\ 0 \\ 1 \\ -1 \\ 0 \end{bmatrix}, \quad \mathbf{u}_5 = \frac{1}{\sqrt{6}}\begin{bmatrix} 0 \\ 0 \\ 0 \\ 1 \\ 1 \\ -2 \end{bmatrix}.$$

$\mathbf{u}_6$, of course, is $\mathbf{1}_6/\sqrt{6}$. The corresponding eigenvalues are $\lambda_1 = 3, \lambda_2 = \lambda_3 = \lambda_4 = \lambda_5 = 3.75$ and $\lambda_6 = 0$. The Moore–Penrose generalized inverse, $\mathbf{C}^+$, is given by

$$\mathbf{C}^+ = \sum_{i=1}^{h} (\mathbf{u}_i\mathbf{u}_i'/\lambda_i), \tag{3.4.5}$$

where

$$\mathbf{C} = \sum_{i=1}^{h} (\lambda_i \mathbf{u}_i \mathbf{u}_i'), \tag{3.4.6}$$

h being the number of non-zero eigenvalues of $\mathbf{C}$, the zero eigenvalues being numbered from $(h+1)$ to v. It has a number of pleasing properties. Thus $\mathbf{C}^+\mathbf{C} = \mathbf{C}\mathbf{C}^+ = \Sigma_{i=1}^{h} (\mathbf{u}_i\mathbf{u}_i')$. Hence, the matrix corresponds to the constraint, $\mathbf{1}'\boldsymbol{\tau} = 0$. Also, $\mathbf{C}^+\mathbf{C}\mathbf{C}^+ = \mathbf{C}^+$.

In the example,

$$\mathbf{C}^+ = \tfrac{1}{90}\begin{bmatrix} 21 & -3 & -3 & -5 & -5 & -5 \\ -3 & 21 & -3 & -5 & -5 & -5 \\ -3 & -3 & 21 & -5 & -5 & -5 \\ -5 & -5 & -5 & 21 & -3 & -3 \\ -5 & -5 & -5 & -3 & 21 & -3 \\ -5 & -5 & -5 & -3 & -3 & 21 \end{bmatrix}.$$

Again, a constant difference separates an element of $\mathbf{C}^+$ from the corresponding value in $\boldsymbol{\Omega}$ or $\boldsymbol{\Xi}$. Of the three methods considered so far, this is the only one that can cope with a confounded contrast, though, as will appear later, $\boldsymbol{\Omega}$ and $\boldsymbol{\Xi}$ can be modified to do so. An example, 3C, will be found at the end of the chapter.

To return for a moment to $\boldsymbol{\Omega}$ and $\boldsymbol{\Xi}$ it may be noted that any matrix of the form

$$(\mathbf{C} + \mathbf{p}\mathbf{p}'/\mathbf{1}_v'\mathbf{p})^{-1} \tag{3.4.7}$$

is a generalized inverse of $\mathbf{C}$, though the only useful ones appear to be those given by $\mathbf{p} = \mathbf{r}$ or $\mathbf{q}$, corresponding respectively to $\boldsymbol{\Omega}$ and $\boldsymbol{\Xi}$. If $\mathbf{p} = \mathbf{1}_v$,

$$(\mathbf{C} + \mathbf{p}\mathbf{p}'/\mathbf{1}_v'\mathbf{p})^{-1} = \mathbf{C}^+ + \mathbf{1}_v\mathbf{1}_v'/v.$$

It is assumed that $\mathbf{1}_v'\mathbf{p} \neq 0$ and that there are no further zero eigenvalues.

Jones (1976) has used the infinite series

$$\boldsymbol{\Xi} = \mathbf{q}^{-\delta} - \sum_{i=1}^{\infty} [\mathbf{1}_v\mathbf{1}_v'/u - (\mathbf{q}^{-\delta}\mathbf{W})^i\mathbf{q}^{-\delta}]. \tag{3.4.8}$$

It does not always converge if some blocks contain only two plots or if there are only two treatments. Catchpole (1981) has suggested that if (3.4.4) is modified to read

$$\boldsymbol{\theta} = (\mathbf{C} + \alpha\mathbf{q}\mathbf{q}'/u)^{-1} \tag{3.4.9}$$

where α is some constant, then

$$\boldsymbol{\theta} = \mathbf{q}^{-\delta} - \sum_{i=1}^{\infty} [\alpha\mathbf{1}_v\mathbf{1}_v'/u - (\mathbf{q}^{-\delta}\mathbf{W})^i\mathbf{q}^{-\delta}] \tag{3.4.10}$$

will always converge given a suitable choice of α and gives another generalized inverse, $\boldsymbol{\theta}$.

3.5 THE KUIPER–CORSTEN ITERATION

The three methods of finding a solution for $\mathbf{C}^-$ just described are mostly used when there is some regularity in the design. With less patterned matrices, or when a general method is needed for a computer, there are advantages in the iterative method introduced by Kuiper (1952) and developed by Corsten (1958) and Caliński (1971).

It depends upon two simple operations. The first is called 'projection' and its purpose is to transform a vector, $\mathbf{v}$, for treatments into a vector, $\mathbf{u}$, for blocks. Each element of $\mathbf{u}$ is found as a weighted mean of the elements of $\mathbf{v}$, the weights being derived from the corresponding column of the incidence matrix. Thus, to continue with the example of the last section,

$$u_1 = \tfrac{1}{4}(v_1 + v_2 + v_3 + v_4)$$
$$u_2 = \tfrac{1}{4}(v_1 + v_2 + v_3 + v_5), \text{ etc.}$$

The multiplier, $\frac{1}{4}$, arises because each block has four plots. The other operation is called 'dual projection' and similarly transforms a vector for blocks into one for treatments, the weights now being the elements of the appropriate row of $\mathbf{N}$. Thus, if $\mathbf{w}$ is the dual projection of $\mathbf{u}$,

$$w_1 = \tfrac{1}{4}(u_1 + u_2 + u_3 + u_4)$$
$$w_2 = \tfrac{1}{4}(u_1 + u_2 + u_3 + u_5), \text{ etc.}$$

The multipliers are now derived from the replications. It will be seen that $\mathbf{u} = \mathbf{k}^{-\delta}\mathbf{N}'\mathbf{v}$ and $\mathbf{w} = \mathbf{r}^{-\delta}\mathbf{N}\mathbf{u}$.

The iteration starts by writing down a vector, $\mathbf{v}_1$, such that

$$\mathbf{r}'\mathbf{v}_1 = 0. \tag{3.5.1}$$

Projection and dual projection follow alternately, i.e.

$$\mathbf{u}_j = \mathbf{k}^{-\delta}\mathbf{N}'\mathbf{v}_j, \qquad \mathbf{v}_{j+1} = \mathbf{r}^{-\delta}\mathbf{N}\mathbf{u}_j \tag{3.5.2}$$

and the sum to infinity of the successive values of $\mathbf{v}_j$ equals $\boldsymbol{\Upsilon}\mathbf{r}^{\delta}\mathbf{v}_1$, where $\boldsymbol{\Upsilon}$ proves to be a generalized inverse of $\mathbf{C}$.

To justify that result it will be helpful to use the matrix, $\mathbf{F}$, such that

$$\mathbf{F} = \mathbf{r}^{-\frac{1}{2}\delta}\mathbf{C}\mathbf{r}^{-\frac{1}{2}\delta} = \mathbf{I}_v - \mathbf{r}^{-\frac{1}{2}\delta}\mathbf{N}\mathbf{k}^{-\delta}\mathbf{N}'\mathbf{r}^{-\frac{1}{2}\delta}. \tag{3.5.3}$$

Its importance lies in its eigenvalues, which will be found from the lemmas that follow. (The notation has been explained in Section 24 of the Appendix.)

Lemma 3.5.A *If a matrix,* $\mathbf{S}$, *can be written in the form* $\mathbf{S} = \mathbf{M}\mathbf{M}'$, *then it has no negative eigenvalues.*

Proof: If $\mathbf{S}\mathbf{p} = \lambda\mathbf{p}$, then $\lambda = \mathbf{p}'\mathbf{S}\mathbf{p} = (\mathbf{p}'\mathbf{M})(\mathbf{p}'\mathbf{M})'$, which is a sum of squares. ■

Lemma 3.5.B $\mathbf{F}$ *has no negative eigenvalues.*

Proof: $\mathbf{F} = \mathbf{r}^{-\frac{1}{2}\delta}\mathbf{C}\mathbf{r}^{\frac{1}{2}\delta} = (\mathbf{r}^{-\frac{1}{2}\delta}\boldsymbol{\Delta}\boldsymbol{\phi})(\mathbf{r}^{-\frac{1}{2}\delta}\boldsymbol{\Delta}\boldsymbol{\phi})'$. ■

Lemma 3.5.C *No eigenvalue of* **F** *exceeds one.*

Proof:
$$(\mathbf{I}_v - \mathbf{F}) = (\mathbf{r}^{-\frac{1}{2}\delta}\mathbf{N}\mathbf{k}^{-\frac{1}{2}\delta})(\mathbf{r}^{-\frac{1}{2}\delta}\mathbf{N}\mathbf{k}^{-\frac{1}{2}\delta})'$$
so no eigenvalue of $(\mathbf{I}_v - \mathbf{F})$ is negative. However, if $\mathbf{F}\mathbf{p} = \varepsilon\mathbf{p}$
$$(\mathbf{I}_v - \mathbf{F})\mathbf{p} = (1-\varepsilon)\mathbf{p}$$
if $\mathbf{F}\mathbf{p} = \varepsilon\mathbf{p}$, so no eigenvalue of **F** exceeds one. ■

For reasons that will appear later after (3.6.8) the matrix **F** will be called the 'efficiency matrix' on account of its eigenvalues, which have some useful properties.

From (3.5.2) it follows that
$$\begin{aligned}\mathbf{v}_{j+1} &= \mathbf{r}^{-\delta}\mathbf{N}\mathbf{k}^{-\delta}\mathbf{N}'\mathbf{v}_j\\ &= \mathbf{r}^{-\frac{1}{2}\delta}(\mathbf{I}_v - \mathbf{F})\mathbf{r}^{\frac{1}{2}\delta}\mathbf{v}_j\\ &= \mathbf{r}^{-\frac{1}{2}\delta}\sum_{i=1}^{v}[(1-\varepsilon_i)\mathbf{p}_i\mathbf{p}_i']\mathbf{r}^{\frac{1}{2}\delta}\mathbf{v}_j.\end{aligned} \tag{3.5.4}$$

It will be noticed that v as a scalar and **v** as a vector are not related. That is unfortunate but convention requires both to be as they are and the reader will no doubt separate the two meanings easily enough.

It appears that there is at least one zero eigenvalue in **F** because there is one in **C**. In fact
$$\mathbf{F}(\mathbf{r}^{\frac{1}{2}\delta}\mathbf{1}_v) = \mathbf{r}^{-\frac{1}{2}\delta}\mathbf{C}\mathbf{1}_v = \mathbf{0} \tag{3.5.5}$$

Consider the eigenvector $\mathbf{p}_v = \mathbf{r}^{\frac{1}{2}\delta}\mathbf{1}_v$; then in (3.5.4), $\mathbf{p}_v'\mathbf{r}^{\frac{1}{2}\delta}\mathbf{v}_1 = \mathbf{1}'\mathbf{r}^{\delta}\mathbf{v}_1 = \mathbf{r}'\mathbf{v}_1$. If $\mathbf{r}'\mathbf{v}_1$ equals 0, the eigenvector $\mathbf{p}_v$ will be having no effect in (3.5.4). Hence, the summation can be limited to the range 1 to $(v-1)$ without harm. For that matter, if there are other zero eigenvalues in **C** and therefore in **F**, they can be regarded in the same way if $\mathbf{v}_1$ is such that $\mathbf{p}_i'\mathbf{r}^{\delta}\mathbf{v}_1 = 0$. If that is so for all eigenvectors with zero eigenvalues the range can be reduced further to be from 1 to h, where h has its familiar meaning as the number of non-zero eigenvalues of **C**. From now on that will be done, but with the mental proviso that h probably equals $(v-1)$. Since squaring a matrix squares its eigenvalues without affecting its eigenvectors,
$$\mathbf{v}_2 = \mathbf{r}^{-\frac{1}{2}\delta}\sum_{i=1}^{h}[(1-\varepsilon_i)\mathbf{p}_i\mathbf{p}_i']\mathbf{r}^{\frac{1}{2}\delta}\mathbf{v}_1$$
$$\mathbf{v}_3 = \mathbf{r}^{-\frac{1}{2}\delta}\sum_{i=1}^{h}[(1-\varepsilon_i)^2\mathbf{p}_i\mathbf{p}_i']\mathbf{r}^{\frac{1}{2}\delta}\mathbf{v}_1, \text{ etc.}$$
and hence
$$\begin{aligned}\sum_{j=1}^{\infty}\mathbf{v}_j &= \mathbf{r}^{-\frac{1}{2}\delta}\sum_{i=1}^{h}\left\{\sum_{j=1}^{\infty}[(1-\varepsilon_i)^{j-1}\mathbf{p}_i\mathbf{p}_i']\right\}\mathbf{r}^{\frac{1}{2}\delta}\mathbf{v}_1\\ &= \mathbf{r}^{-\frac{1}{2}\delta}\sum_{i=1}^{h}(\mathbf{p}_i\mathbf{p}_i'/\varepsilon_i)\mathbf{r}^{\frac{1}{2}\delta}\mathbf{v}_i\end{aligned}$$

because $$\sum_{j=1}^{\infty} (1-\varepsilon_i)^{j-1} = 1/\varepsilon_i.$$

Hence
$$\sum_{j=1}^{\infty} \mathbf{v}_j = \mathbf{r}^{-\frac{1}{2}\delta}\mathbf{F}^{+}\mathbf{r}^{\frac{1}{2}\delta}\mathbf{v}_1$$
$$= \boldsymbol{\Upsilon}\mathbf{r}^{\delta}\mathbf{v}_1,$$
where $$\boldsymbol{\Upsilon} = \mathbf{r}^{-\frac{1}{2}\delta}\mathbf{F}^{+}\mathbf{r}^{-\frac{1}{2}\delta} \tag{3.5.6}$$

Obviously $\boldsymbol{\Upsilon}$ is a generalized inverse of $\mathbf{C}$ $(= \mathbf{r}^{\frac{1}{2}\delta}\mathbf{F}\mathbf{r}^{\frac{1}{2}\delta})$.

The original use of the iteration was that suggested by Kuiper (1952), who put $\mathbf{v}_1 = \mathbf{r}^{-\delta}\mathbf{Q}$ and obtained a valid estimate of $\boldsymbol{\tau}$. Such a calculation will be found in Example 3E at the end of the chapter. The analysis can be completed by noting from (3.1.3a) that

$$\hat{\boldsymbol{\beta}} = \mathbf{k}^{-\delta}\mathbf{B} - \mathbf{k}^{-\delta}\mathbf{N}'\hat{\boldsymbol{\tau}} = \mathbf{k}^{-\delta}\mathbf{B} - \mathbf{k}^{-\delta}\mathbf{N}'\sum_j \mathbf{v}_j$$
$$= \mathbf{k}^{-\delta}\mathbf{B} - \sum_j \mathbf{u}_j \tag{3.5.7}$$

It is then possible to find $\hat{\boldsymbol{\eta}}$, thus:

$$\hat{\boldsymbol{\eta}} = \mathbf{y} - \mathbf{D}'\hat{\boldsymbol{\beta}} - \boldsymbol{\Delta}'\hat{\boldsymbol{\tau}} \tag{3.5.8}$$

and the error sum of squared deviations equals $\hat{\boldsymbol{\eta}}'\hat{\boldsymbol{\eta}}$. Since the treatment sum of squared deviations equals $\mathbf{Q}'\mathbf{C}^{-}\mathbf{Q} = \mathbf{Q}'\boldsymbol{\Upsilon}\mathbf{Q} = \mathbf{Q}'\hat{\boldsymbol{\tau}}$, it is possible to find the error line by difference from $\mathbf{y}'\mathbf{y} - \mathbf{B}'\mathbf{k}^{-\delta}\mathbf{B}$, but there are advantages in knowing $\hat{\boldsymbol{\eta}}$. For one thing there is an effective check. For another it is often useful to have estimates of the residuals.

Later Caliński (1971) showed that there is an easy way of finding $\boldsymbol{\Upsilon}$ and so of obtaining a covariance matrix for the $\hat{\boldsymbol{\tau}}$ already found. He put $\mathbf{v}_1$ equal to a column of $(\mathbf{r}^{-\delta} - \mathbf{1}_v\mathbf{1}_v'/n)$. Now,

$$\boldsymbol{\Upsilon}\mathbf{r}^{\delta}(\mathbf{r}^{-\delta} - \mathbf{1}_v\mathbf{1}_v'/n) = \boldsymbol{\Upsilon} - \boldsymbol{\Upsilon}\mathbf{r}\mathbf{1}_v'/n$$
$$= \boldsymbol{\Upsilon} - \mathbf{r}^{-\frac{1}{2}\delta}\mathbf{F}^{+}\mathbf{r}^{\frac{1}{2}\delta}\mathbf{1}_v\mathbf{1}_v' = \boldsymbol{\Upsilon} \tag{3.5.9}$$

because of (3.5.5) and the fact that a zero eigenvalue in $\mathbf{F}$ corresponds to another in $\mathbf{F}^{+}$. He thus obtained the corresponding column of $\boldsymbol{\Upsilon}$. Example 3D at the end of the chapter illustrates the calculations.

A problem concerns the point at which the iteration should stop, an important matter when programming a computer. First, it may be assumed that ε_i for an unconfounded contrast will scarcely ever fall below $\frac{1}{4}$, which means that $(1-\varepsilon_i)$ will rarely exceed $\frac{3}{4}$. In that case, if $\mathbf{V}$ is the true sum to infinity of the $\mathbf{v}_j$, then for any elements of $\mathbf{V}$, its deviation from the cumulative sum of the corresponding elements in $\mathbf{v}_1, \mathbf{v}_2, \ldots, \mathbf{v}_j$ will be less than three times the element added by $\mathbf{v}_j$. Accordingly, if it is decided that all elements of $\boldsymbol{\Upsilon}$ must be correct to within a limit, α, the iteration can be concluded when no element of the last $\mathbf{v}_j$ is greater than $\alpha/3$, signs being of course ignored.

The introduction of $\mathbf{F}$ does something to explain what happens with zero eigenvalues in $\mathbf{C}$. There is always at least one, i.e. that given by $\mathbf{C}\mathbf{1}_v = \mathbf{0} = \mathbf{F}(\mathbf{r}^{\frac{1}{2}\delta}\mathbf{1}_v)$. In going from $\mathbf{C}$ to $\mathbf{\Omega}^{-1}$, the zero eigenvalue in $\mathbf{F}$ is changed from 0 to 1, i.e.,

$$\mathbf{\Omega}^{-1} = \mathbf{r}^{\frac{1}{2}\delta}(\mathbf{F} + \mathbf{p}_v\mathbf{p}_v')\mathbf{r}^{\frac{1}{2}\delta}, \tag{3.5.10}$$

where $\mathbf{p}_v = \mathbf{r}^{\frac{1}{2}\delta}\mathbf{1}_v/\sqrt{n}$, the divisor being called for to standardize $\mathbf{p}_v$, i.e. to make $\mathbf{p}_v'\mathbf{p}_v = 1$. Other zero eigenvalues can be dealt with in the same way. Let there be $(v-h)$ of them, then it is convenient to write

$$\mathbf{\Omega}^{-1} = \mathbf{r}^{\frac{1}{2}\delta}[\mathbf{F} + \sum_{j=h+1}^{v}(\mathbf{p}_i\mathbf{p}_i')]\mathbf{r}^{\frac{1}{2}\delta}. \tag{3.5.11}$$

The approach can be extended to cover $\mathbf{\Xi}^{-1}$ as well. It may be asked whether this tampering with eigenvalues is valid, but it can do no harm provided no attempt is made to estimate effects that are confounded, i.e., that involve the eigenvectors that correspond to the changed eigenvalues.

There is, however, another way of dealing with the singularities of $\mathbf{C}$. It is to calculate $\mathbf{r}^{-\delta}\mathbf{C}\mathbf{r}^{-\delta}$ and to apply the Kuiper–Corsten iteration twice to each of its columns. The result will be the formation of $\mathbf{\Upsilon}$, because

$$\begin{aligned}
&(\mathbf{r}^{-\frac{1}{2}\delta}\mathbf{F}^{+}\mathbf{r}^{\frac{1}{2}\delta})(\mathbf{r}^{-\frac{1}{2}\delta}\mathbf{F}^{+}\mathbf{r}^{\frac{1}{2}\delta})(\mathbf{r}^{-\delta}\mathbf{C}\mathbf{r}^{-\delta})\\
&= \mathbf{r}^{-\frac{1}{2}\delta}\mathbf{F}^{+}\mathbf{F}^{+}\mathbf{r}^{\frac{1}{2}\delta}(\mathbf{r}^{-\frac{1}{2}\delta}\mathbf{F}\mathbf{r}^{-\frac{1}{2}\delta})\\
&= \mathbf{r}^{-\frac{1}{2}\delta}\mathbf{F}^{+}\mathbf{F}^{+}\mathbf{F}\mathbf{r}^{-\frac{1}{2}\delta}\\
&= \mathbf{r}^{-\frac{1}{2}\delta}\mathbf{F}^{+}\mathbf{r}^{-\frac{1}{2}\delta}\\
&= \mathbf{\Upsilon}
\end{aligned} \tag{3.5.12}$$

3.6 NATURAL AND BASIC CONTRASTS

So far little has been said about independence of estimation of the various contrasts to be studied. Traditionally some importance has been attached to the subject, because it is convenient, to put it no higher, if the degrees of freedom and the sum of squared deviations for treatments can be partitioned in a way that will add up correctly for both quantities. It is also desirable, because no one wants to study an interaction, say, only to find the same information turning up again in a main effect. Nevertheless, independence, though desirable, should not become a fetish because there are situations in which it is unattainable. To take an example, where several treatments are being compared with a control as in Exercises 3A to 3D, the estimated mean of the control will enter into all contrasts of interest and consequently they must be correlated. To take another, someone may be studying the best way to use a fungicide and has decided to try zero, single and double doses applied at a certain stage of development of the fungus. He then has doubts about the double dose. Will it damage the plants? Is it sensible to apply so much at once? In the event he may decide to keep it but to supplement the design with a fourth treatment in which two single doses are applied, say, three days before and three days after the proposed date for the rest. The additional treatment calls for comparison with the double dose, which is itself part of a patterned set. There can be no question of its contrast with the double dose being estimated independently of the linear and quadratic effects of dosage, but does that really matter?

Nevertheless, independence is desirable where practicable and one obvious way of attaining it is to ensure that the eigenvectors of **C** correspond to the contrasts requiring study. The treatment sum of squared deviations, $\mathbf{Q}'\mathbf{C}^{-}\mathbf{Q}$ with h degrees of freedom, then equals

$$\mathbf{Q}'\mathbf{C}^{+}\mathbf{Q} = \sum_{i=1}^{h} \frac{(\mathbf{u}_i'\mathbf{Q})^2}{\lambda_i} \tag{3.6.1}$$

where $\mathbf{C} = \sum_{i=1}^{h} (\lambda_i \mathbf{u}_i \mathbf{u}_i')$. Also, since $\mathbf{Q} = \mathbf{C}\hat{\boldsymbol{\tau}}$ (3.1.4),

$$\mathbf{Q}'\mathbf{C}^{-}\mathbf{Q} = \hat{\boldsymbol{\tau}}'\mathbf{C}\mathbf{C}^{-}\mathbf{C}\hat{\boldsymbol{\tau}} = \hat{\boldsymbol{\tau}}'\mathbf{C}\hat{\boldsymbol{\tau}}$$

$$= \sum_{i=1}^{h} [\lambda_i (\mathbf{u}_i'\hat{\boldsymbol{\tau}})^2] \tag{3.6.2}$$

which could be a more convenient form, especially since the values of $\mathbf{u}_i'\hat{\boldsymbol{\tau}}$, which are the estimates of the various contrasts, are going to be needed anyway. Such contrasts, though obviously important, appear to lack a name and it is here suggested that they be called the 'natural contrasts' of the design, partly because other suitable names appear to have been appropriated for other purposes and partly because they are the ones that immediately and naturally spring to mind.

Where there are multiplicities among the eigenvalues, alternative partitions become possible. If $\lambda_i = \lambda_j$, then $(\theta\mathbf{u}_i + \phi\mathbf{u}_j)/(\theta^2 + \phi^2)^{\frac{1}{2}}$ and $(\phi\mathbf{u}_i - \theta\mathbf{u}_j)/(\theta^2 + \phi^2)^{\frac{1}{2}}$ are just as much natural contrasts as $\mathbf{u}_i$ and $\mathbf{u}_j$. The fact can be useful in seeking a design that will match a problem. Nevertheless, freedom of partition, i.e. an ability to change the contrasts of interest after the experiment has been started, should not be extolled too much since the aims should be clear from the beginning. (It must be admitted, to be honest, that experiments do sometimes produce surprising results and a facility to adjust the partition to meet the unexpected does have its advantages.)

However, the natural contrasts are not the only ones to meet the requirement of independence. The 'basic contrasts' of a design, which derive rather from the eigenvectors of **F**, have the same property (Pearce *et al.*, 1974). Writing

$$\mathbf{F} = \sum_{i=1}^{h} (\varepsilon_i \mathbf{p}_i \mathbf{p}_i').$$

Then, if $\mathbf{z}_i = \mathbf{r}^{\frac{1}{2}\delta}\mathbf{p}_i$, $\mathbf{z}_i$ is a basic contrast. Since $\mathbf{p}_i'\mathbf{p}_i = 1$ and $\mathbf{p}_i'\mathbf{p}_j (i \neq j) = 0$, it follows that

$$\mathbf{z}_i'\mathbf{r}^{-\delta}\mathbf{z}_i = 1, \qquad \mathbf{z}_i'\mathbf{r}^{-\delta}\mathbf{z}_j = 0. \tag{3.6.3}$$

From (3.5.5) $\mathbf{F}\mathbf{p}_v = \mathbf{F}(\mathbf{r}^{\frac{1}{2}\delta}\mathbf{1}_v) = \mathbf{0}$. Writing

$$\mathbf{z}_v = \mathbf{r}^{\frac{1}{2}\delta}\mathbf{p}_v = \mathbf{r}^{\delta}\mathbf{1}_v = \mathbf{r},$$

it follows from (3.6.3) that all other $\mathbf{z}_i$ must represent contrasts, i.e. if $i < v$,

$$\mathbf{z}_i'\mathbf{1}_v = \mathbf{z}_i'\mathbf{r}^{-\delta}\mathbf{z}_v = 0.$$

Further, if $\mathbf{Z}$ is a matrix that holds $\mathbf{z}_i$ in its ith column, then from (3.6.3)

$$\mathbf{Z}'\mathbf{r}^{-\delta}\mathbf{Z} = \mathbf{I}_v \tag{3.6.4a}$$

so

$$\mathbf{Z}\mathbf{Z}' = \mathbf{r}^{\delta}. \tag{3.6.4b}$$

If a contrast is thought to be basic, the matter can readily be checked and the eigenvalue, ε_i, can be found at the same time by using the relationship

$$\mathbf{C}\mathbf{r}^{-\delta}\mathbf{z}_i = \mathbf{r}^{\frac{1}{2}\delta}\mathbf{F}\mathbf{r}^{-\frac{1}{2}\delta}\mathbf{z}_i = \varepsilon_i\mathbf{r}^{\frac{1}{2}\delta}\mathbf{p}_i = \varepsilon_i\mathbf{z}_i. \tag{3.6.5}$$

Similarly

$$\mathbf{C} = \mathbf{r}^{\frac{1}{2}\delta}\sum_i(\varepsilon_i\mathbf{p}_i\mathbf{p}_i')\mathbf{r}^{\frac{1}{2}\delta}$$

$$= \sum_i(\varepsilon_i\mathbf{z}_i\mathbf{z}_i') = \mathbf{Z}\boldsymbol{\varepsilon}^{\delta}\mathbf{Z}'. \tag{3.6.6}$$

It has already been noted that $\mathbf{z}_v = \mathbf{r}$. Hence from (3.6.3), it can more correctly be written $\mathbf{r}/\sqrt{n}$. From (3.4.1), supposing there to be no more zero eigenvalues,

$$\boldsymbol{\Omega}^{-1} = \mathbf{C} + \mathbf{r}\mathbf{r}'/n = \mathbf{Z}\boldsymbol{\varepsilon}_*^{\delta}\mathbf{Z}' \tag{3.6.7}$$

where the elements of $\boldsymbol{\varepsilon}_*$ are the same as those of $\boldsymbol{\varepsilon}$ except that $\varepsilon_{*v} = 1$ instead of zero. Hence, from (3.6.4)

$$\boldsymbol{\Omega} = \mathbf{r}^{-\delta}\mathbf{Z}\boldsymbol{\varepsilon}_*^{-\delta}\mathbf{Z}'\mathbf{r}^{-\delta}. \tag{3.6.8}$$

The relationship to (3.5.10) is obvious.

This is perhaps the place to clear up the mysterious reference to the interesting properties of the eigenvalues of $\mathbf{F}$, which appeared below Lemma 3.5.C. The 'efficiency factor' of a constant, $\mathbf{c}$, is applied to the ratio

$$\mathbf{c}'\mathbf{r}^{-\delta}\mathbf{c}/\mathbf{c}'\mathbf{C}^{-}\mathbf{c}.$$

The numerator indicates the variance of estimation of the contrast if it is found from a completely randomized design, i.e., $\mathbf{c}'\mathbf{r}^{-\delta}\mathbf{c}\sigma_1^2$, while the denominator correspondingly shows the same variance if estimated from some other design, namely, $\mathbf{c}'\mathbf{C}^{-}\mathbf{c}\sigma_2^2$, which is assumed to have the same replications. Unless the blocks or other device to control environmental variation are so effective that σ_2^2/σ_1^2 is less than the efficiency factor of the design, nothing will have been gained. If $\mathbf{c} = \mathbf{z}_i$ and if $\mathbf{c}$ is unconfounded, then from (3.6.3) and (3.6.8) its efficiency factor is

$$1/(1/\varepsilon_{*i}) = \varepsilon_{*i} = \varepsilon_i.$$

That is to say, the eigenvalues of $\mathbf{F}$ are the efficiency factors of such basic contrasts as can be estimated.

The eigenvalues of $\mathbf{C}$ have a similar property. In general if $\mathbf{c}$ is a contrast its 'effective replication' relative to a given design is

$$1/(\mathbf{c}'\mathbf{C}^{-}\mathbf{c})$$

If $\mathbf{c} = \mathbf{u}_i$, an unconfounded natural contrast,

$$\mathbf{c}'\mathbf{C}^-\mathbf{c} = \mathbf{c}'\mathbf{C}^+\mathbf{c} = 1/\lambda_i$$

and λ_i is the effective replication of $\mathbf{c}$. For that reason it will henceforth often be written R_i. Moreover, it is readily found as $\mathbf{c}'\mathbf{C}\mathbf{c}$.

To illustrate the point, let $\mathbf{c}$ represent the difference between the first two treatments, i.e.

$$\mathbf{c}' = (1, -1, 0, \ldots, 0)/\sqrt{2}.$$

Given a completely randomized design

$$R = \frac{1}{\mathbf{c}'\mathbf{r}^{-\delta}\mathbf{c}} = \frac{2r_1r_2}{r_1 + r_2}.$$

If $r_1 = r_2 = r$, then $R = r$. If, however, blocks are introduced and there is a loss of information on their account, $R < r$. In the extreme case, if $\mathbf{c}$ is confounded, $R = 0$ just as $\varepsilon = 0$. For an equireplicate design, the natural and the basic contrasts are the same apart from scaling and $\varepsilon = R/r$.

Continuing from (3.6.8), as at (3.6.2) the sum of squared deviations for treatments may be written as

$$\begin{aligned}\hat{\boldsymbol{\tau}}'\mathbf{C}\hat{\boldsymbol{\tau}} &= \hat{\boldsymbol{\tau}}'\mathbf{r}^{\frac{1}{2}\delta}\mathbf{F}\mathbf{r}^{\frac{1}{2}\delta}\hat{\boldsymbol{\tau}} = \hat{\boldsymbol{\tau}}'\mathbf{r}^{\frac{1}{2}\delta}\sum_{i=1}^{h}(\varepsilon_i\mathbf{p}_i\mathbf{p}_i')\mathbf{r}^{\frac{1}{2}\delta}\boldsymbol{\tau} \\ &= \sum_{i=1}^{h}[\varepsilon_i(\mathbf{z}_i'\hat{\boldsymbol{\tau}})^2].\end{aligned} \tag{3.6.9}$$

There is a further way of writing basic contrasts. Let $\mathbf{s}_i = \mathbf{r}^{-\delta}\mathbf{z}_i = \mathbf{r}^{-\frac{1}{2}\delta}\mathbf{p}_i$; then (3.6.3) becomes

$$\mathbf{s}_i'\mathbf{r}^{\delta}\mathbf{s}_i = 1 \qquad \mathbf{s}_i'\mathbf{r}^{\delta}\mathbf{s}_j = 0 \quad (i \neq j). \tag{3.6.10}$$

Corresponding to (3.6.5), it appears that

$$\mathbf{r}^{-\delta}\mathbf{C}\mathbf{s}_i = \mathbf{r}^{-\frac{1}{2}\delta}\mathbf{F}\mathbf{r}^{\frac{1}{2}\delta}\mathbf{s}_i = \mathbf{r}^{-\frac{1}{2}\delta}\mathbf{F}\mathbf{p}_i = \varepsilon_i\mathbf{r}^{-\frac{1}{2}\delta}\mathbf{p}_i = \varepsilon_i\mathbf{s}_i. \tag{3.6.11}$$

The sum of squared deviations for treatments may now be written

$$\begin{aligned}\mathbf{Q}'\boldsymbol{\Upsilon}\mathbf{Q} &= \mathbf{Q}'\mathbf{r}^{-\frac{1}{2}\delta}\mathbf{F}^+\mathbf{r}^{-\frac{1}{2}\delta}\mathbf{Q} = \mathbf{Q}'\mathbf{r}^{-\frac{1}{2}\delta}\sum_{i=1}^{h}(\mathbf{p}_i\mathbf{p}_i'/\varepsilon_i)\mathbf{r}^{-\frac{1}{2}\delta}\mathbf{Q} \\ &= \sum_{i=1}^{h}[(\mathbf{s}_i'\mathbf{Q})^2/\varepsilon_i]\end{aligned} \tag{3.6.12}$$

analogous to (3.6.9).

It will be noted that though the $\mathbf{s}_i$ have been referred to as contrasts, $\mathbf{1}_v'\mathbf{s}_i$ does not in general equal 0. However, the $\mathbf{s}_i$ are not applied to the estimated treatment parameters, $\hat{\boldsymbol{\tau}}$, but to the adjusted treatment totals, $\mathbf{Q}$. In fact, the relationship $\mathbf{C}\mathbf{1}_v = \mathbf{0}$, which leads to the result $\mathbf{z}_v = \mathbf{r}$, leads also to an $\mathbf{s}$, such that $\mathbf{s}_v = \mathbf{1}_v$. Hence from (3.6.10), for all other $\mathbf{s}_i$

$$\mathbf{r}'\mathbf{s}_i = 0. \tag{3.6.13}$$

Basic contrasts are often studied in the form originally used by Caliński (1971), i.e. as the eigenvectors of $\mathbf{M}_0 = \mathbf{r}^{-\delta}\mathbf{N}\mathbf{k}^{-\delta}\mathbf{N}' - \mathbf{1}\mathbf{r}'/n$. From (3.6.11) and (3.6.13)

$$\mathbf{M}_0\mathbf{s}_i = (1-\varepsilon_i)\mathbf{s}_i. \tag{3.6.14}$$

In many instances the natural and the basic contrasts will be the same. Apart from scaling that will be so for all equireplicate designs. Where the two sets of contrasts differ there is no saying which will fit a particular problem. That depends upon considerations external to the design. With basic contrasts, as with natural ones, multiplicities among the eigenvalues, i.e. among the ε_i, will permit alternative partitions.

To take an example of natural and basic contrasts, in Example 2D it was suggested that if there are three treatments, A, B and C, and 24 plots and the sole purpose is to compare the middle dose with the mean of the others, then in a completely randomized design there should be twice as many plots of B as of A and C. It is here suggested that the situation would be met by having six blocks, each containing the treatments ABBC. In that case

$$\mathbf{C} = \tfrac{3}{2}\begin{bmatrix} 3 & -2 & -1 \\ -2 & 4 & -2 \\ -1 & -2 & 3 \end{bmatrix}, \qquad \mathbf{F} = \tfrac{1}{4}\begin{bmatrix} 3 & -\sqrt{2} & -1 \\ -\sqrt{2} & 3 & -\sqrt{2} \\ -1 & -\sqrt{2} & 3 \end{bmatrix}.$$

The only eigenvectors of $\mathbf{C}$ apart from $(1, 1, 1)'/\sqrt{3}$ i.e. the natural contrasts, are

$$\mathbf{u}_1' = (1, 0, -1)/\sqrt{2}, \qquad \lambda_1 = 6$$
$$\mathbf{u}_2' = (1, -2, 1)/\sqrt{6}, \qquad \lambda_2 = 9$$

The corresponding eigenvectors of $\mathbf{F}$ are

$$\mathbf{p}_1' = (1, 0, -1)/\sqrt{2}, \qquad \varepsilon_1 = 1$$
$$\mathbf{p}_2' = (1, -2, 1)/\sqrt{2}, \qquad \varepsilon_2 = 1$$

Since $\varepsilon_1 = \varepsilon_2$, an alternative solution is given by

$$\mathbf{p}'_{+1} = (\mathbf{p}_1' + \sqrt{2}\mathbf{p}_2')/\sqrt{3} = (\sqrt{2}, -1, 0)/\sqrt{3}, \qquad \varepsilon_{+1} = 1$$
$$\mathbf{p}'_{+2} = (\sqrt{2}\mathbf{p}_1 - \mathbf{p}_2')/\sqrt{3} = (1, \sqrt{2}, -3)/\sqrt{12}, \qquad \varepsilon_{+2} = 1$$

which give rise to basic contrasts

$$\mathbf{z}_1' = 2(1, -1, \ \ 0), \qquad \mathbf{s}_1' = (2, -1, \ \ 0)/6$$
$$\mathbf{z}_2' = \ \ (1, \ \ 2, -3)/\sqrt{2}, \qquad \mathbf{s}_2' = (1, \ \ 1, -3)/\sqrt{72}$$

They are completely meaningful, though, if they had been intended, the design would have been different. Example 3I makes a similar point.

Caliński (1977) has suggested a generalization. Let $\mathbf{X}$ be a $v \times v$, symmetric matrix, all its eigenvalues being positive, and let $\mathbf{w}_i$ be an eigenvector of $\mathbf{X}^{-1}\mathbf{C}$, i.e.

$$\mathbf{X}^{-1}\mathbf{C}\mathbf{w}_i = \kappa_i\mathbf{w}_i \tag{3.6.15}$$

There will be h non-zero eigenvalues, κ_i, corresponding to the h non-zero eigenvalues of $\mathbf{C}$. Then, if $\mathbf{X}^{\frac{1}{2}}$ represents a matrix with eigenvalues equal to the square-roots of those of $\mathbf{X}$ but with the same eigenvectors

$$(\mathbf{X}^{-\frac{1}{2}}\mathbf{C}\mathbf{X}^{-\frac{1}{2}})(\mathbf{X}^{\frac{1}{2}}\mathbf{w}_i) = \kappa(\mathbf{X}^{\frac{1}{2}}\mathbf{w})$$

and it follows that

$$\mathbf{w}_i'\mathbf{X}\mathbf{w}_i = 1, \qquad \mathbf{w}_i'\mathbf{X}\mathbf{w}_j = 0 \quad (i \neq j) \tag{3.6.16}$$

analogous to (3.6.10). In short, Caliński is arguing that it is not necessary to take $\mathbf{X} = \mathbf{r}^{\delta}$ to find basic contrasts; the eigenvectors of $\mathbf{X}^{-1}\mathbf{C}$ and $\mathbf{C}\mathbf{X}^{-1}$ will give analogues of $\mathbf{s}_i$ and $\mathbf{c}_i$ that could well be important also. (The natural contrasts similarly correspond to $\mathbf{X} = \mathbf{I}_v$.) The suggestion is of some importance, though it is fair to comment that the two cases already considered are special, since an eigenvalue of $\mathbf{C}$ indicates the variance of the contrast and one of $\mathbf{F}$ (or $\mathbf{r}^{-\delta}\mathbf{C}$ or $\mathbf{C}\mathbf{r}^{-\delta}$) the efficiency of its estimation, both quantities with an immediate appeal.

3.7 CONFOUNDING AND DISCONNECTION

It appears that zero eigenvalues of $\mathbf{C}$ raise problems. If they rarely arose, the matter could be left, but in fact such eigenvalues occur whenever the design is 'disconnected', i.e. whenever one sub-set of treatments is confined to certain blocks, the complementary sub-set being found only in the others. That situation quite often arises; in fact, it occurs in practice whenever a contrast is confounded, so it calls for examination.

First, let the design be in two disconnected parts, which may themselves be disconnected, and let the treatments be ordered by the two sub-sets, then

$$\mathbf{C} = \begin{bmatrix} \mathbf{C}_1 & \mathbf{0}' \\ \mathbf{0} & \mathbf{C}_2 \end{bmatrix}$$

where $\mathbf{C}_1$ and $\mathbf{C}_2$ are the coefficient matrices of the two parts. Let them have x and y treatments respectively ($v = x + y$). Then $\mathbf{C}_1\mathbf{1}_x = \mathbf{0} = \mathbf{C}_2\mathbf{1}_y$ so

$$\mathbf{C}\begin{bmatrix} y\mathbf{1}_x \\ -x\mathbf{1}_y \end{bmatrix} = \begin{bmatrix} \mathbf{0} \\ \mathbf{0} \end{bmatrix}$$

i.e. the contrast between the two sub-sets is confounded.

Before considering the converse it will be convenient to establish some lemmas.

Lemma 3.7.A *All off-diagonal elements of* $\mathbf{C}$ *are zero or negative.*

Proof:
$$C_{ij}(i \neq j) = -\sum_t (N_{it}N_{jt}/k_t)$$

where N_{it} and N_{jt} cannot be negative and k_t is positive. ■

Lemma 3.7.B *If a sub-matrix,* $\mathbf{C}_0$, *is formed from* $\mathbf{C}$ *by omitting certain rows and the corresponding columns, then* $\mathbf{w}_0'\mathbf{C}_0\mathbf{w}_0 \geqq 0$ *for any real vector,* $\mathbf{w}_0$.

Proof: Expand $\mathbf{w}_0$ to $\mathbf{w}$ by inserting a zero element for each row (or column) omitted, then

$$\mathbf{w}_0{}'\mathbf{C}_0\mathbf{w}_0 = \mathbf{w}'\mathbf{C}\mathbf{w} = (\mathbf{w}'\boldsymbol{\Delta\phi})(\mathbf{w}'\boldsymbol{\Delta\phi})' \geqslant 0. \quad \blacksquare$$

Now, to consider the converse, let $\mathbf{w}$ be a vector such that $\mathbf{Cw} = \mathbf{0}$, i.e. the contrast $\mathbf{w}$ is confounded. Let the treatments be so ordered that $\mathbf{w}$ has three segments, the first with only positive elements, the second in which all are zero and the third in which they are negative. (It is, of course, possible that $\mathbf{w}_2$ will have no elements.) Then,

$$\begin{bmatrix} \mathbf{C}_{11} & \mathbf{C}_{12} & \mathbf{C}_{13} \\ \mathbf{C}'_{12} & \mathbf{C}_{22} & \mathbf{C}_{23} \\ \mathbf{C}'_{13} & \mathbf{C}'_{23} & \mathbf{C}_{33} \end{bmatrix} \begin{bmatrix} \mathbf{w}_1 \\ \mathbf{0} \\ \mathbf{w}_3 \end{bmatrix} = \begin{bmatrix} \mathbf{0} \\ \mathbf{0} \\ \mathbf{0} \end{bmatrix} \tag{3.7.1}$$

From Lemma 3.7.B, $\mathbf{w}'_1\mathbf{C}_{11}\mathbf{w}_1 \geqslant 0$ and $\mathbf{w}'_3\mathbf{C}_{33}\mathbf{w}_3 \geqslant 0$, so $\mathbf{w}'_1\mathbf{C}_{13}\mathbf{w}_3 \leqslant 0$. However, all elements of $\mathbf{w}_1$ are positive, all those of $\mathbf{w}_3$ are negative, while those of $\mathbf{C}_{13}$ cannot be positive, so $\mathbf{w}'_1\mathbf{C}_{13}\mathbf{w}_3 \geqslant 0$. Hence $\mathbf{w}'_1\mathbf{C}_{13}\mathbf{w}_3 = 0$ and $\mathbf{C}_{13}$ is a null-matrix, i.e., treatments of the first sub-set never occur in the same block with any of the third sub-set.

From the first and third rows of (3.7.1), since $\mathbf{C}_{13} = \mathbf{0}$

$$\mathbf{C}_{11}\mathbf{w}_1 = \mathbf{0}, \quad \mathbf{C}_{33}\mathbf{w}_3 = \mathbf{0}. \tag{3.7.2}$$

Also, since $\mathbf{C1}_v = \mathbf{0}$,

$$\mathbf{C}_{11}\mathbf{1}_x + \mathbf{C}_{12}\mathbf{1}_y = \mathbf{0},$$

where the three sub-sets contain x, y and z plots respectively $(x+y+z = n)$. Hence

$$\mathbf{w}'_1\mathbf{C}_{11}\mathbf{1}_x + \mathbf{w}'_1\mathbf{C}_{12}\mathbf{1}_y = 0.$$

Using (3.7.2) it follows that $\mathbf{w}'_1\mathbf{C}_{12}\mathbf{1}_y = 0$. Since all elements of $\mathbf{w}_1$ and $\mathbf{1}_y$ are positive, from Lemma 3.6.A $\mathbf{C}_{12}$ also is a null-matrix and so is $\mathbf{C}_{23}$. Hence the three sub-sets of treatments form three disconnected parts of the design, or two if $\mathbf{w}_2$ is lacking. Hence the converse holds also.

3.8 THE INTER-BLOCK ANALYSIS

So far the block parameters have been regarded as estimates of the effects assigned to blocks, whether at their initial formation on a territorial basis or by subsequent differences brought in to ease administrative problems. In some contexts, however, they may be thought of as being randomly distributed, just as plot residuals are so regarded within blocks. If that is reasonable, it is good policy to find any information that lies between ('inter-block') as well as that already considered, which lies within them ('intra-block').

Of course, it is necessary to consider how much inter-block information is to be found. To take an example, if one block contains Treatments A, C and D and another B, C and D, it is reasonable to regard the difference in their mean performance as due partly to the differing effects of Treatments A and B, but if all blocks are made up in the same way with respect to treatments there is nothing to be found along that path. Again, there may be very few degrees of freedom for estimating an inter-block error, if indeed there are any at all. The design used in Section 3.4 for purposes of illustration makes the point well. It has six blocks and therefore five degrees of freedom altogether inter-block. It also has six treatments, which absorb all the available degrees of freedom, leaving none for error. (The design does in fact afford some inter-block information about treatment differences even though there is no error line with which to judge their precision. The position will be examined in Section 6.4; for the moment it suffices to draw attention to the difficulty.) Nevertheless, there are times when an inter-block analysis is required.

It will be assumed that all blocks contain k plots. It is true that some workers have allowed block sizes to differ (e.g. Patterson and Thompson, 1971) but that is to raise some formidable problems. One concerns randomization. When all blocks are of the same size and there are sub-sets of treatments of the right size to be assigned to them, randomization is feasible, but if some blocks contain this number of plots and some that and if the sub-sets of treatments are the same, no overall randomization procedure is possible. The second problem concerns error structure. For the case of variable block size

$$\mathbf{B} = \mathbf{N}'\boldsymbol{\tau}_0 + \mathbf{k}^{\delta}\boldsymbol{\beta}_0 \tag{3.8.1}$$

the suffix 0 indicating estimates, etc. made inter-block. If now, the error variation derives from that of plots within blocks it is reasonable to write $\mathrm{Cov}(\boldsymbol{\beta}_0) = \mathbf{k}^{-\delta}\sigma_0^2$, but if it derives from differences between blocks $\mathrm{Cov}(\boldsymbol{\beta}_0)$ would be better written as $\mathbf{I}_v\sigma_0^2$. (The σ_0 in the two expressions will not, of course, be the same.) In fact, both sources are to be expected and combined in some indeterminate proportion, so what is $\mathrm{Cov}(\boldsymbol{\beta}_0)$? Only if $\mathbf{k} = k\mathbf{1}_b$ is the difficulty resolved. In that case

$$\mathbf{B} = \mathbf{N}'\boldsymbol{\tau}_0 + k\boldsymbol{\beta}_0. \tag{3.8.2}$$

The $\boldsymbol{\beta}_0$ are now regarded as random independent variates with zero mean and variance σ_0^2/k. (It would be quite possible to call the variance σ_0^2, but the intention is to preserve as much parallelism as possible with the intra-block case and there the variance of a block total would be of the order of σ^2/k.)

There is one important respect in which (3.8.2) differs from (3.1.1). In the intra-block case a general change in the elements of $\boldsymbol{\beta}$ can be compensated by a contrary change in those of $\boldsymbol{\tau}$, but that is not so in the inter-block case. There the fact that all elements of $\boldsymbol{\beta}_0$ have zero expectation implies the constraint that the grand total ($G = \mathbf{1}_b'\mathbf{B}$) shall equal $\mathbf{r}'\boldsymbol{\tau}_0$ ($= \mathbf{1}_b'\mathbf{N}'\boldsymbol{\tau}_0$). The task then is of minimizing $\mathbf{k}\boldsymbol{\beta}_0'\boldsymbol{\beta}_0$ subject to that constraint. Differentiation leads to the conclusion that

$$\mathbf{NN}'\hat{\boldsymbol{\tau}}_0 = \mathbf{NB}. \tag{3.8.3}$$

At this point there appears to be a departure from previous developments and it needs to be examined. Up till now it has been usual to write the total sum of squared deviations as $\mathbf{y}'\boldsymbol{\phi}\mathbf{y}$. Minimizing $\mathbf{k}\boldsymbol{\beta}_0'\boldsymbol{\beta}_0$ in the model

$$\mathbf{B} = \mathbf{1}_b\alpha + \mathbf{k}\boldsymbol{\beta}_0$$

it appears that $\mathbf{y}'\boldsymbol{\phi}_0\mathbf{y}$ should equal $\mathbf{B}'\mathbf{B}/k - G^2/n$, so

$$\boldsymbol{\phi}_0 = \mathbf{D}'\mathbf{D}/k - \mathbf{1}_n\mathbf{1}_n'/n. \tag{3.8.4}$$

By analogy with previous work it is to be expected that

$$\hat{\boldsymbol{\tau}}_0 = \mathbf{C}_0^- \mathbf{Q}_0 \tag{3.8.5}$$

where

$$\mathbf{C}_0 = \boldsymbol{\Delta}\boldsymbol{\phi}_0\boldsymbol{\Delta}' = \mathbf{N}\mathbf{N}'/k - \mathbf{r}\mathbf{r}'/n, \tag{3.8.6a}$$

$$\mathbf{Q}_0 = \boldsymbol{\Delta}\boldsymbol{\phi}_0\mathbf{y} \;= \mathbf{N}\mathbf{B}/k - \mathbf{r}G/n. \tag{3.8.6b}$$

Also, if

$$\boldsymbol{\Omega}_0^{-1} = \mathbf{C}_0 + \mathbf{r}\mathbf{r}'/n = \mathbf{N}\mathbf{N}'/k, \tag{3.8.7}$$

then $\boldsymbol{\Omega}_0$ can be used for $\mathbf{C}_0^-$. (Note that $\boldsymbol{\Omega}_0^{-1}\mathbf{1}_v = \mathbf{r}$ and $\boldsymbol{\Omega}_0\mathbf{r} = \mathbf{1}_v$.) It appears then that the difference between the estimates of $\hat{\boldsymbol{\tau}}_0$ obtained from (3.8.3) and (3.8.5) lies only in a constant difference between corresponding elements equal to the general mean of the data and that will not affect the estimation of differences, i.e.

$$\begin{aligned}\hat{\boldsymbol{\tau}}_0 &= \boldsymbol{\Omega}_0\mathbf{Q}_0 = (\mathbf{N}\mathbf{N}'/k)^{-1}(\mathbf{N}\mathbf{B}/k - \mathbf{r}G/n)\\ &= (\mathbf{N}\mathbf{N}')^{-1}\mathbf{N}\mathbf{B} - \mathbf{1}_vG/n\end{aligned} \tag{3.8.8}$$

because $\boldsymbol{\Omega}_0\mathbf{r} = \mathbf{1}_v$. The point is brought out in Example 3H at the end of the chapter.

From (3.8.5) and (3.8.2)

$$\hat{\boldsymbol{\tau}}_0 = \mathbf{C}_0^-\mathbf{C}_0\boldsymbol{\tau}_0 + \mathbf{C}_0^-(\mathbf{N} - k\mathbf{r}\mathbf{1}_b'/n)\boldsymbol{\beta}_0$$

since $G = \mathbf{1}_b'\mathbf{B}$. Hence $\mathbf{c}'\hat{\boldsymbol{\tau}}_0$ is an unbiassed estimate of $\mathbf{c}'\boldsymbol{\tau}_0$ provided $\mathbf{c}$ is a linear composition of eigenvectors of $\mathbf{C}_0$ (Lemma 3.2.C). Also, since Cov $(\boldsymbol{\beta}_0) = \mathbf{I}_b\sigma_0^2/k$

$$\begin{aligned}\text{Cov}(\hat{\boldsymbol{\tau}}_0) &= \mathbf{C}_0^-(\mathbf{N} - k\mathbf{r}\mathbf{1}_b'/n)(\mathbf{N}' - k\mathbf{1}_b\mathbf{r}'/n)\mathbf{C}_0^-\sigma_0^2\\ &= \mathbf{C}_0^-\mathbf{C}_0\mathbf{C}_0^-\sigma_0^2.\end{aligned} \tag{3.8.9}$$

Although (3.8.9) cannot be used for all purposes, it is permissible to write the variance of $\mathbf{c}'\hat{\boldsymbol{\tau}}$ as

$$\begin{aligned}\text{Var}(\mathbf{c}'\hat{\boldsymbol{\tau}}_0) &= \mathbf{c}'\mathbf{C}_0^-\mathbf{C}_0\mathbf{C}_0^-\mathbf{c}\sigma_0^2\\ &= \mathbf{c}'\mathbf{C}_0^-\mathbf{c}\sigma_0^2\end{aligned} \tag{3.8.10}$$

from Lemma (3.2.C).

From (3.8.2) and (3.8.5)

$$\hat{\boldsymbol{\beta}}_0 = \mathbf{B}/k - \mathbf{N}'\mathbf{C}_0^-\mathbf{Q}_0/k.$$

Hence the minimized value of $k\boldsymbol{\beta}_0'\boldsymbol{\beta}_0$ is

$$\begin{aligned}k\hat{\boldsymbol{\beta}}_0'\hat{\boldsymbol{\beta}}_0 = \mathbf{B}'\mathbf{B}/k &- 2\mathbf{Q}_0'\mathbf{C}_0^-\mathbf{Q}_0 - \mathbf{Q}_0'\mathbf{C}_0^-\mathbf{r}G/n\\ &- \mathbf{r}'\mathbf{C}_0^-\mathbf{Q}_0G/n + \mathbf{Q}_0'\mathbf{C}_0^-\mathbf{C}_0\mathbf{C}_0^-\mathbf{Q}_0\\ &+ \mathbf{Q}_0'\mathbf{C}_0^-\mathbf{r}\mathbf{r}'\mathbf{C}_0^-\mathbf{Q}_0/n = \mathbf{y}'\boldsymbol{\psi}_0\mathbf{y}, \text{ say.}\end{aligned}$$

Here it should be noted that $\mathbf{\Omega}_0$ can be used for $\mathbf{C}_0^-$. Since $\mathbf{\Omega}_0\mathbf{r} = \mathbf{1}_v$ and since Lemma 3.2.D holds for $\boldsymbol{\phi}_0$ as for $\boldsymbol{\phi}$

$$\mathbf{y}'\boldsymbol{\psi}_0\mathbf{y} = k\hat{\boldsymbol{\beta}}_0'\hat{\boldsymbol{\beta}}_0 = \mathbf{B}'\mathbf{B}/k - G^2/n - \mathbf{Q}_0'\mathbf{\Omega}_0\mathbf{Q}_0 = \mathbf{y}'\boldsymbol{\phi}_0\mathbf{y} - \mathbf{Q}_0'\mathbf{\Omega}_0\mathbf{Q}_0. \tag{3.8.11}$$

But Lemma 3.2.E likewise holds for $\boldsymbol{\phi}_0$, so $\mathbf{\Omega}_0$ can be replaced by any other solution for $\mathbf{C}_0^-$. Accordingly the treatment sum of squared deviations may be written as $\mathbf{Q}_0'\mathbf{C}_0^-\mathbf{Q}_0$, analogous to $\mathbf{Q}'\mathbf{C}^-\mathbf{Q}$ for the intra-block analysis.

The Kuiper–Corsten iteration is also available but with the difference that (3.5.3) is replaced by

$$\mathbf{v}_{j+1} = (\mathbf{I}_v - \mathbf{r}^{-\delta}\mathbf{N}\mathbf{N}'/k)\mathbf{v}_j.$$

That is to say, $\mathbf{r}^{-\delta}\mathbf{N}\mathbf{k}^{-\delta}\mathbf{N}'v_j$ is worked out as before, but it is subtracted from $\mathbf{v}_j$ to form $\mathbf{v}_{j+1}$ instead of becoming $\mathbf{v}_{j+1}$ itself.

Accordingly, since $\mathbf{F} = \mathbf{I} - \mathbf{r}^{-\frac{1}{2}\delta}\mathbf{N}\mathbf{k}^{-\delta}\mathbf{N}'\mathbf{r}^{-\frac{1}{2}\delta}$,

$$\begin{aligned}\sum_{i=1}^{\infty} \mathbf{v}_j &= \mathbf{r}^{-\frac{1}{2}\delta}(\mathbf{I} + \mathbf{F} + \mathbf{F}^2 + \mathbf{F}^3 + \cdots)\mathbf{r}^{\frac{1}{2}\delta}\mathbf{v}_1 \\ &= \mathbf{r}^{-\frac{1}{2}\delta}\sum_{i=1}^{\infty}\sum_{j=1}^{h}(\varepsilon_j^{i-1}\,\mathbf{p}_i\mathbf{p}_i')\mathbf{r}^{\frac{1}{2}\delta}\mathbf{v}_1 \\ &= \mathbf{r}^{-\frac{1}{2}\delta}\sum_{j=1}^{h}[\mathbf{p}_i\mathbf{p}_i'/(1-\varepsilon_j)]\mathbf{r}^{\frac{1}{2}\delta}\mathbf{v}_1 \\ &= \mathbf{r}^{-\frac{1}{2}\delta}(\mathbf{I}_v - \mathbf{F})^{+}\,\mathbf{r}^{\frac{1}{2}\delta}\mathbf{v}_i \\ &= \boldsymbol{\Upsilon}_0\mathbf{r}^{\delta}\mathbf{v}_i, \text{ say,}\end{aligned}$$

where

$$\boldsymbol{\Upsilon}_0 = \mathbf{r}^{-\frac{1}{2}\delta}(\mathbf{I}_v - \mathbf{F})^{+}\mathbf{r}^{-\frac{1}{2}\delta}. \tag{3.8.12}$$

It will be seen that $\boldsymbol{\Upsilon}_0$ is a generalized inverse of $\mathbf{C}_0 = \mathbf{r}^{\frac{1}{2}\delta}(\mathbf{I}_v - \mathbf{F})\mathbf{r}^{\frac{1}{2}\delta}$. If $\mathbf{v}_1$ is set equal to $\mathbf{r}^{-\delta}\mathbf{Q}_0$, the iteration will give a valid estimate of $\boldsymbol{\tau}_0$. If, instead, $\mathbf{v}_1$ equals a column of $\mathbf{r}^{-\delta}$, the outcome will be the corresponding column of $\boldsymbol{\Upsilon}_0$. Given contrasts that are confounded inter-block, it is useful to note that

$$\boldsymbol{\Upsilon}_0 = (\boldsymbol{\Upsilon}_0\mathbf{r}^{\delta})(\boldsymbol{\Upsilon}_0\mathbf{r}^{\delta})(\mathbf{r}^{-\delta}\mathbf{C}_0\mathbf{r}^{-\delta}). \tag{3.8.13}$$

That is to say, a column of $\boldsymbol{\Upsilon}_0$ can be found by applying the iteration twice to the corresponding column of $\mathbf{r}^{-\delta}\mathbf{C}_0\mathbf{r}^{-\delta}$.

The algebra above shows clearly the relationship between the inter-block and the intra-block analyses, which depends upon the interchange of roles of $\mathbf{F}$ and $(\mathbf{I}_v - \mathbf{F})$ when passing from one to the other. Hence, if a basic contrast has an efficiency of ε intra-block, its efficiency will be $(1-\varepsilon) = \varepsilon_0$ inter-block. More specifically, if a contrast is of full efficiency in one it will be confounded in the other and *vice versa*.

To consider randomization, it is sufficient to take the groups of treatments to be assigned to blocks and to allocate them at random. On the null-hypothesis the element of $\overset{\circ}{\boldsymbol{\beta}}_0$ corresponding to any field plot is fixed and the error sum of squared deviations can be written as either $k\,\hat{\boldsymbol{\beta}}'_0\,\hat{\boldsymbol{\beta}}_0$ or $k\,\overset{\circ}{\boldsymbol{\beta}}'_0\,\overset{\circ}{\boldsymbol{\beta}}_0$ and equals $(b-1)\sigma_0^2$. Consequently, the expectation of any element squared is $(b-1)\sigma_0^2/n$. Since $\mathbf{1}'_b\,\hat{\boldsymbol{\beta}}_0 = 0$, the expectation of a product of elements is $-\sigma_0^2/n$, i.e.

$$\text{ex}\,(\boldsymbol{\beta}_0\,\boldsymbol{\beta}'_0) = \boldsymbol{\phi}_0\sigma_0^2\,. \tag{3.8.14}$$

That result has conclusions that will be examined further in Section 4.9. For the moment it suffices to note that the randomization argument in Section 3.4 applies inter-block as well as intra-block. As before, the degrees of freedom for treatments number

$$\text{tr}\,(\mathbf{C}_0\,\mathbf{C}_0^-). \tag{3.8.15}$$

3.9 BLOCKS IN CONJUNCTION WITH OTHER FORMS OF LOCAL CONTROL

So far several methods of effecting local control have been considered separately. They can, however, be used in conjunction. Thus blocks and covariance are a common combination, each being used to control different sources of variation, e.g., when blocks are formed to allow for environmental differences and covariates are added to account for provenance or other characteristics of the plant material.

Generally speaking it is best to use whatever method suits the particular sort of variation; if that means combining methods the complication should be accepted. Thus environmental variation may respond to covariance, either upon trends or upon measured characteristics of soil. Alternatively, it may respond rather to adjustment by neighbouring plots, if there is patchiness, or to blocks, if the site suggests them. Even if blocks are unlikely to control much environmental variation they may still be needed for other reasons and must be used whatever methods of local control are required in addition. What is wrong is to try to force a method on to a problem that it does not suit. For example, when long-lived plants, like trees or shrubs, are planted, it is reasonable to grade them and to plant each block with uniform material. It is not reasonable to plant them without regard to the grading and to impose dispersed blocks later, based on initial size, because that could lead to difficulty with environmental variation and would be useless for purposes of administration. If initial grading is not possible, e.g. planting must take place so soon after lifting that detailed measurements are not practicable, it may be sound to use a combination of blocks for environmental and administrative reasons and of covariance for characteristics, like early growth, that are apparent before the treatments are applied. There is no difficulty about that. The matrices $\boldsymbol{\phi}$ and $\boldsymbol{\psi}$ have been re-defined for a block design respectively in (3.1.6) and above (3.3.7). The change in $\boldsymbol{\psi}$ leads to a change in $\mathbf{V}$ (defined in 2.3.3) and of $\boldsymbol{\phi}$ in $\mathbf{V}_0$ (2.3.6), but that is all. The relationships (2.3.8) and (2.3.9) still apply.

Similarly there need be no objection to a combination of blocks with adjustment by neighbouring plots. It should be remembered, however, that the concomitant variate remains the deviations of plot values from treatment means and does not become $\hat{\boldsymbol{\eta}}$ as defined in (3.3.7). The roles of the two quantities are quite different. The one is a measure of local soil effect, whereas the other measures the contribution that the plot is making to the error sum of squared deviations. The fact that they are the same for a completely randomized design is irrelevant.

Examples 3

A. The following experiment was conducted some thirty years ago at the East Malling Research Station (Pearce, 1953) and its data have been used for purposes of demonstration many times since.

It was intended to find out if the application of herbicides, then new-fangled and open to suspicion, would harm the growth of strawberry plants. The treatments comprised four kinds of herbicide, A–D, and an untreated control, O. The data, which follow, represent the total spread in inches of twelve sample plants from each plot:

Block I	Block II	Block III	Block IV
D.107	A.136	C.118	O.173
A.166	O.146	A.117	D. 95
B.133	D.104	O.176	D.109
C.166	C.152	B.132	A.130
O.177	B.119	C.139	B.103
A.163	O.164	O.186	O.185
O.190	B.132	D.103	C.147

Carry out an analysis of variance on the data.

Answer: The method will be by way of (3.3.1), (3.3.5) etc. That will require a solution for $\mathbf{C}^{-}$. Here $\boldsymbol{\Omega}$ will be chosen as defined in (3.4.1).

First it will be helpful to study the design.

$$\mathbf{N} = \begin{bmatrix} 2 & 1 & 1 & 1 \\ 1 & 2 & 1 & 1 \\ 1 & 1 & 2 & 1 \\ 1 & 1 & 1 & 2 \\ 2 & 2 & 2 & 2 \end{bmatrix} \qquad \mathbf{N}\mathbf{k}^{-\delta}\mathbf{N}' = \tfrac{1}{7}\begin{bmatrix} 7 & 6 & 6 & 6 & 10 \\ 6 & 7 & 6 & 6 & 10 \\ 6 & 6 & 7 & 6 & 10 \\ 6 & 6 & 6 & 7 & 10 \\ 10 & 10 & 10 & 10 & 16 \end{bmatrix}$$

$$\mathbf{C} = \tfrac{1}{7}\begin{bmatrix} 28 & -6 & -6 & -6 & -10 \\ -6 & 28 & -6 & -6 & -10 \\ -6 & -6 & 28 & -6 & -10 \\ -6 & -6 & -6 & 28 & -10 \\ -10 & -10 & -10 & -10 & 40 \end{bmatrix} \quad \boldsymbol{\Omega}^{-1} = \tfrac{1}{28}\begin{bmatrix} 137 & 1 & 1 & 1 & 0 \\ 1 & 137 & 1 & 1 & 0 \\ 1 & 1 & 137 & 1 & 0 \\ 1 & 1 & 1 & 137 & 0 \\ 0 & 0 & 0 & 0 & 224 \end{bmatrix}.$$

Check that $\mathbf{N}\mathbf{k}^{-\delta}\mathbf{N}'\mathbf{1}_v = \mathbf{r}$, $\mathbf{C}\mathbf{1}_v = \mathbf{0}$, $\mathbf{\Omega}^{-1}\mathbf{1}_v = \mathbf{r}$.

$$\mathbf{\Omega} = \tfrac{1}{680}\begin{bmatrix} 139 & -1 & -1 & -1 & 0 \\ -1 & 139 & -1 & -1 & 0 \\ -1 & -1 & 139 & -1 & 0 \\ -1 & -1 & -1 & 139 & 0 \\ 0 & 0 & 0 & 0 & 85 \end{bmatrix}.$$

Check that $\mathbf{\Omega}\mathbf{r} = \mathbf{1}_v$.

Next, the adjusted treatment totals are required. The unadjusted totals are A, 712; B, 619; C, 722; D, 518; O, 1397. Written as a column vector these quantities are called $\mathbf{T}$. The corresponding block totals, which are called $\mathbf{B}$, are I, 1102; II, 953; III, 971; IV, 942. The block means ($\mathbf{b} = \mathbf{k}^{-\delta}\mathbf{B}$) are readily found by dividing each total by 7, and the adjusted treatment totals are $\mathbf{Q} = \mathbf{T} - \mathbf{N}\mathbf{k}^{-\delta}\mathbf{B}$, i.e.

$$\begin{aligned} Q_{\mathrm{A}} &= T_{\mathrm{A}} - (2b_1 + b_2 + b_3 + b_4) = -12.29 \\ Q_{\mathrm{B}} &= T_{\mathrm{B}} - (b_1 + 2b_2 + b_3 + b_4) = -84.00 \\ Q_{\mathrm{C}} &= T_{\mathrm{C}} - (b_1 + b_2 + 2b_3 + b_4) = 16.43 \\ Q_{\mathrm{D}} &= T_{\mathrm{D}} - (b_1 + b_2 + b_3 + 2b_2) = -183.43 \\ Q_{\mathrm{O}} &= T_{\mathrm{O}} - (2b_1 + 2b_2 + 2b_3 + 2b_4) = 263.29. \end{aligned}$$

Check that $\mathbf{Q}'\mathbf{1}_v = 0$.

The next step is to find the estimated treatment parameters, $\hat{\boldsymbol{\tau}} = \mathbf{\Omega}\mathbf{Q}$, i.e.,

$$\begin{aligned} \hat{\tau}_{\mathrm{A}} &= (139Q_{\mathrm{A}} - Q_{\mathrm{B}} - Q_{\mathrm{C}} - Q_{\mathrm{D}})/680 = -2.14 \\ \hat{\tau}_{\mathrm{B}} &= (-Q_{\mathrm{A}} + 139Q_{\mathrm{B}} - Q_{\mathrm{C}} - Q_{\mathrm{D}})/680 = -16.91 \\ \hat{\tau}_{\mathrm{C}} &= (-Q_{\mathrm{A}} - Q_{\mathrm{B}} + 139Q_{\mathrm{C}} - Q_{\mathrm{D}})/680 = +3.77 \\ \hat{\tau}_{\mathrm{D}} &= (-Q_{\mathrm{A}} - Q_{\mathrm{B}} - Q_{\mathrm{C}} + 139Q_{\mathrm{D}})/680 = -37.38 \\ \hat{\tau}_{\mathrm{O}} &= 85Q_{\mathrm{O}}/680 = +32.91. \end{aligned}$$

Check that $\mathbf{r}'\hat{\boldsymbol{\tau}} = 0$. (This equation of constraint applies only when $\hat{\boldsymbol{\tau}}$ has been estimated by the use of $\mathbf{\Omega}$.)

Hence the treatment sum of squared deviations ($\mathbf{Q}'\mathbf{\Omega}\mathbf{Q} = \mathbf{Q}'\hat{\boldsymbol{\tau}}$) is

$$\begin{aligned} &(-12.29)(-2.14) + (-84.00)(-16.91) + (16.43)(3.77) \\ &+ (-183.43)(-37.38) + (263.29)(32.91) = 17030. \end{aligned}$$

The corresponding figure for blocks ($\mathbf{B}'\mathbf{k}^{-\delta}\mathbf{B} - G^2/n$) is

$$(1102^2 + 953^2 + 971^2 + 942^2)/7 - 3968^2/28 = 2366.$$

(In calculating this figure no allowance has been made for treatments. Totals are taken as they are.) The total sum of squared deviations ($\mathbf{y}'\mathbf{y} - G^2/n$) is

$$(107^2 + 136^2 + 118^2 + \cdots + 103^2 + 147^2) - 3968^2/12 = 22856.$$

The analysis of variance can therefore be written

Source	*d.f*	*s-s*	*m-s*
Blocks	3	2366	
Treatments	4	17030	4258
Error*	20	3460	$173 = \hat{\sigma}^2$
Total	27	22856	

* by difference

The analysis is needed not for an F-test but to estimate σ^2. From $\mathbf{\Omega}$ the variance of a difference such as $(\tau_A - \tau_O)$ is

$$(139 + 85 - 0 - 0)\sigma^2/680$$
$$= 0.329\sigma^2 = 56.92 = (7.54)^2.$$

The effects of the four herbicides are:

$$\hat{\tau}_A - \hat{\tau}_O = -35.1$$
$$\hat{\tau}_B - \hat{\tau}_O = -49.8$$
$$\hat{\tau}_C - \hat{\tau}_O = -29.2$$
$$\hat{\tau}_D - \hat{\tau}_O = -70.3.$$

From these figures it is clear that an earlier generation was not unreasonable in being suspicious of herbicides. Indeed, since any of the above differences is significant ($P < 0.05$) if it exceeds 15.7 ($= 2.086 \times 7.54$, 2.086 being the critical value of t for 20 degrees of freedom for error), it appears that all the herbicides did harm.

B. Repeat the analysis of data in the last example but using $\mathbf{\Xi}$ instead of $\mathbf{\Omega}$ as a solution for $\mathbf{C}^-$ (3.4.4)

Answer:

$$\mathbf{W} = \tfrac{1}{7}\begin{bmatrix} 0 & 6 & 6 & 6 & 10 \\ 6 & 0 & 6 & 6 & 10 \\ 6 & 6 & 0 & 6 & 10 \\ 6 & 6 & 6 & 0 & 10 \\ 10 & 10 & 10 & 10 & 0 \end{bmatrix},$$

$$\mathbf{q} = \mathbf{W}\mathbf{1}_v, \qquad u = \mathbf{q}'\mathbf{1}_v,$$

$$\mathbf{\Xi}^{-1} = \mathbf{C} + \mathbf{q}\mathbf{q}'/u$$

$$= \tfrac{1}{133}\begin{bmatrix} 630 & -16 & -16 & -16 & -50 \\ -16 & 630 & -16 & -16 & -50 \\ -16 & -16 & 630 & -16 & -50 \\ -16 & -16 & -16 & 630 & -50 \\ -50 & -50 & -50 & -50 & 960 \end{bmatrix}.$$

Check that $\boldsymbol{\Xi}^{-1}\mathbf{1}_v = \mathbf{q}$.

$$\boldsymbol{\Xi} = \frac{1}{245480}\begin{bmatrix} 52185 & 1645 & 1645 & 1645 & 2975 \\ 1645 & 52185 & 1645 & 1645 & 2975 \\ 1645 & 1645 & 52185 & 1645 & 2975 \\ 1645 & 1645 & 1645 & 52185 & 2975 \\ 2975 & 2975 & 2975 & 2975 & 34629 \end{bmatrix}.$$

Check that $\boldsymbol{\Xi}\mathbf{q} = \mathbf{1}_v$.

The example illustrates the awkwardness of $\boldsymbol{\Xi}$ in most arithmetical contexts. Its usefulness lies rather in theoretical studies. From this point $\boldsymbol{\Xi}$ can be used instead of $\boldsymbol{\Omega}$ without affecting any of the results obtained previously.

C. Analyse the data again, this time using $\mathbf{C}^+$ for $\mathbf{C}^-$.
Answer: The difficulty here is that the eigenvectors of $\mathbf{C}$ are not known. However, it is possible to make a good guess. Whenever the elements of a sub-set of treatments have the same status in the design and nothing would be changed if they were permuted, any contrast that involves them alone is usually an eigenvector. An orthogonal set is

$$\begin{aligned} \mathbf{u}_1' &= (1 \quad -1 \quad 0 \quad 0 \quad 0)/\sqrt{2} \\ \mathbf{u}_2' &= (1 \quad 1 \quad -2 \quad 0 \quad 0)/\sqrt{6} \\ \mathbf{u}_3' &= (1 \quad 1 \quad 1 \quad -3 \quad 0)/\sqrt{12}. \end{aligned}$$

That leaves

$$\mathbf{u}_4' = (1 \quad 1 \quad 1 \quad 1 \quad -4)/\sqrt{20}$$

as the only possibility for the last. It appears that all four are in fact eigenvectors, the eigenvalues being

$$\lambda_1 = \lambda_2 = \lambda_3 = \tfrac{34}{7}, \lambda_4 = \tfrac{50}{7}.$$

Hence

$$\mathbf{C}^+ = \tfrac{7}{34}(\mathbf{u}_1\mathbf{u}_1' + \mathbf{u}_2\mathbf{u}_2' + \mathbf{u}_3\mathbf{u}_3') + \tfrac{7}{50}\mathbf{u}_4\mathbf{u}_4'$$

$$= \tfrac{7}{34}\cdot\tfrac{1}{4}\begin{bmatrix} 3 & -1 & -1 & -1 & 0 \\ -1 & 3 & -1 & -1 & 0 \\ -1 & -1 & 3 & -1 & 0 \\ -1 & -1 & -1 & 3 & 0 \\ 0 & 0 & 0 & 0 & 0 \end{bmatrix} + \tfrac{7}{50}\cdot\tfrac{1}{20}\begin{bmatrix} 1 & 1 & 1 & 1 & -4 \\ 1 & 1 & 1 & 1 & -4 \\ 1 & 1 & 1 & 1 & -4 \\ 1 & 1 & 1 & 1 & -4 \\ -4 & -4 & -4 & -4 & 16 \end{bmatrix}$$

$$= \frac{1}{4250}\begin{bmatrix} 686 & -189 & -189 & -189 & -119 \\ -189 & 686 & -189 & -189 & -119 \\ -189 & -189 & 686 & -189 & -119 \\ -189 & -189 & -189 & 686 & -119 \\ -119 & -119 & -119 & -119 & 476 \end{bmatrix}.$$

Check that $\mathbf{C}^+\mathbf{1}_v = \mathbf{0}$. However, the check is not a strong one, because it should work for each matrix, $\mathbf{u}_i\mathbf{u}_i'$, separately and most mistakes concern their coefficients. A further check is afforded by the trace of $\mathbf{C}^+$ being the sum of its intended eigenvalues, i.e.

$$(4 \times \tfrac{686}{4250}) + \tfrac{476}{4250} = (3 \times \tfrac{7}{34}) + \tfrac{7}{50}.$$

It will be found that the same conclusions will be reached as in Examples 3A and 3B.

D. Use the Kuiper–Corsten iteration to find $\boldsymbol{\Upsilon}$ for the design of Example 3A. Elements should be correct to three places of decimals. The method is given in Section 3.5.

Answer: The first task must be to decide the precise form of the projection and dual-projection. If $\mathbf{v}$ is a vector for treatments, then

$$\begin{aligned} u_1 &= (2v_1 + v_2 + v_3 + v_4 + 2v_5)/7 \\ u_2 &= (v_1 + 2v_2 + v_3 + v_4 + 2v_5)/7 \\ u_3 &= (v_1 + v_2 + 2v_3 + v_4 + 2v_5)/7 \\ u_4 &= (v_1 + v_2 + v_3 + 2v_4 + 2v_5)/7, \end{aligned}$$

the coefficients being derived from the appropriate column of $\mathbf{N}$, the incidence matrix, and the divisors being the block sizes. For the dual-projection

$$\begin{aligned} v_1 &= (2u_1 + u_2 + u_3 + u_4)/5 \\ v_2 &= (u_1 + 2u_2 + u_3 + u_4)/5 \\ v_3 &= (u_1 + u_2 + 2u_3 + u_4)/5 \\ v_4 &= (u_1 + u_2 + u_3 + 2u_4)/5 \\ v_5 &= (2u_1 + 2u_2 + 2u_3 + 2u_4)/8. \end{aligned}$$

The coefficients are now derived from the rows of $\mathbf{N}$; the devisors are the treatment replications.

To find the first column of $\boldsymbol{\Upsilon}$ the iteration will start with the first column of $\mathbf{r}^{-\delta} - \mathbf{1}_v\mathbf{1}_v'/n$, i.e.

$$\mathbf{v}_1' = (0.1643,\ -0.0357,\ -0.0357,\ -0.0357,\ -0.0357).$$

The iteration will cease when no element of $\mathbf{v}_j$ lies outside the range $\pm$ 0.0003

$\mathbf{v}_1$	$\mathbf{v}_2$	$\mathbf{v}_3$	$\mathbf{u}_1$	$\mathbf{u}_2$
0.1643	0.0043	0.0001	0.0214	0.0006
−0.0357	−0.0014	0.0000	−0.0071	−0.0002
−0.0357	−0.0014	0.0000	−0.0071	−0.0002
−0.0357	−0.0014	0.0000	−0.0071	−0.0002
−0.0357	0.0000	0.0000		

Hence

$$\Sigma \mathbf{v}'_j = (0.169, -0.037, -0.037, -0.037, -0.036).$$

A danger comes from the required relationship, $\mathbf{r}'\mathbf{v}_j = 0$, breaking down in some cycle on account of rounding errors. To avoid that danger, in each cycle, j, the elements of $\mathbf{v}_j$ should be adjusted by subtracting $\mathbf{r}'\mathbf{v}_j/n$ before proceeding. Otherwise the iteration may give a false result and may even diverge.

The second, third and fourth columns can be found similarly, but there is no need to repeat the iteration. The only requirement is a permutation of the elements of the first. The fifth column presents no difficulty, because

$$\mathbf{v}'_1 = (-0.0357, -0.0357, -0.0357, -0.0357, 0.0893)$$

gives $\mathbf{u}_1 = \mathbf{0}$, so

$$\Upsilon = \begin{bmatrix} 0.169 & -0.037 & -0.037 & -0.037 & -0.036 \\ -0.037 & 0.169 & -0.037 & -0.037 & -0.036 \\ -0.037 & -0.037 & 0.169 & -0.037 & -0.036 \\ -0.037 & -0.037 & -0.037 & 0.169 & -0.036 \\ -0.036 & -0.036 & -0.036 & -0.036 & 0.089 \end{bmatrix}.$$

E. Analyse the data in Example 3A using the Kuiper–Corsten iteration, as described in Section 3.5.

Answer: Although Υ has been found in the last example, the calculation would more usually proceed thus:

$$\mathbf{v}_1 = \mathbf{r}^{-\delta}\mathbf{Q} = (-2.458, -16.800, +3.286, -36.686, +32.91)'.$$

Hence by the iteration

$$\hat{\boldsymbol{\tau}} = \sum_j \mathbf{v}_j = (-2.14, -16.91, +3.77, -37.38, +32.91)'.$$

The treatment sum of squares can now be found directly as $\mathbf{Q}'\Upsilon^{-}\mathbf{Q} = \mathbf{Q}'\Upsilon\mathbf{Q} = \mathbf{Q}'\hat{\boldsymbol{\tau}} = 17\,030$, as in Example 3A.

The error sum of squared deviations can now be found by difference but it is better to find it from the block parameters even if only for the check. Now,

$$\hat{\boldsymbol{\beta}} = \mathbf{k}^{-\delta}\mathbf{B} - \sum_j \mathbf{u}_j.$$

The first term represents the block means.

$$\hat{\boldsymbol{\beta}} = (155.86, 136.68, 136.30, 138.03)'.$$

It is now possible to find the residual for each plot. Thus, for the first it is

$$107 - 155.86 - (-37.38) = -11.5$$

and for the last it is

$$147 - 138.03 - 3.76 = +5.2.$$

Squaring the residuals and summing provides a check on the error sum of squared deviations.

F. In the design introduced in Section 3.4 to illustrate the chapter, i.e.,

Block I	A	B	C	D	Block IV	A	D	E	F
II	A	B	C	E	V	B	D	E	F
III	A	B	C	F	VI	C	D	E	F

let the treatments represent three fertilizer substances, X, Y and Z, applied at two levels, 1 and 2, i.e.

$$A = X1 \qquad B = Y1$$
$$C = Z1 \qquad D = X2$$
$$E = Y2 \qquad F = Z2.$$

Now consider the following contrasts:

$$\mathbf{u}_1 = (1 \;\; -1 \;\; 0 \;\; 1 \;\; -1 \;\; 0)'/2,$$
$$\mathbf{u}_2 = (1 \;\; 1 \;\; -2 \;\; 1 \;\; 1 \;\; -2)'/\sqrt{12}.$$

Together they represent the main effect of substances, because the first refers to the difference between X and Y, while the second relates Z to the means of the two levels considered. Also,

$$\mathbf{u}_3 = (1 \;\; 1 \;\; 1 \;\; -1 \;\; -1 \;\; -1)'/\sqrt{6}$$

represents the main effect of doses. The interaction of substance and dose is given by $\mathbf{u}_4$ and $\mathbf{u}_5$, derived respectively from $\mathbf{u}_1$ and $\mathbf{u}_3$ and from $\mathbf{u}_2$ and $\mathbf{u}_3$, i.e.

$$\mathbf{u}_4 = (1 \;\; -1 \;\; 0 \;\; -1 \;\; 1 \;\; 0)'/2,$$
$$\mathbf{u}_5 = (1 \;\; 1 \;\; -2 \;\; -1 \;\; -1 \;\; 2)'/\sqrt{12}.$$

What are the variances of these contrasts?

Answer: The variance of a contrast, $\mathbf{c}$, is $\mathbf{c}'\mathbf{C}^-\mathbf{c}\sigma^2$. The evaluation of σ^2 must derive from the data, but it is immaterial what matrix is used for $\mathbf{C}^-$, whether $\boldsymbol{\Omega}$, $\boldsymbol{\Xi}$, or any other generalized inverse. It will be found that

$$\mathbf{u}_1'\mathbf{C}^-\mathbf{u}_1 = \mathbf{u}_2'\mathbf{C}^-\mathbf{u}_2 = \tfrac{4}{15}$$
$$\mathbf{u}_3'\mathbf{C}^-\mathbf{u}_3 = \tfrac{1}{3}$$
$$\mathbf{u}_4'\mathbf{C}^-\mathbf{u}_4 = \mathbf{u}_5'\mathbf{C}^-\mathbf{u}_5 = \tfrac{4}{15}.$$

G. A visitor comments that the above design was obviously intended to effect a good comparison between X, Y and Z. (He has noticed the care in keeping A, B and C together in blocks and similarly D, E and F.) He asks if that was intended. Would it not have been better, he asks, to have studied the main effects of levels better by keeping A and D, B and E, C and F together? The design could then have been like this:

Block	I	B	C	E	F
	II	A	C	D	F
	III	A	B	D	E
	IV	B	C	E	F
	V	A	C	D	F
	VI	A	B	D	E

(Design α)

He also suggests trying this design to see what happens:

Block	I	A	B	E	F
	II	A	B	D	F
	III	A	C	D	E
	IV	A	C	E	F
	V	B	C	D	E
	VI	B	C	D	F

(Design β)

Examine these proposals by working out $\mathbf{c}_j'\mathbf{C}^-\mathbf{c}_j$ for all j in all designs.
Answer: Possible generalized inverses are:

$$\boldsymbol{\Omega}_\alpha = \tfrac{1}{72}\begin{bmatrix} 20 & -1 & -1 & 2 & -1 & -1 \\ -1 & 20 & -1 & -1 & 2 & -1 \\ -1 & -1 & 20 & -1 & -1 & 2 \\ 2 & -1 & -1 & 20 & -1 & -1 \\ -1 & 2 & -1 & -1 & 20 & -1 \\ -1 & -1 & 2 & -1 & -1 & 20 \end{bmatrix}$$

$$\boldsymbol{\Omega}_\beta = \tfrac{1}{2340}\begin{bmatrix} 643 & -29 & -29 & -32 & 16 & 16 \\ -29 & 643 & -29 & 16 & -32 & 16 \\ -29 & -29 & 643 & 16 & 16 & -32 \\ -32 & 16 & 16 & 643 & -29 & -29 \\ 16 & -32 & 16 & -29 & 643 & -29 \\ 16 & 16 & -32 & -29 & -29 & 643 \end{bmatrix}.$$

Hence, for Design α,

$$\mathbf{c}_1'\boldsymbol{\Omega}\mathbf{c}_1 = \mathbf{c}_2'\boldsymbol{\Omega}\mathbf{c}_2 = \tfrac{1}{3}, \quad \mathbf{c}_3'\boldsymbol{\Omega}\mathbf{c}_3 = \mathbf{c}_4'\boldsymbol{\Omega}\mathbf{c}_4 = \mathbf{c}_5'\boldsymbol{\Omega}\mathbf{c}_5 = \tfrac{1}{4}$$

and for Design β,

$$\mathbf{c}_1'\boldsymbol{\Omega}\mathbf{c}_1 = \mathbf{c}_2'\boldsymbol{\Omega}\mathbf{c}_2 = \tfrac{4}{15}, \quad \mathbf{c}_3'\boldsymbol{\Omega}\mathbf{c}_3 = \tfrac{1}{4}, \quad \mathbf{c}_4'\boldsymbol{\Omega}\mathbf{c}_4 = \mathbf{c}_5'\boldsymbol{\Omega}\mathbf{c}_5 = \tfrac{4}{13}.$$

Comment: The visitor has correctly noticed the salient feature of the adopted design and has successfully transferred its good feature to the other factor, i.e. that of levels, so that it is now estimated with full efficiency. (That is to be expected since A and D always come together, as do B and E and also C and F, so that the particular effects of levels for each substance are well estimated. For the same reason, in Design α the interaction also is found with full efficiency.) In Design β the visitor has associated each treatment as far as possible with others that have nothing in common with it, either substance or level, e.g. A with E and F, B with D and F, etc. The effect has not been to estimate the interaction well. (In that respect Design α is unassailable.) In fact he has contrived to effect a compromise between the other two, one of which was good for the main effect of substances and the other for that of levels, in each case at the expense of information elsewhere.

It will be noticed that the contrasts have been standardized. That would not have been done in practice but was done here to allow a comparison between contrasts as well as between designs. It will be noticed also that they have been estimated independently, i.e., for all designs $\mathbf{c}_i'\mathbf{C}^-\mathbf{c}_j = 0$ if $i \neq j$. Such a feature is certainly desirable and arises here because the contrasts of interest are all eigenvectors of the $\mathbf{C}$ for the various designs.

H. Examine the inter-block analysis of the design used to illustrate this chapter, i.e. that in Example 3F. (The fact that there are no degrees of freedom for the inter-block error should be noted, though it is of no immediate importance.)
Answer: First

$$\boldsymbol{\Omega}_0^{-1} = \mathbf{NN}'/k = \tfrac{1}{4}\begin{bmatrix} 4 & 3 & 3 & 2 & 2 & 2 \\ 3 & 4 & 3 & 2 & 2 & 2 \\ 3 & 3 & 4 & 2 & 2 & 2 \\ 2 & 2 & 2 & 4 & 3 & 3 \\ 2 & 2 & 2 & 3 & 4 & 3 \\ 2 & 2 & 2 & 3 & 3 & 4 \end{bmatrix},$$

so

$$\boldsymbol{\Omega}_0 = \tfrac{1}{8}\begin{bmatrix} 23 & -9 & -9 & -1 & -1 & -1 \\ -9 & 23 & -9 & -1 & -1 & -1 \\ -9 & -9 & 23 & -1 & -1 & -1 \\ -1 & -1 & -1 & 23 & -9 & -9 \\ -1 & -1 & -1 & -9 & 23 & -9 \\ -1 & -1 & -1 & -9 & -9 & 23 \end{bmatrix}.$$

The treatment parameters may be estimated as $\boldsymbol{\Omega}_0\,\mathbf{Q}_0 = \boldsymbol{\Omega}_0\,(\mathbf{NB}/k - \mathbf{r}G/n)$. Alternatively,

$$\mathbf{C}_0 = \mathbf{NN}'/k - \mathbf{rr}'/n = \tfrac{1}{12}\begin{bmatrix} 4 & 1 & 1 & -2 & -2 & -2 \\ 1 & 4 & 1 & -2 & -2 & -2 \\ 1 & 1 & 4 & -2 & -2 & -2 \\ -2 & -2 & -2 & 4 & 1 & 1 \\ -2 & -2 & -2 & 1 & 4 & 1 \\ -2 & -2 & -2 & 1 & 1 & 4 \end{bmatrix},$$

$$\mathbf{C}_0^{+} = \tfrac{1}{6}\begin{bmatrix} 17 & -7 & -7 & -1 & -1 & -1 \\ -7 & 17 & -7 & -1 & -1 & -1 \\ -7 & -7 & 17 & -1 & -1 & -1 \\ -1 & -1 & -1 & 17 & -7 & -7 \\ -1 & -1 & -1 & -7 & 17 & -7 \\ -1 & -1 & -1 & -7 & -7 & 17 \end{bmatrix}.$$

Another estimator for $\boldsymbol{\tau}_0$ is now available, namely $\mathbf{C}_0^+\mathbf{Q}_0 = \mathbf{C}_0^+(\mathbf{NB}/k - \mathbf{r}G/n)$.

The alternatives can be illustrated by supposing that each plot of A gives a datum of 1, each of B gives 2 and so on, then

$$\mathbf{B} = (10, 11, 12, 16, 17, 18)'$$
$$\mathbf{NB}/k = \tfrac{1}{4}(49, 50, 51, 61, 62, 63)'$$
$$\mathbf{Q}_0 = \tfrac{1}{4}(-7, -6, -5, 5, 6, 7)'.$$

Hence

$$(\mathbf{NN}')^{-1}\mathbf{NB} = \boldsymbol{\Omega}_0\mathbf{NB}/k = \begin{bmatrix} 1 \\ 2 \\ 3 \\ 4 \\ 5 \\ 6 \end{bmatrix}, \qquad \boldsymbol{\Omega}_0\mathbf{Q}_0 = \mathbf{C}_0^+\mathbf{Q}_0 = \tfrac{1}{2}\begin{bmatrix} -5 \\ -3 \\ -1 \\ 1 \\ 3 \\ 5 \end{bmatrix}.$$

For purposes of estimating contrasts the two solutions are equivalent. The first gives the same general mean as the data, because

$$\mathbf{r}'\boldsymbol{\Omega}_0\mathbf{NB}/k = \mathbf{1}_v'\mathbf{NB}/k = \mathbf{1}_v'\mathbf{B} = G,$$

whereas

$$\mathbf{r}'\mathbf{C}_0^+\mathbf{Q}_0 = \mathbf{r}'\mathbf{r}^{-\frac{1}{2}\delta}\mathbf{F}^+\mathbf{r}^{-\frac{1}{2}\delta}\mathbf{Q}_0 = (\mathbf{1}_v'\mathbf{r}^{\frac{1}{2}})\mathbf{F}^+\mathbf{r}^{-\frac{1}{2}\delta}\mathbf{Q}_0 = 0$$

because $\mathbf{r}^{\frac{1}{2}\delta}\mathbf{1}_v$ is the eigenvector of $\mathbf{F}$, and hence of $\mathbf{F}^+$, with the zero eigenvalue.

Since the Kuiper–Corsten iteration converged quickly for the intra-block analysis it is unlikely to do so for the inter-block and a few cycles will confirm that suspicion. In general,

$$\mathbf{v}_{j+1} = (\mathbf{I}_v - \mathbf{r}^{-\delta}\mathbf{NN}'/k)\mathbf{v}_j.$$

Here

$$\mathbf{v}_{j+1} = \frac{1}{16}\begin{bmatrix} 12 & -3 & -3 & -2 & -2 & -2 \\ -3 & 12 & -3 & -2 & -2 & -2 \\ -3 & -3 & 12 & -2 & -2 & -2 \\ -2 & -2 & -2 & 12 & -3 & -3 \\ -2 & -2 & -2 & -3 & 12 & -3 \\ -2 & -2 & -2 & -3 & -3 & 12 \end{bmatrix}\mathbf{v}_j.$$

To find the treatment parameters, the iteration begins

$$\mathbf{v}_1 = \mathbf{r}^{-\delta}\mathbf{Q}_0 = \begin{bmatrix} -0.4375 \\ -0.3750 \\ -0.3125 \\ 0.3125 \\ 0.3750 \\ 0.4375 \end{bmatrix}, \qquad \mathbf{v}_2 = \begin{bmatrix} -0.3398 \\ -0.2813 \\ -0.2227 \\ 0.2227 \\ 0.2813 \\ 0.3398 \end{bmatrix}, \qquad \mathbf{v}_3 = \begin{bmatrix} -0.2658 \\ -0.2110 \\ -0.1560 \\ 0.1560 \\ 0.2110 \\ 0.2658 \end{bmatrix}.$$

From either $\boldsymbol{\Omega}_0$ or $\mathbf{C}_0^+$ it will be found that for the contrast of (*A − *B) the variance is $8\sigma_0^2$, so $\varepsilon_0 = \frac{1}{16}$. For the intra-block analysis $\varepsilon = \frac{15}{16}$, so $\varepsilon + \varepsilon_0 = 1$, but that is a basic contrast. For (*A − *D), however, $\varepsilon_0 = \frac{1}{12}$ and $\varepsilon = \frac{45}{62}$, so $\varepsilon + \varepsilon_0 < 1$.

I. Find a generalized inverse for this design:

Block I	A	B	C		Block V	A	C	H
II	D	E	F		VI	D	F	G
III	E	F	G		VII	A	B	H
IV	B	C	H		VIII	D	E	G

Hint: For a disconnected design $\mathbf{C}^-$ can be found from the relationship

$$\begin{bmatrix} \mathbf{C}_1 & \mathbf{0} \\ \mathbf{0} & \mathbf{C}_2 \end{bmatrix}^- = \begin{bmatrix} \mathbf{C}_1^- & \mathbf{0} \\ \mathbf{0} & \mathbf{C}_2^- \end{bmatrix}.$$

Chapter 4

Some useful design concepts

4.1 ORTHOGONALITY

The commonest way of designing an experiment in practice is to use 'randomized blocks', introduced by Fisher. In that design each block contains as many plots as there are treatments and each treatment occurs once in each block i.e. $\mathbf{N} = \mathbf{1}_v\mathbf{1}'_b$, $r = b$, $k = v$, $n = bv$, $\mathbf{C} = b\mathbf{I}_v - (b/v)\mathbf{1}_v\mathbf{1}'_v$, $\mathbf{\Omega}^{-1} = b\mathbf{I}_v$, $\mathbf{\Omega} = \mathbf{I}_v/b$. That is to say, the covariance matrix of treatment means is $\mathbf{r}^{-\delta}\sigma^2$ as for a completely randomized design and all contrasts are estimated with full efficiency. This property, together with its simplicity, justifies its widespread use.

Indeed, it may be asked why anyone ever uses anything else. In general it is a most useful design but there are times when it raises problems. For example it may not be possible to form blocks of the size required. If it is possible, there could still be advantages in some other blocking system. Again, as was explained at the end of Section 2.1, there are experimental situations when the treatments are better not made equi-replicate. Nevertheless, most people called upon to design an experiment think first of randomized blocks and they are quite right to do so.

Randomized blocks are an obvious example of an 'orthogonal' design. The term was first used (Yates, 1933a) to indicate that the treatment means are unaffected by differences between blocks. These days when it is recognized that the intra-block analysis does not estimate treatment means anyway, it is necessary to modernize the definition and say that an orthogonal design is one in which all contrasts are estimated with full efficiency, i.e. in $\mathbf{F}$ all eigenvalues except the last equal one. Since the last eigenvalue is zero and corresponds to the eigenvector, $\mathbf{r}^{\frac{1}{2}\delta}\mathbf{1}_v/\sqrt{n}$, then

$$\mathbf{F} = \sum_{i=1}^{v}\left(\varepsilon_i\mathbf{p}_i\mathbf{p}'_i\right) = \mathbf{I}_v - \mathbf{r}^{\frac{1}{2}\delta}\mathbf{1}_v\mathbf{1}'_v\mathbf{r}^{\frac{1}{2}\delta}/n \tag{4.1.1}$$

and

$$\mathbf{C} = \mathbf{r}^{\delta} - \mathbf{r}\mathbf{r}'/n \qquad (= \mathbf{r}^{\delta} - \mathbf{N}\mathbf{k}^{-\delta}\mathbf{N}'). \tag{4.1.2}$$

Orthogonality then depends upon

$$\mathbf{N}\mathbf{k}^{-\delta}\mathbf{N}' - \mathbf{r}\mathbf{r}'/n$$

being a null-matrix. Its ith diagonal element equals

$$\sum_{j=1}^{b} (N_{ij}^2/k_j) - r_i^2/n = \sum_j [k_j (N_{ij}/k_j - r_i/n)^2].$$

If all such elements equal zero, then all N_{ij} must equal $r_i k_j/n$, i.e.

$$\mathbf{N} = \mathbf{r}\mathbf{k}'/n. \tag{4.1.3}$$

If that is so, $\mathbf{\Omega}^{-1} = \mathbf{r}^{\delta}$ and $\mathbf{\Omega} = \mathbf{r}^{-\delta}$. Example 4B at the end of the chapter takes the argument a little further. An equivalent definition of orthogonality can thus be given in combinatorial terms by saying that an orthogonal design is one in which each block is made up proportionately in the same way with respect to treatments. Thus, the following design is orthogonal:

Block I	A	A	B	C				
II	A	A	A	A	B	B	C	C

Here half the plots of each block are given over to Treatment A and a quarter to each of Treatments B and C. More usual is the case in which each block contains an equal number of plots of all treatments. Thus, someone might start out with the idea of using four blocks, each of three treatments. It then occurs to him that he has left only six degrees of freedom for error, so he decides to use two blocks instead, thus:

Block I	A	A	B	B	C	C
II	A	A	B	B	C	C

which leaves eight. There need be no objection if, as may well be the case, there is no call for particularly small blocks. (Of course, if he decided that an experiment with only twelve plots did not need blocks at all, he might well go to the extreme and use a completely randomized design.)

In an orthogonal design, treatment parameters can be estimated by the corresponding actual means because

$$\hat{\tau} = \mathbf{\Omega}\mathbf{Q} = \mathbf{r}^{-\delta}(\mathbf{T} - \mathbf{r}G/n) \tag{4.1.4}$$

and the treatment sum of squared deviations is

$$\mathbf{Q}'\mathbf{\Omega}\mathbf{Q} = \mathbf{T}'\mathbf{r}^{-\delta}\mathbf{T} - G^2/n. \tag{4.1.5}$$

Hence, the corresponding value for error is

$$\mathbf{y}'\mathbf{y} - \mathbf{B}'\mathbf{k}^{-\delta}\mathbf{B} - \mathbf{T}'\mathbf{r}^{-\delta}\mathbf{T} + G^2/n. \tag{4.1.6}$$

Example 4A shows some actual calculations.

That leads to a particularly simple way of computing the analysis of variance for an orthogonal block design using four 'summation terms'. The first, the total term, is the sum of data squared, i.e.

$$S = \mathbf{y}'\mathbf{y} = \sum y_i^2.$$

The second, S_b, the block term, is found from

$$S_b = \mathbf{B}'\mathbf{k}^{-\delta}\mathbf{B} = \sum(B_j^2/k_j).$$

The third, the treatment term, is

$$S_t = \mathbf{T}'\mathbf{r}^{-\delta}\mathbf{T} = \sum(T_i^2/r_i),$$

and the fourth, the correction term, is

$$S_0 = G^2/n.$$

From what has been said, the sum of squared deviations for blocks is $(S_b - S_0)$ with $(b-1)$ degrees of freedom, that for treatments is $(S_t - S_0)$ with $(v-1)$ and that for error is $(S - S_b - S_t + S_0)$ with $(b-1)(v-1)$. The whole sums to $(S - S_0)$ with $(bv-1)$ degrees of freedom.

On account of the development of these alternative orthogonal designs difficulty has arisen about nomenclature. Thus some prefer to speak of the original design as being in 'randomized replicate blocks' to distinguish designs where each block contains exactly one replicate from those in which it contains two or more. There need be no objection provided they do not presume to correct those who use the traditional terminology. Again some are careful always to get the word 'complete' in somewhere, arguing that randomization is required of any block design and the distinguishing feature implied is that each block shall contain a complete replicate. It is better not to become involved in such disputes. The name 'randomized blocks' now has an established and honoured place in statistical nomenclature and it would be a pity to alter it. (When it is considered how many statistical terms arose in the first place from ignorance or misapprehension but are now hallowed by use, a change here would be absurd.) However, to explain the usage in this monograph, the word 'orthogonal' will indicate the most general case in which $\mathbf{N} = \mathbf{r}\mathbf{k}'/n$, the term 'randomized blocks' being confined to designs in which $\mathbf{N} = \alpha \mathbf{1}_v \mathbf{1}_b'$, where α is some integer. If it is necessary to specify the original case in which $\alpha = 1$, the term 'randomized replicate blocks' certainly has merit.

The term 'orthogonal' itself has acquired a variety of meanings and Preece (1977) has helpfully examined the terminological confusion.

4.2 PARTIAL ORTHOGONALITY

It is possible for the orthogonality relationship given by (4.1.3) to apply for some treatments but not for others. In that case

$$\mathbf{N} = \frac{1}{n}\begin{bmatrix} \rho \mathbf{k}' \\ \mathbf{M} \end{bmatrix}, \qquad \mathbf{r} = \mathbf{N}\mathbf{1}_b = \begin{bmatrix} \rho \\ \rho_0 \end{bmatrix}, \text{ say,}$$

where ρ is the vector of replications of the treatments disposed orthogonally and $\mathbf{M}$ is arbitrary $(\mathbf{M}\mathbf{1}_b/n = \rho_0)$. Then

$$\mathbf{\Omega}^{-1} = \begin{bmatrix} \rho^\delta & \mathbf{0}' \\ \mathbf{0} & \rho_0^\delta - \mathbf{M}\mathbf{k}^{-\delta}\mathbf{M}'/n^2 + \rho_0\rho_0'/n \end{bmatrix}. \tag{4.2.1.}$$

It appears that a contrast will be estimated with full efficiency if it involves only treatments for which the orthogonality law holds.

This little piece of algebra, trivial though it may appear, provides a further argument for using orthogonal designs. If mistakes do occur in the application of treatments, the arbitrariness of **M** will not affect contrasts between treatments correctly applied. Also, if a complete solution is required, it is necessary to invert only that sub-matrix of $\mathbf{\Omega}^{-1}$ that relates to the treatments involved in the muddle. If there are only two such treatments, a simple method exists for finding the variance of the contrast between them, as the next section shows.

In some contexts it is helpful to isolate the non-orthogonal component of the design. Writing $\mathbf{\Omega}^{-1}$ as in (3.4.1), i.e.

$$\mathbf{\Omega}^{-1} = \mathbf{r}^{\delta} - \mathbf{N}\mathbf{k}^{-\delta}\mathbf{N}' + \mathbf{r}\mathbf{r}'/n,$$

then

$$\mathbf{\Omega}^{-1} = \mathbf{r}^{\delta} - \mathbf{A}\mathbf{k}^{-\delta}\mathbf{A}', \tag{4.2.2}$$

where

$$\mathbf{A} = \mathbf{N} - \mathbf{r}\mathbf{k}'/n. \tag{4.2.3}$$

In (4.2.2), $\mathbf{\Omega}^{-1}$ is written in two terms, the first of which corresponds to the orthogonal case and the second to the deviation brought about by any non-orthogonality. The matrix, **A**, is called the 'non-orthogonality matrix' (Pearce, 1970a) and will be used in various contexts later. If any treatment conforms to (4.1.3) the row and column in **A** for that treatment is void. In any case $\mathbf{A}\mathbf{1}_v = \mathbf{0}$.

It may be noted in connection with the Kuiper–Corsten iteration (Section 3.5) that

$$\mathbf{r}^{-\delta}\mathbf{A}\mathbf{k}^{-\delta}\mathbf{A}' = \mathbf{r}^{-\delta}\mathbf{N}\mathbf{k}^{-\delta}\mathbf{N}' - \mathbf{1}_v\mathbf{r}'/n.$$

Since in each cycle $\mathbf{r}'\mathbf{v}_i = 0$, it follows that **A** can be used instead of **N**. Where **A** contains several void rows and columns, that can be a saving; on the other hand, if it does not, the labour of calculating it is not repaid.

4.3 CONCORDANCE AND DISCORDANCE

Although 'concordance' is not of itself necessarily a desirable relationship between two treatments, an understanding of what is implied by the term and its opposite, 'discordance', can be helpful. The difference will be illustrated by reference to the design used in Section 3.4. First, it may be recalled that

$$\mathbf{\Omega} = \tfrac{1}{360}\begin{bmatrix} 99 & 3 & 3 & -5 & -5 & -5 \\ 3 & 99 & 3 & -5 & -5 & -5 \\ 3 & 3 & 99 & -5 & -5 & -5 \\ -5 & -5 & -5 & 99 & 3 & 3 \\ -5 & -5 & -5 & 3 & 99 & 3 \\ -5 & -5 & -5 & 3 & 3 & 99 \end{bmatrix},$$

the order of the treatments being A, B, C, D, E, F. Suppose now that Treatments A and B turn out to be the same. That may happen because of a mishap, some chemical spray arriving too late, its defection forcing a change of plan, or because the specification of the treatments failed to distinguish them. Thus it may be that irrigation was to be applied to A whenever the soil moisture deficit exceeded five centimetres, B being left unirrigated for purposes of comparison, but the season was a wet one and the deficit never reached such levels. Whatever the reason, A and B have become merged and the experiment has only five treatments instead of six, namely, AB, C, D, E, F. Using the suffix, *m*, to indicate the merged case,

$$\mathbf{\Omega}_m = \frac{1}{360}\begin{bmatrix} 51 & 3 & -5 & -5 & -5 \\ 3 & 99 & -5 & -5 & -5 \\ -5 & -5 & 99 & 3 & 3 \\ -5 & -5 & 3 & 99 & 3 \\ -5 & -5 & 3 & 3 & 99 \end{bmatrix}.$$

It will be seen that contrasts that do not involve either of the merged treatments are estimated with the same precision as before. That is to say, A and B have merged without there being repercussions elsewhere in the experiment. For that reason they will be said to be 'concordant'.

The situation is different if the merged treatments are A and D. The treatments are now AD, B, C, E, F and will be taken in that order. For the new design

$$\mathbf{\Omega}_m = \frac{1}{4680}\begin{bmatrix} 611 & -13 & -13 & -13 & -13 \\ -13 & 1283 & 35 & -61 & -61 \\ -13 & 35 & 1283 & -61 & -61 \\ -13 & -61 & -61 & 1283 & 35 \\ -13 & -61 & -61 & 35 & 1283 \end{bmatrix}.$$

It will be seen that some of the contrasts between B, C, E and F now have different precisions. For (*B – *C) and (*E – *F) the variance of estimation remains $\frac{8}{15}\sigma^2$, but for those like (*B – *E) it is $\frac{112}{195}\sigma^2$ where before it was $\frac{26}{45}\sigma^2$. It is true that the change is for the better, but the present point is that treatments A and D in merging have had effects elsewhere. Accordingly they will be said to be 'discordant'.

To consider the phenomenon further, let the matrix **E** be either **C** or $\mathbf{\Omega}^{-1}$. (The final choice will be made later and could depend upon the actual design, but for the moment that can be left.) Whichever **E** may be chosen it will be convenient to order the treatments so that the two to be merged stand first. Then, **E** may be written in partitioned form, thus, **X** referring to the merged treatments,

$$\mathbf{E} = \begin{bmatrix} \mathbf{X} & \mathbf{Y} \\ \mathbf{Y}' & \mathbf{Z} \end{bmatrix} \tag{4.3.1}$$

Some lemmas will now give properties of partitioned matrices.

Lemma 4.3.A *If a matrix,* **E**, *can be written as in (4.3.1), then*

$$\mathbf{E}^{-1} = \begin{bmatrix} \mathbf{X}^{-1} + \mathbf{X}^{-1}\mathbf{Y}\mathbf{W}^{-1}\mathbf{Y}'\mathbf{X}^{-1} & -\mathbf{X}^{-1}\mathbf{Y}\mathbf{W}^{-1} \\ -\mathbf{W}^{-1}\mathbf{Y}'\mathbf{X}^{-1} & \mathbf{W}^{-1} \end{bmatrix},$$

where $\mathbf{W} = \mathbf{Z} - \mathbf{Y}'\mathbf{X}^{-1}\mathbf{Y}$. *The lemma is readily proved by calculating* $\mathbf{EE}^{-1}$. *This result is, of course, well known.*

Lemma 4.3.B *If* $\mathbf{E}$ *in (4.3.1) is singular, so also is* $\mathbf{W}$.

Proof: If

$$\begin{bmatrix} \mathbf{X} & \mathbf{Y} \\ \mathbf{Y}' & \mathbf{Z} \end{bmatrix} \begin{bmatrix} \mathbf{u} \\ \mathbf{v} \end{bmatrix} = \begin{bmatrix} \mathbf{0} \\ \mathbf{0} \end{bmatrix},$$

then $\mathbf{Xu} + \mathbf{Yv} = \mathbf{0}$, $\mathbf{Y}'\mathbf{u} + \mathbf{Zv} = \mathbf{0}$. Hence

$$\begin{aligned} \mathbf{Wv} &= \mathbf{Zv} - \mathbf{Y}'\mathbf{X}^{-1}\mathbf{Yv} = \mathbf{Zv} + \mathbf{Y}'\mathbf{X}^{-1}\mathbf{Xu} \\ &= \mathbf{Zv} + \mathbf{Y}'\mathbf{u} = \mathbf{0}. \quad \blacksquare \end{aligned}$$

Lemma 4.3.C *If* $\mathbf{E}$ *and therefore* $\mathbf{W}$ *are singular,* $\mathbf{E}^-$ *may be written for* $\mathbf{E}^{-1}$ *in Lemma 4.3.A, using* $\mathbf{W}^-$ *instead of* $\mathbf{W}^{-1}$.

Proof: The result is proved by calculating $\mathbf{EE}^-\mathbf{E}$. ■

If the first two treatments are merged, $\mathbf{E}$ becomes

$$\mathbf{E}_m = \begin{bmatrix} T & \mathbf{1}_2'\mathbf{Y} \\ \mathbf{Y}'\mathbf{1}_2 & \mathbf{Z} \end{bmatrix},$$

where T is the sum of the elements in $\mathbf{X}$. Now

$$\mathbf{W}_m = \mathbf{Z} - \mathbf{Y}'\mathbf{1}_2\mathbf{1}_2'\mathbf{Y}/T.$$

If

$$\mathbf{Y}'\mathbf{X}^{-1}\mathbf{Y} = \mathbf{Y}'\mathbf{1}_2\mathbf{1}_2'\mathbf{Y}/\mathbf{T}, \tag{4.3.2}$$

then $\mathbf{W} = \mathbf{W}_m$ and the merged treatments are concordant. It is not difficult to show that (4.3.2) is equivalent to the condition that

$$(X_{11} + X_{12})Y_{2j} = (X_{12} + X_{22})Y_{1j} \tag{4.3.3}$$

for all j. If then a matrix $\mathbf{E}$ exists ($\mathbf{E} = \mathbf{C}$ or $\mathbf{\Omega}^{-1}$) such that (4.3.3) holds, then the first two treatments, as now ordered, are concordant.

Further, (4.3.2) implies that $\mathbf{Y}$ can be written

$$\mathbf{Y} = \begin{bmatrix} (X_{11} + X_{12})\boldsymbol{\alpha}' \\ (X_{12} + X_{22})\boldsymbol{\alpha}' \end{bmatrix},$$

where $\boldsymbol{\alpha}$ is some vector with $(v-2)$ elements. Hence

$$\mathbf{X}^{-1}\mathbf{Y} = \frac{1}{D}\begin{bmatrix} X_{22} & -X_{12} \\ -X_{12} & X_{11} \end{bmatrix} \begin{bmatrix} (X_{11} + X_{12})\boldsymbol{\alpha}' \\ (X_{12} + X_{22})\boldsymbol{\alpha}' \end{bmatrix} = \begin{bmatrix} \boldsymbol{\alpha}' \\ \boldsymbol{\alpha}' \end{bmatrix}, \tag{4.3.4}$$

where $D = X_{11}X_{22} - (X_{12})^2$, the determinant of $\mathbf{X}$.

From Lemma 4.3.A the contrast between the parameters of the first two treatments has a variance of

$$(\mathbf{c}'\mathbf{X}^{-1}\mathbf{c} + \mathbf{c}'\mathbf{X}^{-1}\mathbf{YW}^{-1}\mathbf{YX}^{-1}\mathbf{c})\sigma^2 = \mathbf{c}'\mathbf{X}^{-1}\mathbf{c}\sigma^2,$$

where $\mathbf{c}' = (1 \quad -1 \quad \mathbf{0}')$, because from (4.3.4) $\mathbf{c}'\mathbf{X}^{-1}\mathbf{Y} = \mathbf{0}$. Writing T for the sum of elements of $\mathbf{X}$, it follows that the variance is

$$\frac{T\sigma^2}{D}. \tag{4.3.5}$$

This result is of some importance. It shows that if two treatments are concordant, the variance of the contrast between them can be derived entirely from characteristics of $\mathbf{X}$, the rest of $\mathbf{E}$ being of no importance. Further, two treatments are concordant if $\mathbf{E}$ can be found to satisfy (4.3.3).

To take the example of the design already considered, for purposes of merging Treatments A and B, $\mathbf{E}$ can be taken as either $\mathbf{C}$ or $\boldsymbol{\Omega}^{-1}$; both comply with the condition in (4.3.3). Using $\mathbf{C}$,

$$\mathbf{X} = \tfrac{1}{4}\begin{bmatrix} 12 & -3 \\ -3 & 12 \end{bmatrix} \qquad \mathbf{Y} = \tfrac{1}{4}\begin{bmatrix} -3 & -2 & -2 & -2 \\ -3 & -2 & -2 & -2 \end{bmatrix}$$

Hence, from (4.3.3) the two treatments are concordant. Further, from (4.3.5) the variance of their contrast is

$$\frac{T\sigma^2}{D} = \frac{8\sigma^2}{15},$$

because

$$T = \tfrac{1}{4}(12 - 3 - 3 + 12) = \tfrac{9}{2},$$
$$D = \tfrac{1}{16}(12^2 - 3^2) = \tfrac{135}{16}.$$

If, however, Treatments A and D are merged,

$$\mathbf{X} = \tfrac{1}{4}\begin{bmatrix} 12 & -2 \\ -2 & 12 \end{bmatrix} \qquad \mathbf{Y} = \begin{bmatrix} -3 & -3 & -2 & -2 \\ -2 & -2 & -3 & -3 \end{bmatrix}$$

and (4.3.3) does not hold, nor does it do so for any alternative $\mathbf{E}$, i.e. $\boldsymbol{\Omega}^{-1}$.

To take another example, for the design

Block I	A	A	B	C
II	A	A	B	D
III	A	B	C	D

$$\mathbf{C} = \tfrac{1}{4}\begin{bmatrix} 11 & -5 & -3 & -3 \\ -5 & 9 & -2 & -2 \\ -3 & -2 & 6 & -1 \\ -3 & -3 & -1 & 6 \end{bmatrix}$$

There is no need to look far to establish the concordance of Treatments A and B, because from $\mathbf{C}$

$$\mathbf{X} = \tfrac{1}{4}\begin{bmatrix} 11 & -5 \\ -5 & 9 \end{bmatrix}, \qquad \mathbf{Y} = \tfrac{1}{4}\begin{bmatrix} -3 & -3 \\ -2 & -2 \end{bmatrix}$$

and (4.3.3) holds, though it does not do so for $\boldsymbol{\Omega}^{-1}$.

Finally, if a design is orthogonal and two treatments are involved in a muddle, $\mathbf{Y}$ will be a null-matrix on account of (4.2.1) and (4.3.3) will necessarily hold for $\mathbf{\Omega}^{-1}$. Hence the variance of the contrast between the two wrongly assigned treatments can be found by putting $\mathbf{X}$ equal to $\rho_0^\delta - \mathbf{MK}^{-\delta}\mathbf{M}'/n^2 + \rho_0\rho_0'/n$ and calculating T and D for use in (4.3.5).

4.4 PROPORTIONATE TREATMENTS

An important special case of treatments being concordant arises when they are 'proportionate'. Section 4.2 dealt with treatments such that their occurrence in any block bears a constant ratio to the size of that block; the condition now is that the occurrence of one should be in a constant ratio to the occurrence of the other, as happens with A and B in this example:

Block	I	A	B	B	C	D	
	II	A	A	B	B	B	B
	III	C	C	D	D		

As a practical design it is nearly useless, but it does illustrate proportionality, because, whenever A occurs in a block, B occurs twice as often. In general,

$$\mathbf{N} = \begin{bmatrix} \boldsymbol{\alpha}' \\ \theta\boldsymbol{\alpha}' \\ \mathbf{M} \end{bmatrix}, \qquad \mathbf{r} = \begin{bmatrix} e_1 \\ \theta r_1 \\ \rho \end{bmatrix},$$

where $\rho = \mathbf{M1}_{v-2}$. Hence

$$\mathbf{C} = \begin{bmatrix} r_1 - x & -\theta x & -\mathbf{z}' \\ -\theta x & \theta r_1 - \theta^2 x & -\theta\mathbf{z}' \\ -\mathbf{z} & -\theta\mathbf{z} & \rho^\delta - \mathbf{Mk}^{-\delta}\mathbf{M}' \end{bmatrix},$$

where $x = \boldsymbol{\alpha}'\mathbf{k}^{-\delta}\boldsymbol{\alpha}$ and $\mathbf{z} = \mathbf{Mk}^{-\delta}\boldsymbol{\alpha}$.

It follows from (4.3.3) that the first two treatments are concordant. Hence, from (4.3.5), the variance of their contrast is

$$\frac{(1+\theta)r_1 - (1+\theta)^2 x}{\theta r_1^2 - \theta(1+\theta)r_1 x}\sigma^2 = \left(\frac{1}{r_1} + \frac{1}{\theta r_1}\right)\sigma^2.$$

That is to say their contrast is estimated with full efficiency regardless of the rest of the design.

Again, the result could be used as an argument for orthogonal designs where initially all treatments taken in pairs are proportionate. If a disaster ensues leaving some treatments undamaged, some of the pairs will remain proportionate and contrasts between them will still be estimated with full efficiency. (A similar argument could, of course, be used, though with reduced effect, in support of other designs in which some pairs are initially proportionate.)

However, the main usefulness of the property becomes apparent not when plots are lost but when additional ones are proposed. It sometimes happens that the field is not rectangular and a conventionally designed experiment will leave areas round itself that could form additional plots. In such cases someone will almost certainly propose that they be used for supplementary treatments that were omitted from the main experiment only with reluctance. (Of course, the additional plots should be associated with the nearest block and randomized with its other plots. The supplementary treatments should not be left round the perimeter.) If the design is in randomized blocks there need be no objection on narrowly statistical grounds, always provided that the arbitrary enlargement of some blocks is not going to cause trouble with the error variance, bearing in mind the assumption that all plots, regardless of block, should be equally subject to error. A design like

Block	I	A	B	C	D		
	II	A	B	C	D	E	
	III	A	B	C	D	E	F
	IV	A	B	C	D	F	
	V	A	B	C	D	E	F
	VI	A	B	C	D		

is not very likeable, but the main treatments, A, B, C and D, remain proportionate, so contrasts between them remain of full efficiency. Federer (1956) called such designs 'augmented' and the name is very appropriate. A colleague, however, had suggested calling them 'penthouse designs' and that would be appropriate too, because it could well express the reaction of a biometrician, who found that his plans for a superb cathedral had been spoiled by the unthinking addition of bicycle sheds for the clergy.

4.5 EQUIVALENT TREATMENTS

Two treatments are said to be 'equivalent' if interchanging them has no effect on **C**. Thus, to take the design that has already been used for purposes of illustration, i.e.

Block	I	A	B	C	D
	II	A	B	C	E
	III	A	B	C	F
	IV	A	D	E	F
	V	B	D	E	F
	VI	C	D	E	F

it would make no difference to either **C** or **r** if A, B, and C were permuted. They therefore form an 'equivalence set', as do D, E and F. Further, it would make no difference if A, B and C as a whole were interchanged with D, E and F, which is a third equivalence property in the design.

If permutation has no effect on **C** or **r**, it cannot affect $\mathbf{\Omega}^{-1}$ either or, consequently $\mathbf{\Omega}$. First of all, that helps in the matrix inversion, because it makes clear that $\Omega_{AA} = \Omega_{BB} = \Omega_{CC}$ and $\Omega_{BC} = \Omega_{AC} = \Omega_{AD}$. Taking the other equivalence properties into account, it follows that $\boldsymbol{\Omega}$ has the form

$$\begin{bmatrix} a & b & b & c & c & c \\ b & a & b & c & c & c \\ b & b & a & c & c & c \\ c & c & c & a & b & b \\ c & c & c & b & a & b \\ c & c & c & b & b & a \end{bmatrix}.$$

From a knowledge of $\boldsymbol{\Omega}^{-1}$ it appears that

$$\begin{aligned} 44a - 2b + 6c &= 12 \\ -a + 43b + 6c &= 0 \\ 2a + 4b + 42c &= 0 \end{aligned} \qquad (4.5.1)$$

and the inversion is easy.

More important, however, are the statistical consequences. If the first two treatments are equivalent, i.e. if

$$\mathbf{X} = \begin{bmatrix} a & b \\ b & a \end{bmatrix} \quad \text{and} \quad Y_{1j} = Y_{2j} \qquad \text{for all } j \geqslant 3,$$

then from (4.3.3) the two treatments are concordant. The variance of their contrast can therefore be derived from (4.3.5). It is

$$\frac{(2a+2b)\sigma^2}{(a-b)} = \frac{2\sigma^2}{a-b}. \qquad (4.5.2)$$

Thus, in the example, the difference between the first two equations of (4.5.1) gives $45(a-b) = 12$, so the variance is $\frac{24}{45}\sigma^2 = \frac{8}{15}\sigma^2$, a result already known but by less simple methods. A similar situation is presented in Example 4B.

In fact, where treatments are of similar logical status in the objectives of the experiment, they should ideally be equivalent in the design also. Thus, in Example 3A, there was no reason to distinguish between the four herbicides because the same questions were being asked about all of them. It was therefore satisfactory that they could form an equivalence set. If that were not practicable, and one of them had to be compared with the control in a less precise manner than the others, which one should it be? A logical but foolish answer would be to select for that unfavoured position the one most likely to have a large effect, but imagine the reproaches of the experimenter if it did indeed perform well but did not show significance on account of 'a statistical trick', the others, which had had less effect receiving the accolade of significance! As a matter of diplomacy, which should be in the armoury of every good biometrician, it would be better to deal frankly with the experimenter and to explain that one of his treatments will have to be less precisely compared than the others. Which can he best spare? Sometimes, where his treatments are not intended for immediate practical use but form part of a logical sequence, he may insist that they must be treated alike. If he does, the biometrician should think again with this additional requirement in mind, possibly sacrificing some other property that he had hoped to incorporate.

4.6 DUALITY

Although equation (3.1.1) is symmetrical with respect to blocks and treatments all the attention so far has been directed to the study of treatment effects eliminating those due to blocks, but there is no reason in principle against studying the effect of blocks eliminating those of treatments. It is just that mostly no one is interested. However, it is wise at times to enquire whether blocks are indeed removing much variation and there are occasions when the blocks represent not spatial distinctions but, say, the source of material or the residual effects of past treatments. It is then convenient to use the 'dual' of the original design, blocks and treatments having exchanged roles. (Of course, the blocks have not been randomized.)

There is no difficulty about that. Using an asterisk to denote quantities in the dual design,

$$\mathbf{N}_* = \mathbf{N}', \qquad \mathbf{r}_* = \mathbf{k}, \qquad \mathbf{k}_* = \mathbf{r} \tag{4.6.1}$$

and

$$\mathbf{C}_* = \mathbf{k}^\delta - \mathbf{N}'\mathbf{r}^{-\delta}\mathbf{N}. \tag{4.6.2}$$

For example, to take the design in Exercises 3A to 3D,

$$\mathbf{C}_* = \tfrac{17}{10}\begin{bmatrix} 3 & -1 & -1 & -1 \\ -1 & 3 & -1 & -1 \\ -1 & -1 & 3 & -1 \\ -1 & -1 & -1 & 3 \end{bmatrix},$$

which is easier to deal with than $\mathbf{C}$.

Further, if $\mathbf{F}$ is written as $\mathbf{I}_v - \boldsymbol{\xi}\boldsymbol{\xi}'$, where $\boldsymbol{\xi} = \mathbf{r}^{-\frac{1}{2}\delta}\mathbf{N}\,\mathbf{k}^{-\frac{1}{2}\delta}$, then

$$\mathbf{F}_* = \mathbf{I}_b - \boldsymbol{\xi}'\boldsymbol{\xi} \tag{4.6.3}$$

Consequently, if $\mathbf{F}\mathbf{p}_i = \varepsilon_i\,\mathbf{p}_i$

$$\mathbf{F}_*(\boldsymbol{\xi}'\mathbf{p}_i) = \varepsilon_i(\boldsymbol{\xi}'\mathbf{p}_i) \tag{4.6.4}$$

i.e., $\boldsymbol{\xi}'\mathbf{p}_i$ is a corresponding eigenvector of $\mathbf{F}_*$ with the same efficiency.

It is, however, of interest to trace the relationship between the two designs. The following argument is derived from that of Corsten (1976) and starts from the completely neutral position of asking if the two design matrices, $\mathbf{D}$ for blocks and $\boldsymbol{\Delta}$ for treatments, are independent, i.e. if there are perhaps vectors $\mathbf{a}$ and $\mathbf{b}$ such that the elements of $\mathbf{D}'\mathbf{a}$ are correlated with those of $\boldsymbol{\Delta}'\mathbf{b}$. Since a correlation coefficient between two sets of quantities is unaffected by a change of origin, there will be no loss of generality in referring both $\mathbf{a}$ and $\mathbf{b}$ to origins equal to the respective means of the elements, i.e. in writing

$$\mathbf{k}'\mathbf{a} = 0 = \mathbf{r}'\mathbf{b}. \tag{4.6.5}$$

The correlation coefficient, ρ, is then such that

$$\rho = \frac{\mathbf{b}'\boldsymbol{\Delta}\mathbf{D}'\mathbf{a}}{(\mathbf{b}'\boldsymbol{\Delta}\boldsymbol{\Delta}'\mathbf{b}\,\mathbf{a}'\mathbf{D}\mathbf{D}'\mathbf{a})^{1/2}} = \frac{\mathbf{b}'\mathbf{N}\mathbf{a}}{(\mathbf{b}'\mathbf{r}^\delta\mathbf{b}\,\mathbf{a}'\mathbf{k}^\delta\mathbf{a})^{1/2}}.$$

Scaling **a** and **b** so that

$$\mathbf{a}'\mathbf{k}^{\delta}\mathbf{a} = 1 = \mathbf{b}'\mathbf{r}^{\delta}\mathbf{b}, \tag{4.6.6}$$

$\rho = \mathbf{b}'\mathbf{N}\mathbf{a}$. If now **b** is chosen and the question asked what solution $\mathbf{a}_0$ of **a** will be most highly correlated with it, the problem reduces to the maximization of $\mathbf{b}'\mathbf{N}\mathbf{a}$ subject to the constraint that $\mathbf{a}'\mathbf{k}^{\delta}\mathbf{a} = 1$. (If that appears to neglect the possibility of a high negative correlation, it should be recalled that a reversal of the sign of **a** would give a high positive one.) In fact,

$$\mathbf{N}'\mathbf{b} = 2\lambda\mathbf{k}^{\delta}\mathbf{a}_0,$$

where λ is the Lagrangian multiplier, so

$$\mathbf{a}_0 = \mathbf{k}^{-\delta}\mathbf{N}'\mathbf{b}/(2\lambda). \tag{4.6.7}$$

Hence, since $\mathbf{a}'\mathbf{k}^{\delta}\mathbf{a}_0 = 1$, from (4.6.7),

$$\mathbf{b}'\mathbf{N}\mathbf{k}^{-\delta}\mathbf{N}'\mathbf{b} = 4\lambda^2$$

and

$$\max(\rho) = \mathbf{b}'\mathbf{N}\mathbf{a}_0 = \mathbf{b}'\mathbf{N}\mathbf{k}^{-\delta}\mathbf{N}'\mathbf{b}/(2\lambda). \tag{4.6.8}$$

If the design is orthogonal, i.e., if $\mathbf{N} = \mathbf{r}\mathbf{k}'/n$ (see 4.1.3), then $\mathbf{N}\mathbf{k}^{-\delta}\mathbf{N}' = \mathbf{r}\mathbf{r}'/n$ and from (4.6.5), $\mathbf{N}\mathbf{k}^{-\delta}\mathbf{N}'\mathbf{b} = 0$ regardless of the choice of **b**. For such a design, then, the two matrices are quite independent.

Special interest attaches the case when **b** is chosen to equal a basic contrast, i.e., $\mathbf{b} = \mathbf{s}_i$, say, $= \mathbf{r}^{-\frac{1}{2}\delta}\mathbf{p}_i$, where

$$\mathbf{F}\mathbf{p}_i = (\mathbf{I}_v - \mathbf{r}^{\frac{1}{2}\delta}\mathbf{N}\mathbf{k}^{-\delta}\mathbf{N}'\mathbf{r}^{-\frac{1}{2}\delta})\mathbf{p}_i = \varepsilon_i\mathbf{p}_i.$$

In that case

$$\rho^2 = \mathbf{s}_i'\mathbf{N}\mathbf{k}^{-\delta}\mathbf{N}'\mathbf{s}_i = \mathbf{p}_i'\mathbf{r}^{-\frac{1}{2}\delta}\mathbf{N}\mathbf{k}^{-\delta}\mathbf{N}'\mathbf{r}^{-\frac{1}{2}\delta}\mathbf{p}_i = 1 - \varepsilon_i. \tag{4.6.9}$$

The orthogonal design has already been noted as having all $\varepsilon_i = 1$ so $\rho^2 = 0$. If, however, the design is confounded and some $\varepsilon_i = 0$, then $\rho^2 = 1$. The conclusion is indeed obvious; the disconnection can be represented equally well by a contrast between treatments or by one between blocks. In short, as the two matrices become more related, so it becomes more feasible to express the same contrast in terms of either blocks or treatments. Further, if $\mathbf{b} = \mathbf{s}_i = \mathbf{r}^{-\frac{1}{2}\delta}\mathbf{p}_i$, then from (4.6.7) $\mathbf{a}_0 = \mathbf{k}^{-\delta}\mathbf{N}'\mathbf{r}^{-\frac{1}{2}\delta}\mathbf{p}_i = \mathbf{k}^{-\delta}\mathbf{N}'\mathbf{s}_i$, which is a basic contrast of the dual design, conforming as it does to (3.6.7), i.e. $\mathbf{k}^{-\delta}\mathbf{z}_*\mathbf{a}_0 = \varepsilon_i\mathbf{a}_0$. Indeed, $\mathbf{a}_0$ could be written $\mathbf{s}_{*i}$. It will be seen that $\mathbf{s}_{*i}$ is the projection of $\mathbf{s}_i$, as defined for the Kuiper–Corsten iteration in Section 3.5.

To return to the designs in Exercises 3A to 3D, $\mathbf{F}_* = \frac{1}{7}\mathbf{C}_*$. In the original design

$$\mathbf{F} = \frac{1}{70}\begin{bmatrix} 50 & -5\sqrt{10} & -5\sqrt{10} & -5\sqrt{10} & -5\sqrt{10} \\ -5\sqrt{10} & 56 & -12 & -12 & -12 \\ -5\sqrt{10} & -12 & 56 & -12 & -12 \\ -5\sqrt{10} & -12 & -12 & 56 & -12 \\ -5\sqrt{10} & -12 & -12 & -12 & 56 \end{bmatrix}$$

It has an eigenvector, $\mathbf{p}_1$, such that

$$\mathbf{p}'_1 = (-1, -1, -1, -1, \sqrt{10})/\sqrt{14}$$

with $\varepsilon_1 = 1$. It may be written equally well as

$$\mathbf{z}'_1 = \mathbf{p}'_1 \mathbf{r}^{\frac{1}{2}\delta} = (-1, -1, -1, -1, 4)/\sqrt{14/5}$$
$$\mathbf{s}'_1 = \mathbf{p}'_1 \mathbf{r}^{-\frac{1}{2}\delta} = (-2, -2, -2, -2, 5)/\sqrt{280}$$

Hence $\mathbf{s}'_{*1} = \mathbf{a}'_0 = \mathbf{s}'_1 \mathbf{N}\mathbf{k}^{-\delta} = \mathbf{0}'$ with an undefined eigenvalue. However, to take another eigenvector,

$$\mathbf{p}'_2 = (3, -1, -1, -1, 0)/\sqrt{12}$$

with $\varepsilon_2 = \frac{34}{35}$. It will be found that $\mathbf{z}'_2$, $\mathbf{s}'_2$, $\mathbf{a}'_0 = \mathbf{s}'_{*2}$ all equal to

$$(3, -1, -1, -1, 0)$$

with varying divisors and that the efficiency, ε_{*2}, does in fact equal $\varepsilon_2 = \frac{34}{35}$. The association of the basic contrasts in the two designs has in fact been maintained, the difficulty of there being four such contrasts in the original design and only three in the dual having been met by $\mathbf{s}_1$ having only a formal analogue in $\mathbf{s}_{*1}$.

4.7 DISTANCES BETWEEN TREATMENTS

It has become clear that treatments often exhibit a relationship to one another that arises from the characteristics of the design, quite apart from any relationships that will appear when the data come to be examined. The work of Jaccottet and Tomassone (1976) illustrates that very well and leads to a useful means of representing a design diagrammatically. First, they define the 'profile' of a treatment as its proportional allocation between the blocks. Thus, for the design introduced in Section 3.4 with six treatments in six blocks of four plots each, the profile of Treatment A is

$$(\tfrac{1}{4}, \tfrac{1}{4}, \tfrac{1}{4}, \tfrac{1}{4}, 0, 0)$$

and that of Treatment D is

$$(\tfrac{1}{4}, 0, 0, \tfrac{1}{4}, \tfrac{1}{4}, \tfrac{1}{4}).$$

In short, the profile of a treatment is its row in $\mathbf{r}^{-\delta}\mathbf{N}$. As a prelude to what follows it may be remarked that if two treatments both exhibit the orthogonality relationship (4.1.3) or if they are proportionate (Section 4.4), then in both cases their contrast will be estimated with full efficiency; also, they will have the same profile.

The elements of a profile can be regarded as the co-ordinates of a point in a b-dimensional space—the 'block space'—but first it will be necessary to decide on scales. Jaccottet and Tomassone adopt the rule that the unit of measurement along any axis is $(k_j/n)^{1/2}$, where k_j is the number of plots in the corresponding block, but here it will be taken to be $k_j{}^{1/2}$. As a result, the profile, $\mathbf{r}^{-\delta}\mathbf{N}$, becomes a set of co-ordinates, $\mathbf{r}^{-\delta}\mathbf{N}\mathbf{k}^{-\frac{1}{2}\delta}$. That being so the distance between any two treatment points, say $d(e, f)$ for Treatments e and f, is given by

$$d^2(e,f) = \sum_j \left[\frac{1}{k_j} \left(\frac{N_{ej}}{r_e} - \frac{N_{fj}}{r_f} \right)^2 \right]$$

$$= \mathbf{c}'\mathbf{r}^{-\delta}\mathbf{N}\mathbf{k}^{-\delta}\mathbf{N}'\mathbf{r}^{-\delta}\mathbf{c} = \mathbf{c}'\mathbf{r}^{-\delta}\mathbf{c} - \mathbf{c}'\mathbf{r}^{-\delta}\mathbf{C}\mathbf{r}^{-\delta}\mathbf{c}, \qquad (4.7.1)$$

where $\mathbf{c}$ is the unscaled elementary contrast between the two treatments, i.e. $c_e = 1$, $c_f = -1$, $c_g = 0$ if $g \neq e, f$.

Equation (4.7.1) may equally well be written as

$$d^2(e,f) = \mathbf{c}'\mathbf{r}^{-\delta}\mathbf{A}\mathbf{k}^{-\delta}\mathbf{A}'\mathbf{r}^{-\delta}\mathbf{c} \qquad (4.7.2)$$

where $\mathbf{A}$ is the non-orthogonality matrix defined in (4.2.3). If $\mathbf{c}$ corresponds to a basic contrast, it should be scaled to accord with (3.6.3) and then related to its eigenvector of $\mathbf{F}$, i.e.

$$\mathbf{c} = \left[\frac{1}{r_e} + \frac{1}{r_f} \right]^{\frac{1}{2}} \mathbf{r}^{\frac{1}{2}\delta}\mathbf{p}_i$$

Then, from (4.7.1)

$$d^2(e,f) = \left[\frac{1}{r_e} + \frac{1}{r_f} \right] \mathbf{p}_i'(\mathbf{I}_v - \mathbf{F})\mathbf{p}_i$$

$$= \left[\frac{1}{r_e} + \frac{1}{r_f} \right] (1 - \varepsilon_i) \qquad (4.7.3)$$

i.e., if $\varepsilon_i = 1$, $d(e, f) = 0$. On the other hand, if the two treatments lie in different disconnected parts of the design, $\varepsilon_i = 0$ and the distance attains a maximum value of

$$\left[\frac{1}{r_e} + \frac{1}{r_f} \right]^{1/2}.$$

In general, if the elementary contrast of two treatments is basic, its variance of estimation is

$$\frac{(r_e + r_f)^2 \sigma^2}{r_e r_f (r_e + r_f - d^2 r_e r_f)} \qquad (4.7.4)$$

Jaccottet and Tomassone then set themselves the task of examining the cloud of points generated by the treatments, i.e. in the block space and scaling as indicated. They obtained a mean for each co-ordinate that could serve as an origin. In that way $\mathbf{r}^{-\delta}\mathbf{N}\mathbf{k}^{-\frac{1}{2}\delta}$ reduces to $\mathbf{r}^{-\delta}\mathbf{A}\mathbf{k}^{-\frac{1}{2}\delta}$ and the matrix of squares and products is seen to be $\mathbf{r}^{-\delta}\mathbf{A}\mathbf{k}^{-\delta}\mathbf{A}'\mathbf{r}^{-\delta}$.

If $\mathbf{z}_i = \mathbf{r}^{\frac{1}{2}\delta}\mathbf{p}_i$ is a basic contrast of the design, i.e.

$$(1 - \varepsilon_i)\mathbf{p}_i = \mathbf{r}^{-\frac{1}{2}\delta}\mathbf{N}\mathbf{k}^{-\delta}\mathbf{N}'\mathbf{r}^{-\frac{1}{2}\delta}\mathbf{p}_i$$

$$= \mathbf{r}^{-\frac{1}{2}\delta}\mathbf{A}\mathbf{k}^{-\delta}\mathbf{A}'\mathbf{r}^{-\frac{1}{2}\delta}\mathbf{p}_i + \mathbf{r}^{\frac{1}{2}\delta}\mathbf{1}_v\mathbf{1}_v'\mathbf{r}^{\frac{1}{2}\delta}\mathbf{p}_i/n$$

$$= \mathbf{r}^{-\frac{1}{2}\delta}\mathbf{A}\mathbf{k}^{-\delta}\mathbf{A}'\mathbf{r}^{-\delta}\mathbf{z}_i + \mathbf{r}^{\frac{1}{2}\delta}\mathbf{1}_v\mathbf{1}_v'\mathbf{z}_i/n.$$

Since $\mathbf{1}'_v \mathbf{z}_i = 0$, it follows that

$$(1-\varepsilon_i)\mathbf{z}_i = \mathbf{r}^{-\delta}\mathbf{A}\mathbf{k}^{-\delta}\mathbf{A}'\mathbf{r}^{-\delta}\mathbf{z}_i, \tag{4.7.5}$$

i.e. regarding the cloud of points as a hyper-ellipsoid, its axes are represented by the basic contrasts of the design.

To return to the example, the co-ordinates of the six treatment points are given by the elements of $\mathbf{r}^{-\delta}\mathbf{A}\mathbf{k}^{-\frac{1}{2}\delta}$, i.e.

$$\begin{array}{ll} \text{A} & (\;\;1,\;\;\;1,\;\;\;1,\;\;\;1,\;-2,\;-2)/24 \\ \text{B} & (\;\;1,\;\;\;1,\;\;\;1,\;-2,\;\;\;1,\;-2)/24 \\ \text{C} & (\;\;1,\;\;\;1,\;\;\;1,\;-2,\;-2,\;\;\;1)/24 \\ \text{D} & (\;\;1,\;-2,\;-2,\;\;\;1,\;\;\;1,\;\;\;1)/24 \\ \text{E} & (-2,\;\;\;1,\;-2,\;\;\;1,\;\;\;1,\;\;\;1)/24 \\ \text{F} & (-2,\;-2,\;\;\;1,\;\;\;1,\;\;\;1,\;\;\;1)/24. \end{array} \tag{4.7.6}$$

The difference in co-ordinates between two points like A and B are respectively

$$(0, 0, 0, 3, -3, 0)/24$$

and $d^2(\text{A, B}) = [\frac{3}{24}]^2 + [\frac{3}{24}]^2 = \frac{1}{32}$. Since the elementary contrast of the two treatments is also basic, from (4.7.3) $\varepsilon = \frac{15}{16}$, a result that can be verified by previous work. Thus the variance is known to be $\frac{8}{15}\sigma^2$, whereas in an orthogonal design it would have been $\frac{1}{2}\sigma^2$. To take another kind of contrast, e.g. that of A with D, $d^2(\text{A, D}) = \frac{1}{16}$. Since, however, that contrast is not basic, (4.7.3) does not apply.

It is perhaps difficult to visualize these points in a six-dimensional space. However, the design being equi-replicate its natural and basic contrasts are the same apart from scaling. Thus, two of the basic contrasts represent the main effect of the kind of fertilizer (see Exercises 3F and 3G), i.e.

$$\mathbf{z}'_1 = (1, -1, \;\; 0, 1, -1, \;\; 0)$$
$$\mathbf{z}'_2 = (1, \;\; 1, -2, 1, \;\; 1, -2)/\sqrt{3}.$$

Specifying these vectors by the co-ordinates of the treatment points set out in (4.7.6) gives

Block	I	II	III	IV	V	VI
$\mathbf{z}_1$	$\frac{1}{8}$	$-\frac{1}{8}$	0	$\frac{1}{8}$	$-\frac{1}{8}$	0
$\mathbf{z}_2$	$\frac{1}{8\sqrt{3}}$	$\frac{1}{8\sqrt{3}}$	$\frac{1}{4\sqrt{3}}$	$\frac{1}{8\sqrt{3}}$	$\frac{1}{8\sqrt{3}}$	$\frac{1}{4\sqrt{3}}$

It appears then that when the six treatment points are projected on to the plane defined by the two chosen vectors, they form an equilateral triangle, the points for A, B, and C being respectively the same as those for D, E, and F. If, however, the treatments are represented on a plane defined by the two basic contrasts for the interaction, i.e.,

$$\mathbf{z}'_4 = (1, -1, \;\; 0, -1, \;\; 1, 0)$$
$$\mathbf{z}'_5 = (1, \;\; 1, -2, -1, -1, 2)/\sqrt{3}$$

they form a regular hexagon, the order of the vertices being A E C D B F, i.e., each of the points A, B and C is now opposite its previous partner. Such diagrams are often provocative and often informative, as Example 4F shows.

4.8 GENERAL BALANCE

The property of general balance, descrbed by Nelder (1965a, 1965b), is primarily of importance in the analysis of data, though not exclusively so. Certainly it was not advanced in the first place as a desirable design feature. Rather Nelder was pointing out that many, if not most, designs used in practice possess the property and that it leads to simplification in computing the analysis of variance. Nevertheless, the fact that so many useful designs have general balance argues that it represents some sort of norm of good practice and so it does, though there are occasions when other designs are to be preferred.

First it is necessary to explain what is meant by 'strata'. In the design used to illustrate previous sections there are three strata, namely (1) plots within blocks, (2) blocks within the total area and (3) the total area. Ignoring treatments, the residuals matrices are respectively

$$\boldsymbol{\phi}_1 = \mathbf{I} - \mathbf{D}'\mathbf{k}^{-\delta}\mathbf{D}, \quad \boldsymbol{\phi}_2 = \mathbf{D}'\mathbf{k}^{-\delta}\mathbf{D} - \mathbf{1}_n\mathbf{1}_n'/n, \quad \boldsymbol{\phi}_3 = \mathbf{1}_n\mathbf{1}_n'/n \tag{4.8.1}$$

The data can now be regarded as having three components, $\mathbf{y}_i = \boldsymbol{\phi}_i\mathbf{y}$ $(i = 1, 2, 3)$ each appropriate to its own stratum. It will be seen that

$$\mathbf{I}_n = \boldsymbol{\phi}_1 + \boldsymbol{\phi}_2 + \boldsymbol{\phi}_3$$

so

$$\mathbf{y} = \mathbf{y}_1 + \mathbf{y}_2 + \mathbf{y}_3. \tag{4.8.2}$$

Also

$$\boldsymbol{\phi}_1 = \boldsymbol{\phi}_1', \qquad \boldsymbol{\phi}_i\boldsymbol{\phi}_i = \boldsymbol{\phi}_i, \qquad \boldsymbol{\phi}_i\boldsymbol{\phi}_j = \mathbf{0} \quad (i \neq j). \tag{4.8.3}$$

It follows that

$$\mathbf{y}'\mathbf{y} = \mathbf{y}'\boldsymbol{\phi}_1\mathbf{y} + \mathbf{y}'\boldsymbol{\phi}_2\mathbf{y} + \mathbf{y}'\boldsymbol{\phi}_3\mathbf{y} \tag{4.8.4}$$

$$= \mathbf{y}_1'\mathbf{y}_1 + \mathbf{y}_2'\mathbf{y}_2 + \mathbf{y}_3'\mathbf{y}_3 \tag{4.8.5}$$

$$= (\mathbf{y}'\mathbf{y} - \mathbf{B}'\mathbf{k}^{-\delta}\mathbf{B}) + (\mathbf{B}'\mathbf{k}^{-\delta}\mathbf{B} - G^2/n) + G^2/n. \tag{4.8.6}$$

The three terms represent sums of squared deviations appropriate to the three strata.

The above equations may be said to show the 'block structure' but there is 'treatment structure' as well. It depends upon the purpose of the experiment. It is possible for the same 'design', as ordinarily understood, to be used on two occasions but with different treatment structures thus, quite rightly, calling for two different forms of the analyses of variance. The sum of squared deviations for error might well be calculated in the same way on both occasions, but differences would come in the partition of the corresponding value for treatments. These considerations enter largely into the concept of general balance, which is concerned specifically with the way in which various groups of contrasts, chosen on account of their relevance to the questions under study, relate to the various strata of the design. It is not, therefore, an absolute property like orthogonality, which exists independently of the specification of the treatments. Rather it is problem-based and depends upon the treatment structure.

Before continuing it will therefore be necessary to specify the treatments hitherto referred to mysteriously as A to F. For purposes of illustration, it will be taken that there are three alternative forms of nitrogeneous fertilizer (X, Y, and Z) each applied at two levels (1 and 2), as was done in Example 3F, i.e. A = X1, B = Y1, C = Z1, D = X2, E = Y2, and F = Z2. With this specification of treatments, it is possible to write down a matrix, **H**, setting out the treatment structure. It has indeed effectively been done in Section 4.7. First there are the two degrees of freedom between, X, Y and Z, which may be related to the contrast vectors,

$$\mathbf{u}_1' = (1, -1, \quad 0, 1, -1, \quad 0)/2$$
$$\mathbf{u}_2' = (1, \quad 1, -2, 1, \quad 1, -2)/\sqrt{12}$$

so the first component of **H** (i.e. $\mathbf{H}_1$) will equal $\mathbf{u}_1\mathbf{u}_1' + \mathbf{u}_2\mathbf{u}_2'$. Then there is the degree of freedom for the main effect of quantity, which may be written

$$\mathbf{u}_3' = (1, 1, 1, -1, -1, -1)/\sqrt{6}$$

giving a second component of **H**, namely, $\mathbf{H}_2 = \mathbf{u}_3\mathbf{u}_3'$. With the main effects defined, the two degrees of freedom for the interaction remain.

Here

$$\mathbf{u}_4' = (1, -1, \quad 0, -1, \quad 1, 0)/2$$
$$\mathbf{u}_5' = (1, \quad 1, -2, -1, -1, 2)/\sqrt{12}$$

making $\mathbf{H}_3 = \mathbf{u}_4\mathbf{u}_4' + \mathbf{u}_5\mathbf{u}_5'$. That leaves the general mean

$$\mathbf{u}_6' = (1, 1, 1, 1, 1, 1)/\sqrt{6}.$$

Then,

$$\mathbf{H} = \sum_j \mathbf{H}_j$$

In this instance, $\mathbf{H} = \mathbf{I}_v$, but that does not hold in all cases, because there could be treatment contrasts that are to be ignored. It will be noticed that the six vectors are mutually orthogonal. Also

$$\mathbf{H}_j = \mathbf{H}_j', \qquad \mathbf{H}_j\mathbf{H}_j = \mathbf{H}_j, \qquad \mathbf{H}_i\mathbf{H}_j = \mathbf{0} \quad (i \neq j), \tag{4.8.7}$$

as for residual matrices.

General balance can now be defined as the property whereby

$$\mathbf{H}_j\boldsymbol{\Delta}\boldsymbol{\phi}_i\boldsymbol{\Delta}'\mathbf{H}_j = R_{ij}\mathbf{H}_j \tag{4.8.8a}$$

for all i and j, the constant, R_{ij}, being the effective replication for that group of contrasts, j, relative to that stratum, i. That result follows from the passage above

(3.6.9). The property defined in (4.8.8a) is equivalent to

$$\mathbf{u}_i'\boldsymbol{\Delta}\boldsymbol{\phi}_i\boldsymbol{\Delta}'\mathbf{u}_i = R_{ij} \tag{4.8.8b}$$

for all contrasts, i, because $\mathbf{u}_i'\mathbf{H} = \mathbf{u}_i'$.

For the example, $\boldsymbol{\Delta}\boldsymbol{\phi}_1\boldsymbol{\Delta}' = \mathbf{C}$, $\boldsymbol{\Delta}\boldsymbol{\phi}_2\boldsymbol{\Delta}' = \mathbf{C}_0$ and $\boldsymbol{\Delta}\boldsymbol{\phi}_3\boldsymbol{\Delta}' = \mathbf{rr}'/n$, it will readily be found that the design is generally balanced, the effective replications being:

$$\begin{array}{llll} R_{11} = 3\tfrac{3}{4} & R_{12} = 3 & R_{13} = 3\tfrac{3}{4} & R_{14} = 0 \\ R_{21} = \tfrac{1}{4} & R_{22} = 1 & R_{23} = \tfrac{1}{4} & R_{24} = 0 \\ R_{31} = 0 & R_{32} = 0 & R_{33} = 0 & R_{34} = 4. \end{array}$$

As would be expected from previous studies, information about treatment contrasts comes mainly from the first stratum (the intra-block analysis) but partly from the second (the inter-block analysis). To illustrate the use of effective replications, it may be noted that the contrast between the means of fertilizers X and Y should be $2\sigma^2/R_{11}$, when estimated intra-block, i.e. $8\sigma^2/15$. This figure can be confirmed from any of the generalized inverses, i.e. $\boldsymbol{\Omega}$, $\boldsymbol{\Xi}$, $\mathbf{C}^+$, or $\boldsymbol{\Upsilon}$. The method works because the vectors used to form $\mathbf{H}_1$, $\mathbf{H}_2$, etc. are all eigenvectors of $\mathbf{C}$, $\mathbf{C}_0$, etc. and the effective replications are the eigenvalues. The device of ignoring an eigenvalue if it equals zero and taking a reciprocal otherwise, is effectively a derivation of the Moore–Penrose generalized inverse, but more about that when analysis comes to be considered. Some further designs are studied in Example 4E.

The strength of this approach lies in its recognition that blocks and treatments have structure and basing everything thereon. It does therefore relate to the needs of an experimenter, who knows more or less how he ought to form blocks and who is interested only in certain specified contrasts between treatments.

It is therefore a little disturbing that some recent writers have shown a tendency to extend the definition given above and to do so in a way that is not acceptable to practical men. There is indeed a wider class of design in which the coefficient matrices, $\mathbf{C}_i$, for the various strata all have the same eigenvectors. It could well be that the mathematical study of such designs will amply repay the effort, but they should not be regarded as necessarily in general balance. That property should be firmly related to the treatment structure. If there are six treatments and they form a 2×3 factorial set, then the eigenvectors found in each stratum must correspond to the contrasts indicated by the structure of the treatments, i.e. two main effects, one with a single degree of freedom and the other with two, together with an interaction. Further, it could well be that the contrasts of a group all need to have the same eigenvalue and for that to be so in each stratum separately. For example, if the factor with three levels represented three varieties, there being no particular reason to regard any one contrast between them as pre-eminent, then it must be possible to study all contrasts with equal facility and, however the two contrasts are formed, the two eigenvalues must be the same. In effect, the three varieties must be equivalent in the sense of Section 4.5.

On the matter of nomenclature, however, there can be doubts. Nelder uses the term 'orthogonal' to indicate that for each group of contrasts all information comes from a single stratum. That is to say, a 2^3 factorial design with the three-factor interaction confounded is 'orthogonal', because everything is estimated with full efficiency intra-block, except the confounded interaction, and that is estimated with full efficiency inter-block. One must comment that most people would not regard that as a valid use of the familiar word. Also, it is only the contrasts of interest that are thus considered. In a design like randomized blocks all contrasts, whether of interest or not, are found with full efficiency, i.e. the property of orthogonality does not depend upon the treatment structure, and that is another difference.

4.9 DESIGNS WITH SIMPLE RANDOMIZATION PROCEDURES

So far randomization has been considered with reference to two sorts of design, namely those that are completely randomized and those in blocks, the latter giving both intra-block and inter-block information. For all cases a randomization procedure has been suggested that makes the expectation of $\hat{\boldsymbol{\eta}}\hat{\boldsymbol{\eta}}'$ equal to $\boldsymbol{\phi}\sigma^2$, where $\boldsymbol{\phi}$ has the following properties.

(1) It is idempotent, i.e. $\boldsymbol{\phi}\boldsymbol{\phi} = \boldsymbol{\phi}$.
(2) The sum of squared deviations for error and treatments together is $\mathbf{y}'\boldsymbol{\phi}\mathbf{y}$.
(3) The coefficient matrix, $\mathbf{C}$, equals $\boldsymbol{\Delta}\boldsymbol{\phi}\boldsymbol{\Delta}'$.
(4) The treatment parameters are estimated by $\mathbf{C}^-\mathbf{Q}$, where $\mathbf{Q} = \boldsymbol{\Delta}\boldsymbol{\phi}\mathbf{y} = \boldsymbol{\Delta}\boldsymbol{\phi}\hat{\boldsymbol{\eta}}$, making the treatment sum of squared deviations to be $\mathbf{Q}'\mathbf{C}^-\mathbf{Q}$.

In such circumstances, the weak criterion of randomization is satisfied, i.e. given suitable degrees of freedom the F-test is unbiassed. For $\mathbf{y}'\boldsymbol{\phi}\mathbf{y}$ there are tr $(\boldsymbol{\phi})$ degrees of freedom and for $\mathbf{Q}'\mathbf{C}^-\mathbf{Q}$ there are tr $(\mathbf{C}\mathbf{C}^-)$. All of that follows from the argument in Section 2.1, which is readily generalized.

There remains, however, the strong criterion that the variance of any given contrast can be estimated without bias. Assuming that $\mathbf{c}'\boldsymbol{\tau}$ can be estimated with a variance of $\mathbf{c}'\mathbf{C}^-\mathbf{c}\hat{\sigma}^2$, that is satisfied also. Since $\mathbf{c}'\boldsymbol{\tau}$ can be estimated $\mathbf{c}$ must be a linear combination of eigenvectors of $\mathbf{C}$ with non-zero eigenvalues. (Again, the argument is readily generalized.) It is therefore possible to write

$$\mathbf{c} = \sum_{i=1}^{h} (\alpha_i \mathbf{u}_i)$$

where $\mathbf{C}\mathbf{u}_i = \lambda_i \mathbf{u}_i$, $\lambda_i \neq 0$. Hence $\mathbf{C}^+\mathbf{u}_i = \mathbf{u}_i/\lambda_i$.

Now

$$\begin{aligned}\mathbf{c}'\hat{\boldsymbol{\tau}} &= \mathbf{c}'\mathbf{C}^-\boldsymbol{\Delta}\boldsymbol{\phi}\hat{\boldsymbol{\eta}} \\ &= \sum (\alpha_i \mathbf{u}_i'/\lambda_i)\boldsymbol{\Delta}\boldsymbol{\phi}\hat{\boldsymbol{\eta}}. \end{aligned} \tag{4.9.1}$$

Therefore

$$(\mathbf{c}'\hat{\boldsymbol{\tau}})^2 = \sum (\alpha_i \mathbf{u}_i'/\lambda_i)\boldsymbol{\Delta}\boldsymbol{\phi}\hat{\boldsymbol{\eta}}\hat{\boldsymbol{\eta}}'\boldsymbol{\phi}\boldsymbol{\Delta}' \sum (\alpha_i \mathbf{u}_i/\lambda_i)$$

so

$$\begin{aligned}\text{ex } (\mathbf{c}'\hat{\boldsymbol{\tau}})^2 &= \sum (\alpha_i \mathbf{u}_i'/\lambda_i)\mathbf{C}\sum (\alpha_i \mathbf{u}_i/\lambda_i)\hat{\sigma}^2 \\ &= \sum (\alpha_i^2/\lambda_i)\hat{\sigma}^2. \end{aligned} \tag{4.9.2}$$

If the variance of $\mathbf{c}'\hat{\boldsymbol{\tau}}$ is $\mathbf{c}'\mathbf{C}^{-}\mathbf{c}\hat{\sigma}^2$, that equals

$$\sum(\alpha_i \mathbf{u}_i')\mathbf{C}^{+}\sum(\alpha_i \mathbf{u}_i)\sigma^2 = \sum(\alpha_i^2/\lambda_i)\hat{\sigma}^2 \tag{4.9.3}$$

as before, so the test of significance of any given contrast is unbiassed.

The problem often takes the form of partitioning the treatment sum of squared deviations into independent components, each estimated without bias. Two solutions have already been suggested, one based on (3.6.1), which uses natural contrasts, and the other on (3.6.9), which has basic contrasts.

It is not to be supposed that all designs fulfil the requirements stated. Most do, but some do not. In awkward cases it may be necessary to find some restrictive schemes to achieve the desired result, e.g. that of Grundy and Healy (1950). As each design comes to be considered, it will be necessary to enquire what randomization procedure is called for. The work of Preece *et al.* (1978) is useful in that connection. There is an important paper by Bailey (1981) in which she argues that randomization is an integral part of the whole process of design and analysis.

4.10 TOWARDS A THEORY OF INFORMATION

The central task confronting anyone who sets out to design an experiment is to obtain good estimates, possibly independent ones, of the contrasts designated as important. If he fails, it is no good pleading that his proposals exemplify some class of balance or provide an interesting application of someone's theorem; his work will be judged by standards that are purely pragmatic. At this point some may conclude that the arcane subject of experimental design has little application to the practical task of designing an experiment. Certainly there is not a lot in the literature of direct assistance. Most of the useful work has been presented earlier in this chapter.

When Ξ was introduced in Section 3.4, it was pointed out that the precision of an experiment depends upon the off-diagonal elements of $\mathbf{N}\mathbf{k}^{-\delta}\mathbf{N}'$. The more that two treatments can be brought together in blocks, the better will be the estimate of the contrast between them. The argument can be taken further. Suppose that a design is proposed with a matrix, $\mathbf{W}$, (3.4.3) and suppose further that a modification or augmentation leads to an increase of α in the elements, W_{ij} and W_{ji}. Using the suffix + to show the improved design and writing $\mathbf{u}_i$ for a column vector with all elements zero except for the ith, which equals one, then

$$\begin{aligned}\mathbf{W}_{+} &= \mathbf{W} + \alpha(\mathbf{u}_i\mathbf{u}_j' + \mathbf{u}_j\mathbf{u}_i') \\ \mathbf{q}_{+}^{\delta} &= \mathbf{q}^{\delta} + \alpha(\mathbf{u}_i\mathbf{u}_i' + \mathbf{u}_j\mathbf{u}_j') \\ \Xi^{-1} &= \Xi_{+}^{-1} - \alpha(\mathbf{u}_i - \mathbf{u}_j)(\mathbf{u}_i' - \mathbf{u}_j') - \mathbf{q}_{+}\mathbf{q}_{+}'/u + \mathbf{q}\mathbf{q}'/u\end{aligned} \tag{4.10.1}$$

Premultiplying by Ξ, and postmultiplying by Ξ_{+}, and noting that $\Xi\mathbf{q} = \mathbf{1}_v = \Xi_{+}\mathbf{q}_{+}$, gives

$$\Xi_{+} = \Xi - \alpha\Xi\mathbf{c}_1\mathbf{c}_1'\Xi_{+} - \Xi_{q+}\mathbf{1}'/u_{+} + \mathbf{1}_v\mathbf{q}'\Xi_{+}/u \tag{4.10.2}$$

where $\mathbf{c}_1$ $(= \mathbf{u}_i - \mathbf{u}_j)$ is a contrast between treatments i and j. Writing $\theta_{st} = \mathbf{c}_s'\Xi\mathbf{c}_t$ and $\theta_{+st} = \mathbf{c}_s'\Xi_+\mathbf{c}_t$,

$$\theta_{+11} = \theta_{11} - \alpha\theta_{11}\theta_{+11}.$$

Thinking in terms of information, i.e. Inf $= 1/\theta$,

$$\text{Inf}_{+11} = \text{Inf}_{11} + \alpha. \tag{4.10.3}$$

That is to say, the information about the contrast between the two treatments has been increased exactly by the quantity added to their relevant elements in **W**. There are two things to note here. One is that $\mathbf{c}_1$ has been used in an unstandardized form. The other is that the derivation of (4.10.3) from (4.10.2) breaks down if θ_{11} is infinite, i.e., if initially there was no information at all about the contrast. The usefulness of the approach is illustrated in Example 4D.

To take the matter further, if $\mathbf{c}_2$ is some other contrast then, by taking all further combinations of pre- and post-multiplication of (4.10.1) by $\mathbf{c}_1$ and $\mathbf{c}_2$,

$$\begin{aligned}\theta_{12} &= \theta_{+12} + \alpha\theta_{11}\theta_{+12}\\ \theta_{12} &= \theta_{+12} + \alpha\theta_{12}\theta_{+11}\\ \theta_{22} &= \theta_{+22} + \alpha\theta_{12}\theta_{+12}.\end{aligned}$$

From the first two of these equations, since θ_{11} and θ_{+11} are both positive, it follows that θ_{12} and θ_{+12} are either both zero or both of the same sign. Hence, from the third,

$$\theta_{+22} \leqq \theta_{22}. \tag{4.10.4}$$

Accordingly, in addition to the direct advantage shown by (4.10.3) which arises from improvements in the elements of **W** immediately relevant to the contrast, there could be indirect advantages brought about by improvements elsewhere. Welcome as the additional information is, it does confuse the issue.

The indirect information so obtained can be quite considerable. To take the familiar example, A and D come together in two blocks, namely, I and IV, so it is to be expected that the variance between them will not exceed $(\frac{1}{2} + \frac{1}{2})\sigma^2$, i.e. there will be at least unit information about their contrast. The variance does in fact equal $\frac{17}{30}\sigma^2$, so the information is 1.76, the addition coming indirectly, i.e. from information concerning the comparisons of both treatments with the other four. Here it may be noted that when the two treatments under study do come together in a block, it may not matter much what other treatments are present. Proportionate treatments provide an example. Indirect information appears to come from blocks in which only one of the two treatments is present along with others, which complete the contrast by occurring also in some block or blocks where they effect a comparison with the second treatment, the first being absent. At the moment the whole subject is rather mysterious, which is a pity when more knowledge would be useful.

Examples 4

A. Campbell (1967) gives the following yields per plot of an experiment to study the effect of changing plant-density in maize. The treatments, 20, 25, 30, 35, and 40, represent plants per unit area and there were five randomized replicate blocks. Data show kilogrammes of dry matter per plot.

	Block					
Treatment	I	II	III	IV	V	Total
20	21.1	16.7	14.9	15.5	19.7	87.9
25	26.8	23.8	21.4	22.6	23.6	118.2
30	30.4	25.5	27.1	26.3	26.6	135.9
35	28.4	28.2	25.3	26.5	32.6	141.0
40	27.6	24.5	26.5	27.0	30.1	135.7
Total	134.3	118.7	115.2	117.9	132.6	618.7

Calculate an appropriate analysis of variance.

Answer: Since the design is orthogonal the analysis can readily be calculated by the use of summation terms, as described in Section 4.1.

S, the data term $= \mathbf{y}'\mathbf{y} = 21.1^2 + 16.7^2 + \cdots + 30.1^2 = 15\,799.0500$

S_b, the block term $= \mathbf{B}'\mathbf{k}^{-\delta}\mathbf{B} = \mathbf{B}'\mathbf{B}/k$ in this instance where all block sizes are the same $= \frac{1}{5}(134.3^2 + 118.7^2 + \cdots + 132.6^2) = 15\,376.0780$

S_t, the treatment term $= \mathbf{T}'\mathbf{r}^{-\delta}\mathbf{T} = \mathbf{T}'\mathbf{T}/r$ in this instance since treatments are equireplicate $= \frac{1}{5}(87.9^2 + 118.2^2 + \ldots + 135.7^2) = 15\,692.3900$

S_0, the correction term $= G^2/n = (618.7)^2/25 = 15\,311.5876$

It will be seen that terms have been calculated to two more decimal places than arose from the squaring of the data. The above data have been recorded with some precision and the usual rule has perhaps given more decimal places than are needed. Nevertheless, it is better to err by having too many rather than too few.

Three sums of squared deviations are required.

Treatments, i.e. $S_t - S_0 = 380.8024$

Error, i.e. $S - S_b - S_t + S_0 = 42.1696$

Total, i.e. $S - S_b = 422.9720$

They give the following analysis of variance

Source	*d.f.*	*s.s.*	*m.s.*	*F*
Treatments	4	380.8024	95.2006	36.12
Error	16	42.1696	2.6356	
Total	20	422.9720		

Clearly the treatments have had a marked effect. Since the density levels are in arithmetic progression there is something to be said for partitioning the treatment effects using the orthogonal polynomials given by Fisher and

Yates (1938), thus:

Linear effect

$$[2(135.7)+(141.0)-(118.2)-2(87.9)]^2/(5\times 10) = 280.3712$$

Quadratic effect

$$[2(135.7)-(141.0)-2(135.9)-(118.2)+2(87.9)]^2/(5\times 14) = 100.3206$$

Cubic effect

$$[-(135.7)+2(141.0)-2(118.2)+(87.9)]^2/(5\times 10) = 0.0968$$

Quartic effect

$$[(135.7)-4(141.0)+6(135.9)-4(118.2)+(87.9)]^2/(5\times 70) = 0.0138$$

In effect that gives a partition of the treatment sum of squared deviations into four effects, each with one degree of freedom, and it shows that only the linear and quadratic effects need be considered, i.e. the response curve is effectively a parabola, at least over the range of densities studied.

Following the thinking of Section 1.6 that amounts to a consideration of the contrasts,

$$\begin{aligned}
\mathbf{c}_1' &= (\ \ 2,\ \ \ 1,\ \ \ 0,\ -1,\ -2)/\sqrt{10}\\
\mathbf{c}_2' &= (\ \ 2,\ -1,\ -2,\ -1,\ \ \ 2)/\sqrt{14}\\
\mathbf{c}_3' &= (-1,\ \ \ 2,\ \ \ 0,\ -2,\ \ \ 1)/\sqrt{10}\\
\mathbf{c}_4' &= (\ \ 1,\ -4,\ \ \ 6,\ -4,\ \ \ 1)/\sqrt{70}
\end{aligned}$$

Further, they are orthogonal in the sense that $\mathbf{c}_i'\mathbf{c}_j = 0$ for all $i \neq j$. It is that property that ensures the separate sums of squared deviations adding up correctly.

For present purposes it is enough to show how the analysis of variance can be calculated using summation terms. The arithmetic would have been more complicated if, for example, the treatments had not all had the same replications, but that has been illustrated in Example 2A. Similar modifications would have been needed if the blocks had not all been of the same size. (It is, of course, assumed that the design is nonetheless orthogonal.)

Convention requires that the analysis should be presented with a line for blocks. In this example it would have had four degrees of freedom and a sum of squared deviations of $(S_b - S_0) = 64.4904$, the total thereby being increased to $(S - S_0) = 487.4624$ with 24 degrees of freedom. The addition does not help much except in showing whether the blocking system has or has not been effective and it would complicate thinking when strata come to be considered in Chapter 6. For that reason, it has not been made here.

B. Show that the Kuiper–Corsten iteration (Section 3.5) applied to an orthogonal design (Section 4.1) converges in the first cycle.

Answer: If $\mathbf{v}_1 = \mathbf{r}^{-\delta}\mathbf{Q}$ and $\mathbf{1}_v'\mathbf{Q} = 0$, then

$$\begin{aligned}
\mathbf{u}_1 &= \mathbf{k}^{-\delta}\mathbf{N}'\mathbf{v}_1 = \mathbf{k}^{-\delta}\mathbf{k}\mathbf{r}'\mathbf{r}^{-\delta}\mathbf{Q}/n\\
&= \mathbf{1}_b\mathbf{1}_v'\mathbf{Q} = \mathbf{0}.
\end{aligned}$$

Comment: Hence $\mathbf{r}^{-\delta}\mathbf{Q}$ is a valid estimate of $\boldsymbol{\tau}$. In fact $\mathbf{r}^{-\delta}$ is a generalized inverse of $\mathbf{C} = \mathbf{r}^{\delta} - \mathbf{rr}'/n$.

C. In the following design, which pairs of treatments are concordant and what are the variances of contrasts between them?

Block	I	A	B	C	D
	II	A	B	C	E
	III	D	E	F	
	IV	D	E	F	

Answer: The pairs BC, AC and AB are all proportionate (Section 4.4). They are therefore concordant and the contrasts between members of those pairs are estimated with full efficiency. Also, D and E are equivalent (Section 4.5), so they also are concordant. In (4.5.2) $a = \frac{29}{12}$ and $b = \frac{-4}{12}$, so the variance of their difference is

$$\frac{2\sigma^2}{a-b} = \frac{8\sigma^2}{11},$$

where σ^2 is the error variance. To take an alternative approach, in (4.3.5.) $T = \frac{17}{6}$ and $D = \frac{187}{48}$, so the variance is $T\sigma^2/D$, which comes to the same thing.

D. An experiment is designed according to the following scheme.

Block	I	A*	B*	C	D	E	
	II	A	B	C	D§		
	III	A	B	C	D	E	E
	IV	A	B	C	D	E	
	V	A	B	C§	D		

What is the standard error of the difference between the estimated parameters of treatments A and B if the plots marked * are omitted?

What is the standard error of the difference between the estimated parameters of treatments C and D if additionally the treatments applied to the plots marked § are accidentally interchanged?

Assuming that neither of these mishaps occurs, what difference would it make to the precision of estimation of contrasts between the treatments A, B, C and D if the plots with treatment E were omitted?

(University of Kent at Canterbury, Diploma/M.Sc. 1979)

Answer: If the plots marked * are omitted (or even if they are not) Treatments A and B are proportionate and their contrast is therefore estimated with full efficiency. Since each has four-fold replication, the variance of (*A − *B) is

$$[\tfrac{1}{4} + \tfrac{1}{4}]\sigma^2 = \frac{\sigma^2}{2}.$$

If the plots marked § are interchanged, Treatments C and D are no longer proportionate, but they are still equivalent. The sub-matrix of Ω^{-1}

($= \mathbf{C}+\mathbf{rr}'/n$) that relates to the two treatments (i.e. $\mathbf{X}$) is

$$\tfrac{1}{30}\begin{bmatrix} 103 & -17 \\ -17 & 103 \end{bmatrix} + \tfrac{1}{24}\begin{bmatrix} 25 & 25 \\ 25 & 25 \end{bmatrix} = \begin{bmatrix} A & B \\ B & A \end{bmatrix}.$$

From (4.5.2) the variance of (*C − *D) is

$$\frac{2\sigma^2}{A-B} = \frac{\sigma^2}{2}.$$

Since Treatments A, B, C and D are proprotionate, all contrasts between them are of full efficiency (i.e. effective replication = 5) whether E is included or not.

Corrollary: It is, of course, conceivable that some of these events would happen in conjunction. If all plots of Treatment E were omitted as well as those marked *, the comparison of Treatments A and B would still be of full efficiency. If Treatment E were omitted and the plots marked § were interchanged, $\mathbf{X}$ would become

$$\tfrac{1}{4}\begin{bmatrix} 13 & -3 \\ -3 & 13 \end{bmatrix} + \tfrac{1}{20}\begin{bmatrix} 25 & 25 \\ 25 & 25 \end{bmatrix}$$

so $2\sigma^2/(A-B)$ would still be $\sigma^2/2$.

To take the further case in which plots marked * are omitted and those marked § are interchanged, Treatments A and B are still proportionate. For Treatments C and D, $\mathbf{X}$ becomes

$$\tfrac{1}{10}\begin{bmatrix} 33 & -7 \\ -7 & 33 \end{bmatrix} + \tfrac{1}{21}\begin{bmatrix} 25 & 25 \\ 25 & 25 \end{bmatrix}$$

and the variance of (*C − *D) is still $\frac{1}{2}\sigma^2$. The further omission of E does not alter the variances.

It is of some interest to examine what happens to the variance of a contrast like (*A − *C). In the full design the treatments are proportionate and remain so if E is removed. Other accidents, however, whether the omission of plots marked * or the interchange of plots marked §, increase the variance.

E. General balance was described in Section 4.8. Are the Designs α and β of Example 3G in general balance? If so, what are the effective replications?
Answer: Both designs are in general balance relative to the declared contrasts.

Design α

$$\begin{array}{llll} R_{11}=3 & R_{12}=4 & R_{13}=4 & R_{14}=0 \\ R_{21}=1 & R_{22}=0 & R_{23}=0 & R_{24}=0 \\ R_{31}=0 & R_{32}=0 & R_{33}=0 & R_{34}=4 \end{array}$$

Design β

$$\begin{array}{llll} R_{11}=3\tfrac{3}{4} & R_{12}=4 & R_{13}=3\tfrac{1}{4} & R_{14}=0 \\ R_{21}=\tfrac{1}{4} & R_{22}=0 & R_{31}=\tfrac{3}{4} & R_{24}=0 \\ R_{31}=0 & R_{32}=0 & R_{33}=0 & R_{34}=4 \end{array}$$

Intra-block generalized inverses have already been found. It will be seen that the intra-block variances of the contrasts that make up $\mathbf{H}_1$, $\mathbf{H}_2$ and $\mathbf{H}_3$ could have been found by using $\mathbf{I}_v/R_{1i}$ instead, i taking a value appropriate to the contrast.

F. Compare the two following designs:

Block					Block			
I	A	B	C		I	A	B	C
II	A	B	C		II	A	B	C
III	A	B	B		III	A	B	C
IV	A	C	C		IV	A	B	C

The second is, of course, in randomized blocks. The first was intended to be, but a mistake was made at planting.

Does the result suggest an expression for the general case in which such a mistake occurs in planting, the design being in randomized replicate blocks? *Answer:* In the sense of (4.10.1) the correctly planted design is an improvement on the first design, α being $2/k$, where k is the block size. That being so, an incorrectly planted pair of treatments has that much less information, i.e. $r/2-2/k = (n-4)/(2k)$, where r is the replication of each treatment and $n = rk$. Hence the variance is $2k\sigma^2/(n-4)$.

Alternatively, using the methods of Section 4.7,

$$\mathbf{A} = \mathbf{N} - \mathbf{rk}'/n = \begin{bmatrix} \mathbf{0} & \mathbf{0} \\ \mathbf{0} & \mathbf{K} \end{bmatrix}, \quad \text{where} \quad \mathbf{K} = \begin{bmatrix} 1 & -1 \\ -1 & 1 \end{bmatrix}.$$ (Note that $\mathbf{KK} = 2\mathbf{K}$).

Hence

$$\mathbf{A}\mathbf{k}^{-\delta}\mathbf{A}' = \frac{2}{k}\begin{bmatrix} \mathbf{0} & \mathbf{0} \\ \mathbf{0} & \mathbf{K} \end{bmatrix}$$

and

$$\mathbf{r}^{-\delta}\mathbf{A}\mathbf{k}^{-\delta}\mathbf{A}'\mathbf{r}^{-\delta} = \frac{2}{(r^2k)}\begin{bmatrix} \mathbf{0} & \mathbf{0} \\ \mathbf{0} & \mathbf{K} \end{bmatrix}.$$

The distance between two correctly planted treatments is therefore zero. Between one that is correctly planted and one that is not, the distance squared is $(2/r^2k)$. It is four times as much if both were incorrectly planted. Hence, plotting the treatments in the block space puts the two incorrectly planted treatments a distance $2\sqrt{2}/(r\sqrt{k})$ apart with all the others halfway between.

The elementary contrast between the two incorrectly planted treatments is basic, so (4.7.3) applies and

$$(1-\varepsilon) = 4/n.$$

Hence, the variance of their contrast is

$$\frac{2\sigma^2}{r\varepsilon} = \frac{2k\sigma^2}{n-4}.$$

as before.

Chapter 5

Classes of design

5.1. THE ROLE OF STANDARD DESIGNS

In a text that emphasizes the need to relate designs to the problems it may seem strange to find a chapter on classes of design. It is as if a tailor were to write about the importance of every article of clothing being made to fit its wearer perfectly and then go on to praise ready-made garments. Nevertheless, there are people with measurements so standard that they do not need to have clothes made specially and there are some problems that arise so frequently that the same sort of design is used again and again. When that happens it may become worth-while to consider a group of designs that possess some desired property, because there could be ways of generating its members or of finding simple ways of analysing the resultant data. It is true that some standard computer program will be used in any practical application. Nevertheless, in choosing a particular design from a class it is helpful to have formulae for variances, etc. However, a class is useful only in so far as it possesses some statistical property, e.g. it affords a precise study of a certain kind of contrast. There is not the same interest, if any at all, in classes based upon some combinatorial property.

In all this it should be remembered that there is one sort of design, namely, those that are orthogonal, which have such admirable properties that they are used whenever possible. First, they estimate all contrasts with full efficiency. Next, since all pairs of treatments are initially orthogonal to blocks and proportionate, the designs are robust in the event of mistakes and accidents. Not least, data from them are easily analysed. Only if there is some objection to their use, e.g. they do not fit any sensible system of blocks or their inflexibility with respect to replication causes problems or there are contrasts that could well be confounded, should anything else be used.

When an experimenter is forced to adopt a non-orthogonal design for practical reasons the difficulty arises as to which of those admirable qualities is to be sacrificed. After Section 1.6 the reader will not be surprised that the author's first priority is the effective estimation of the contrasts of interest, if possible with full efficiency and good replication. It is often also reasonable to ask that they be estimated independently. Also, there are considerable advantages in forming equivalence sets, as described in Section 4.5, because that conduces to general balance, which is always a convenience. Accepting that any non-orthogonal design is going to fall short in some respect, it is necessary to have a clear view of objectives and to seek a design in which the defects are irrelevant to the purposes in hand. Continual use of orthogonal designs can make people lazy because there is no need for them to ask how far a loss of efficiency would be compensated by the achievement of independence and there are many like questions that can be avoided. In a non-orthogonal context they can be awkward and pressing, especially to someone who has avoided facing them before.

In general, to speak of a design as 'balanced' implies 'variance-balance', i.e. the eigenvalues of $\mathbf{C}$ are such that, usually on account of multiplicities, there are groups of important contrasts all estimated with the same variance. Latterly attention has been given to 'efficiency-balance' in which groups of contrasts are estimated with the same efficiency, i.e., the emphasis is on the eigenvalues of $\mathbf{F}$ rather than on those of $\mathbf{C}$ (Puri and Nigam, 1975ab, 1976, 1977ab, 1978; Puri, Nigam and Narain, 1977). With an equi-replicate design one kind of balance implies the other, but not in other cases. It is, however, not clear what practical advantages are conferred by efficiency-balance.

The case may be argued thus. If an experiment is to be conducted, there are two main statistical requirements concerning the design. The first is that certain contrasts should be estimated with maximum precision and possibly equal precision. The other is that the overall efficiency should be high. Although the contrasts of interest may need to be scrutinized individually to ensure that they all have the same variance, any requirement about efficiencies concerns an average figure. For example, a design is open to criticism if it uses 40 plots to achieve what a rival could have done with 36, but no one is going to examine it, contrast by contrast, and complain that the loss of efficiency is unevenly spread. The only obvious advantage of efficiency-balance is the freedom of partition it allows among basic contrasts (Section 3.6), just as multiplicities among the eigenvalues of $\mathbf{C}$ permit freedom of partition among the natural contrasts.

By an extension of (4.1.1), if all contrasts are to be estimated with a constant efficiency, ε, then

$$\mathbf{F} = \varepsilon(I_v - \mathbf{r}^{\frac{1}{2}\delta}\mathbf{1}_v\mathbf{1}_v'\,\mathbf{r}^{\frac{1}{2}\delta}), \tag{5.1.1}$$

$$\mathbf{C} = \varepsilon\mathbf{r}^{\delta} - \varepsilon\mathbf{r}\mathbf{r}'/n, \tag{5.1.2}$$

$$\mathbf{N}\mathbf{k}^{-\delta}\mathbf{N}' = \mathbf{r}^{\delta} - \mathbf{C} = (1-\varepsilon)\,\mathbf{r}^{\delta} + \varepsilon\mathbf{r}\mathbf{r}'/n.$$

That is to say, the off-diagonal elements of $\mathbf{N}\mathbf{k}^{-\delta}\mathbf{N}'$ must be such that each element bears a constant ratio to the product of the replications of the treatments to which it refers. Conversely, if that property holds,

$$\mathbf{C} = \mathbf{p}^{\delta} - \alpha\mathbf{r}\mathbf{r}'$$

where $\mathbf{p}^{\delta}$ is some diagonal matrix and α is some constant. Because $\mathbf{C}\mathbf{1}_v = \mathbf{0}$, it follows that $\mathbf{p} = n\alpha\mathbf{r}$ and

$$\mathbf{C} = \alpha(nr^{\delta} - \mathbf{r}\mathbf{r}')$$

which leads back to (5.1.2) with $\varepsilon = n\alpha$.

The following design will serve as an example:

Block	I	A	A	A	C
	II	B	C		
	III	A	A	B	B
	IV	A	B		

Here

$$\mathbf{C} = \tfrac{1}{4}\begin{bmatrix} 9 & -6 & -3 \\ -6 & 8 & -2 \\ -3 & -2 & 5 \end{bmatrix}, \qquad \mathbf{\Omega} = \tfrac{1}{36}\begin{bmatrix} 7 & -1 & -1 \\ -1 & 11 & -1 \\ -1 & -1 & 23 \end{bmatrix}.$$

The contrasts between B and C, A and C, A and B would have had variances of $\frac{3}{4}\sigma^2$, $\frac{2}{3}\sigma^2$, and $\frac{5}{12}\sigma^2$ respectively had they been estimated with full efficiency. The respective variances are in fact σ^2, $\frac{8}{9}\sigma^2$, and $\frac{5}{9}\sigma^2$, i.e. all contrasts are estimated with an efficiency of $\frac{3}{4}$.

As has been said, some importance attaches to mean efficiency and a certain amount of work has been done on the subject (Pearce, 1968, 1970a; Conniffe and Stone, 1974, 1975; Gnot, 1976; Jarrett, 1977; Williams and Patterson, 1977).

Caliński (1977) has argued that variance-balance and efficiency-balance are more closely related than appears at first sight, and it is true that a suitable choice of **X** in (3.6.13) can give rise to multiple eigenvalues in any design.

There is an immense amount of work that could be considered in this chapter but for various reasons will be ignored. Those who would like to pursue matters further are referred to the admirable bibliographies that have appeared from Cornell University due to the exertions of Professor Federer and his associates (Federer and Balaam, 1972; Federer, 1980, 1981ab).

5.2. TOTAL BALANCE

The oldest class of non-orthogonal designs is that of 'balanced incomplete blocks' (Yates, 1933a), in which all blocks have the same number of plots, k, all treatments have the same degree of replication, r, and all pairs of treatments concur, i.e. come together in a block, equally often, i.e. λ times. It is also required that k should be less than the number of treatments, v, and that the elements of the incidence matrix should all be idempotent, i.e. either 0 or 1. An example is the following:

Block	I	ABC	Block	V	CEG
	II	ADE		VI	BEF
	III	AFG		VII	CDF
	IV	BDG			

Here $b = v = 7$, $r = k = 3$, and $\lambda = 1$. In general $\lambda = r(k-1)/(v-1)$ because the number of pairs of plots that can be formed within blocks equals both $\frac{1}{2}k(k-1)b$

and $\frac{1}{2}v(v-1)\lambda$. Also $bk = rv$. For such a design

$$\mathbf{C} = r\mathbf{I}_v - \frac{1}{k}[(r-\lambda)\mathbf{I}_v + \lambda \mathbf{1}_v \mathbf{1}_v']$$

$$= \frac{r(k-1)}{k(v-1)}\mathbf{L}, \tag{5.2.1}$$

where $\mathbf{L} = v\mathbf{I}_v - \mathbf{1}_v\mathbf{1}_v'$. However, all eigenvalues of $\mathbf{L}$ equal either v or 0, so $\mathbf{L}^+ = \mathbf{L}/v^2$. Hence

$$\mathbf{C}^+ = \frac{k(v-1)\mathbf{L}}{v^2 r(k-1)} = \frac{(v-1)\mathbf{L}}{bv(k-1)}. \tag{5.2.2}$$

It follows that for any contrast, $\mathbf{c}$,

$$\mathbf{c}'\mathbf{C}^+\mathbf{c} = \frac{(v-1)}{b(k-1)}\mathbf{c}'\mathbf{c} = \frac{\mathbf{c}'\mathbf{c}}{R}, \text{ say,} \tag{5.2.3}$$

i.e. all contrasts are estimated with the same precision, the effective replication being $R = b(k-1)/(v-1)$.

A useful catalogue of available designs of this class has been prepared and has been kept up to date in successive editions (Fisher and Yates, 1938, Tables XVII to XIX). Quite apart from the value of these designs in practice, a catalogue is useful as providing a basis for more complex configurations of treatments.

Where $k < v$ and there are $v!/[k!(v-k)!]$ blocks, one for each combination of the v treatments taken k at a time, the design is said to be 'unreduced'.

Balanced incomplete blocks are not only the oldest class of non-orthogonal designs; they are also widely used, their obvious advantages including reduction of block size and simplicity of analysis. It is also urged in their favour that all contrasts are estimated with equal precision, i.e., they have 'total balance'. However that is not necessarily an advantage if the problem under study calls for concentration on certain specified contrasts. Nevertheless, there are many occasions when the argument can reasonably be used.

On the other hand, there are many occasions when totally balanced designs are used for no satisfactory reason at all. A good scientist knows why each treatment was included. With a little prompting he will provide an explanation that enables the biometrician to write down the contrasts of interest. A bad one does not think out clearly enough what he is trying to do and, if pressed, is likely to take refuge in high-sounding phrases as if 'lack of prejudice' were the same as lack of thought or 'an open mind' the same as an empty one. Unless the biometrician is to make his decisions for him, which is a temptation to be avoided, the only possible course may be to recommend the silly fellow to use a design that at least keeps all the options open. Similar problems attend the giving of advice to a committee that has not resolved its differences. The easiest and sometimes the only course is to propose a design that permits the study of any hypothesis whatsoever. Then, if someone asks if it will be possible to compare this with that, whatever the question the biometrician can reply that it will be possible. There is, however, an ethical problem here. If such difficulties arise only rarely, the biometrician may be content to deal diplomatically with them as they come; if they are common, he may well decide that he would do better to seek employment where scientific standards are higher.

If equality of precision is desired, whether for good reasons or bad, it may be noted that designs with total balance form a much wider class than just balanced incomplete blocks (Rao, 1952b; Pearce, 1964; Ragavarao, 1971). Since no contrasts are confounded, **C** has only one zero eigenvalue corresponding to the eigenvector, $\mathbf{1}_v/\sqrt{v}$. Further, if $\mathbf{c}'\mathbf{C}^+\mathbf{c}$ is always to equal $\mathbf{c}'\mathbf{c}/R$, as in (5.2.3), where R is some constant, then all non-zero eigenvalues of $\mathbf{C}^+$ must equal $1/R$, which means that the corresponding eigenvalues of **C** must equal R. Hence,

$$\mathbf{C} = R(\mathbf{I}_v - \mathbf{1}_v\mathbf{1}_v'/v) \tag{5.2.4}$$

and all diagonal elements of $\mathbf{N}\mathbf{k}^{-\delta}\mathbf{N}'$, i.e., all weighted concurrences, must equal R/v. If all blocks contain k plots, then

$$R/v = \lambda/k. \tag{5.2.5}$$

Given a constant R, some very simple conclusions follow. Thus

$$\hat{\mathbf{t}} = \mathbf{C}^+\mathbf{Q} = \mathbf{Q}/R \tag{5.2.6}$$

and the treatment sum of squared deviations is

$$\mathbf{Q}'\mathbf{C}^+\mathbf{Q} = \mathbf{Q}'\mathbf{Q}/R = \sum_i Q_i^2/R. \tag{5.2.7}$$

In fact, everything is as for an orthogonal design except that **Q**, the vector of adjusted treatment totals, is used instead of **T**, in which the totals are unadjusted, and R instead of r. Actually (5.2.6) and (5.2.7) are simpler than their analogues (4.1.4) and (4.1.5) partly because $\mathbf{Q}'\mathbf{1}_v = 0$ and partly because effective replications are necessarily the same for all treatments. Examples 5A and 5B both have a bearing on designs in total balance.

An example of a totally balanced block design that is not in balanced incomplete blocks is the following:

Block	I	A	A	B	C
	II	A	A	B	C
	III	B	C		

Here

$$\mathbf{C} = \begin{bmatrix} 2 & -1 & -1 \\ -1 & 2 & -1 \\ -1 & -1 & 2 \end{bmatrix}$$

and $R = 3$.

It will be seen that in totally balanced designs all treatments are equivalent. That provides an alternative way of deriving some of the results already presented, e.g. by the use of (4.3.5).

Totally balanced designs have an obvious relationship to general balance as set out in Section 4.8. Since all contrasts have the same effective replication all groups of them (main effects, interactions, etc.) fulfil the requirement of (4.8.8ab).

Various attempts have been made to approximate to total balance when an exact solution is not possible. They will be considered later in Section 5.8.

5.3 SUPPLEMENTATION AND REINFORCEMENT

When contrasts are of unequal importance, the reason often lies in one treatment having a special status. Sometimes it is a standard with which each of the others is to be compared. In that case the interest will centre on elementary contrasts that involve it, others being of less importance or of none. Alternatively, there may be a treatment, e.g., taking no action at all, that has been included only as a base-line for the rest. To take an example, in a spraying experiment it is often a good idea to leave a few plots unsprayed as a demonstration that the pest is indeed present. If that is not done there is the danger of concluding that all the sprays are markedly and equally effective when really there has been nothing for them to control. It would be a mistake, however, to have too many such plots. Indeed, they can become a nuisance by providing foci of infection or infestation. (If unsprayed plots are used, it is necessary to keep an eye on them because drastic action may be needed quickly if the pest does begin to build up. Provided that all are treated alike at the same time and then recorded, that will not affect the design, which will continue as before but with one treatment altered. In any case, the data can be analysed even if some plots have to be omitted, though there may be a loss of precision in consequence.)

To cope with a design situation in which one treatment has a special status, two suggestions have been made. One is 'supplementation' (Hoblyn *et al.*, 1954, Pearce, 1960a, 1965); the other is 'reinforcement' (Das, 1958). In the first the designer contrives that the special treatment ('the supplementing treatment') shall make the same weighted concurrences, w_0, with all the others, which are disposed so as to make the same weighted concurrences, w, with one another. Then in (3.4.3)

$$\mathbf{W} = \begin{bmatrix} 0 & w_0 & w_0 & \dots & w_0 \\ w_0 & 0 & w & \dots & w \\ w_0 & w & 0 & \dots & w \\ \vdots & \vdots & \vdots & & \vdots \\ w_0 & w & w & \dots & 0 \end{bmatrix}, \qquad \mathbf{q} = \begin{bmatrix} (v-1)w_0 \\ w_0 + (v-2)w \\ w_0 + (v-2)w \\ \vdots \\ w_0 + (v-2)w \end{bmatrix},$$

$$\mathbf{C} = \mathbf{q}^{\delta} - \mathbf{W} = \left[\begin{array}{c|c} (v-1)w_0 & -w_0 \mathbf{1}'_{v-1} \\ \hline -w_0 \mathbf{1}_{v-1} & [w_0 + (v-1)w]\mathbf{I}_{v-1} - w\mathbf{1}_{v-1}\mathbf{1}'_{v-1'} \end{array}\right]. \tag{5.3.1}$$

Here it may be noted that

$$\mathbf{C}^{-} = \frac{1}{Rw_0}\left[\begin{array}{c|c} w & \mathbf{0}' \\ \hline \mathbf{0} & w_0 \mathbf{I}_{v-1} \end{array}\right] \tag{5.3.2}$$

is a generalized inverse of **C**, where $R[=w_0+(v-1)w]$ is the effective replication of any non-supplementing treatment and $R_0 = Rw_0/w$ is the corresponding value for the supplementing one. (Incidentally, it is possible for R_0 to exceed r_0, the actual replication.) To continue

$$\hat{\tau}_0 = Q_0/R_0, \qquad \hat{\tau}_i = Q_i/R \quad (i \neq 0), \tag{5.3.3}$$

where the supplementing treatment is numbered 0. Also,

Treatment sum of squared deviations

$$= \mathbf{Q}'\mathbf{C}^{-}\mathbf{Q} = \frac{Q_0^2}{R_0} + \sum_{i \neq 0} \frac{Q_i^2}{R}. \tag{5.3.4}$$

The variance of the contrast between the estimated mean of the supplementing treatment and that of any other is

$$\left(\frac{1}{R_0} + \frac{1}{R}\right)\sigma^2. \tag{5.3.5}$$

Between the means of any two non-supplementing treatments it is

$$\frac{2\sigma^2}{R}. \tag{5.3.6}$$

A design in supplemented balance can be regarded as having general balance also if the contrasts of interest are (a) those within the equivalence set of non-supplementing treatments and (b) that between the supplementing treatment and the rest. More usually, however, the enquiry concerns the elementary contrasts between the supplementing treatment and each of the others. Since they are not independent they do not qualify for purposes of general balance.

The design in Example 3A is in supplemented balance, i.e.,

Block	I	O	O	A	A	B	C	D	Block	III	O	O	A	B	C	C	D
	II	O	O	A	B	B	C	D		IV	O	O	A	B	C	D	D

In fact, it was the first of its kind. As initially planned, the design was intended to be orthogonal with each block made up of two plots of O and one each of A, B, C, D, and E. At the last moment supplies of substance E still had not arrived and a decision had to be made quickly. At that time non-orthogonal designs were little understood and the first reaction was to double the number of plots assigned to one of the other substances, but sufficient supplies were not available for any of them, so the final design was adopted in a spirit of desperation. It was decided that, if no valid analysis could be contrived, the four plots originally assigned to E would have to be omitted. In the event a useful class of designs was discovered.

There is no great difficulty about writing down supplemented designs for a particular task provided the biometrician is skilled at producing totally balanced designs for all occasions. For example, let there be three equivalent treatments, A, B, and C, and a fourth, O, with special status and let the most reasonable blocks each contain three plots. Various 'bricks' are available. It is unlikely that two plots of O will be needed in any block so the first brick will not be very useful. It is:

O O A
O O B
O O C

In making a brick the need is to secure balance for the non-supplementing treatments both among themselves and with the remaining one. A second possible brick is:

O B C
O A C
O A B

and, of course, it would be possible to add blocks of A B C, which provides a third brick. Any combination of bricks will give a design in supplemented balance. The particular combination to be chosen will depend upon the status of the supplementing treatment, whether comparisons that involve it are to be more or less precisely determined than the others. Since weighted concurrences are additive, it is enough to work out w and w_0 for each brick and to add values as may be required. Thus, in the example above,

Brick 1	$w = 0,$	$w_0 = \frac{2}{3}$
2	$w = \frac{1}{3},$	$w_0 = \frac{2}{3}$
3	$w = \frac{1}{3},$	$w_0 = 0$

Accordingly, for the design formed from using Brick 2 once and Brick 3 twice, i.e.

Block I	O	B	C	Block IV	A	B	C
II	O	A	C	V	A	B	C
III	O	A	B				

$w = 1$, $w_0 = \frac{2}{3}$, so $R = w_0 + 2w = \frac{11}{3}$ and $R_0 = Rw_0/w = \frac{22}{9}$. Even if no inter-block analysis is intended, it is still advisable to allocate the five sets of treatments to the blocks at random. (In the above design as it stands Treatment O is excluded from Blocks IV and V. Since it is possibly markedly different from A, B and C and the one most likely to behave differently in other conditions, it would be most unfortunate if Blocks I and II, which supposedly lie at one end of the area, were on slightly different soil. Indeed, with all non-orthogonal designs, if blocks are of equal size, it does no harm to randomize in that way and it can avoid awkward questions later.) Examples 4C and 4D also relate to designs in supplemented balance.

Incidentally, there need be no objection to mixing blocks of different size. Thus, the following is a perfectly good design, if it happens to fit the problem:

Block I	O	B	C	Block IV	O	O	A	B	C
II	O	A	C	V	O	O	A	B	C
III	O	A	B	VI	O	O	A	B	C

(It is, of course, being assumed that the variation is much the same within blocks of the two sizes.) Although it is unusual for a design in supplemented balance to be equi-replicate, it is quite possible as the following example shows,

Block I	O	O	A	A	B	C	Block III	O	O	A	B	C	C
II	O	O	A	B	B	C	IV	A	A	B	B	C	C

It comes from a row-and-column design (Pearce, 1963). The better distribution of the non-supplementing treatments leads to R being greater than R_0 ($w = \frac{3}{2}$, $w_0 = \frac{4}{3}$, $R = \frac{35}{6}$, $R_0 = \frac{140}{27}$).

In reinforcement any proper design is taken and a further treatment is added equally often to each block. That definition is a little wider than that given by Das (1958), but it will serve. Thus, the design in Example 3A can be regarded as a double reinforcement of the following, which is itself in total balance

Block I	A	A	B	C	D	Block III	A	B	C	C	D
II	A	B	B	C	D	IV	A	B	C	D	D

In general, if a design has an incidence matrix, $\mathbf{N}$, then a g-fold reinforcement gives

$$\mathbf{N}_r = \begin{bmatrix} g\mathbf{1}'_b \\ \mathbf{N} \end{bmatrix} \tag{5.3.7}$$

where the suffix r indicates the reinforced design. Accordingly,

$$\boldsymbol{\Omega}_r^{-1} = \begin{bmatrix} gb & \mathbf{0}' \\ \mathbf{0} & \mathbf{M}^{-1} \end{bmatrix} \qquad \boldsymbol{\Omega}_r = \begin{bmatrix} 1/(gb) & \mathbf{0}' \\ \mathbf{0} & \mathbf{M} \end{bmatrix} \tag{5.3.8}$$

where $\mathbf{M}^{-1} = (k\boldsymbol{\Omega}^{-1} + g\mathbf{r}^{\delta})/(k+g)$ and k is the constant block size of the original design. Mostly the inversion of $\mathbf{M}^{-1}$ is no more difficult than that of $\boldsymbol{\Omega}^{-1}$, which is just as well.

To consider the design in Example 3A, as has already been said, it can be obtained by reinforcing a totally balanced design. Here,

$$\boldsymbol{\Omega}^{-1} = \frac{1}{20}\begin{bmatrix} 97 & 1 & 1 & 1 \\ 1 & 97 & 1 & 1 \\ 1 & 1 & 97 & 1 \\ 1 & 1 & 1 & 97 \end{bmatrix}, \qquad \mathbf{M}^{-1} = \frac{1}{28}\begin{bmatrix} 137 & 1 & 1 & 1 \\ 1 & 137 & 1 & 1 \\ 1 & 1 & 137 & 1 \\ 1 & 1 & 1 & 137 \end{bmatrix},$$

$$\mathbf{M} = \frac{1}{680}\begin{bmatrix} 139 & -1 & -1 & -1 \\ -1 & 139 & -1 & -1 \\ -1 & -1 & 139 & -1 \\ -1 & -1 & -1 & 139 \end{bmatrix}, \qquad \boldsymbol{\Omega}_r = \frac{1}{680}\begin{bmatrix} 85 & 0 & 0 & 0 & 0 \\ 0 & 139 & -1 & -1 & -1 \\ 0 & -1 & 139 & -1 & -1 \\ 0 & -1 & -1 & 139 & -1 \\ 0 & -1 & -1 & -1 & 139 \end{bmatrix},$$

which is correct.

It will be seen that the classes of supplemented designs and of reinforced designs are not mutually exclusive. Also, some useful designs possess one property but not the other. To sum up, in supplemented designs the non-supplementing component must be in total balance; in reinforced designs the special treatment must be orthogonal to blocks. Are there perhaps useful designs that have neither property? The answer must be that there are, but they are usually more complicated and are often better considered as individuals rather than as members of a class, unless it is some very general one. Often a satisfactory solution is found in a multipartite design like those to be considered in Section 5.6. The following is a possibility:

Block I	O	A	B	C		Block IV	A	B	C	Q
II	O	A	B	C		V	O	A	P	Q
III	A	B	C	D		VI	O	B	P	Q
						Block VII	O	C	P	Q

Here the two equivalence groups, A B C and P Q, are primarily compared within themselves, but some attention is given to comparisons between groups and with O.

Caliński and Ceranka (1974) have considered a class of design, which they advanced as a generalization of those with supplemented balance, but they are perhaps better considered as a special case of the bipartite designs to be considered later. There is in fact real need for a generalization that will allow for two treatments to occupy special positions, one with a higher status, because the main set of treatments is to be compared with it, and one with lower to provide a base-line. The best suggestion at the moment is to start with a design in randomized blocks and to substitute one treatment for another in a range of blocks (Pearce, 1948). For example, the following is possible:

Block I	S	S	A	B	C	D	E	Block IV	S	O	A	B	C	D	E
II	S	S	A	B	C	D	E	V	S	O	A	B	C	D	E
III	S	O	A	B	C	D	E	VI	S	O	A	B	C	D	E

where S is the standard method, A B C D and E are modifications of it and O is an untreated control. Section 4.2, especially (4.2.1), indicates a relevant approach, as does Section 5.6 below. The matter is examined further in Example 5E.

5.4 GROUP-DIVISIBLE DESIGNS

The design introduced in Section 3.4, namely

Block I	A	B	C	D	Block IV	A	D	E	F
II	A	B	C	E	V	B	D	E	F
III	A	B	C	F	VI	C	D	E	F

is representative of a useful class, known as 'group-divisible designs'. The distinctive characteristic lies in the treatments forming μ groups, each of v members, such that any treatment makes w weighted concurrences with any other treatment of its own group and w' with any treatment in another group. In the example the groups are A B C and D E F, so $\mu = 2$, $v = 3$, $w = \frac{3}{4}$ and $w' = \frac{1}{2}$. As in Section 4.7 the treatments will be written as a factorial set, i.e., A = X1, B = Y1, C = Z1, D = X2, E = Y2 and F = Z2.

Historically the group-divisible designs have been regarded as a sub-class of those with partial balance (see Section 5.8), but there is no need to restrict them in that way. For example, there is no necessity for blocks to be of the same size or for treatments to be equally replicated, though those properties usually obtain. Neither is there any need for the elements of the incidence matrix to be restricted to values of 0 and 1.

However, for the definition all treatments must have the same quasi-replications q, such that

$$q = (v-1)w + (\mu-1)vw' \tag{5.4.1}$$

Hence, from (3.4.3)

$$\mathbf{C} = \begin{bmatrix} \mathbf{S} & \mathbf{S}_* & \dots & \mathbf{S}_* \\ \mathbf{S}_* & \mathbf{S} & & \mathbf{S}_* \\ \vdots & \vdots & & \vdots \\ \mathbf{S}_* & \mathbf{S}_* & \dots & \mathbf{S} \end{bmatrix} \tag{5.4.2}$$

where $\mathbf{S} = (q+w)\mathbf{I}_v - w\mathbf{1}_v\mathbf{1}'_v$ and $\mathbf{S}_* = -w'\mathbf{1}_v\mathbf{1}'_v$. Since $\mathbf{C}$ and hence Ξ^{-1} are unaltered by permutation either of treatments within groups or of groups as a whole, the same must be true of Ξ, which therefore has the same pattern as $\mathbf{C}$. Further, any matrix of the form $\Xi + \theta\mathbf{1}_v\mathbf{1}'_v$ must also be a generalized inverse of $\mathbf{C}$, so there is a solution of the form

$$\mathbf{C}^- = \begin{bmatrix} \mathbf{M} & \mathbf{0} & \dots & \mathbf{0} \\ \mathbf{0} & \mathbf{M} & \dots & \mathbf{0} \\ \vdots & \vdots & & \vdots \\ \mathbf{0} & \mathbf{0} & \dots & \mathbf{M} \end{bmatrix}, \tag{5.4.3}$$

where $\mathbf{M}$ is some $v \times v$ matrix. Since $\mathbf{CC}^-\mathbf{C} = \mathbf{C}$, from (5.4.2) and (5.4.3) it appears that

$$\mathbf{SMS} + (\mu-1)\mathbf{S}_*\mathbf{MS}_* = \mathbf{S}$$
$$\mathbf{S}_*\mathbf{MS} + \mathbf{SMS}_* + (\mu-2)\mathbf{S}_*\mathbf{MS}_* = \mathbf{S}_*,$$

so, by subtraction

$$(\mathbf{S}-\mathbf{S}_*)\mathbf{M}(\mathbf{S}-\mathbf{S}_*) = \mathbf{S}-\mathbf{S}_*$$

and

$$\mathbf{M} = (\mathbf{S}-\mathbf{S}_*)^{-1}$$
$$= \frac{1}{q+w}\mathbf{I}_v + \frac{w-w'}{\mu v(q+w)w'}\mathbf{1}_v\mathbf{1}'_v. \tag{5.4.4}$$

Hence, the contrast of any two treatments in the same group must have a variance of

$$\frac{2\sigma^2}{q+w} = \frac{2\sigma^2}{R}, \tag{5.4.5}$$

where $R = v[w+(\mu-1)w']$.

If a mean value is taken over the treatments in one group and compared with the mean of all those in another, the variance is

$$\frac{2}{v^2}\,\mathbf{1}'_v\mathbf{M}\mathbf{1}_v\sigma^2 = \frac{2\sigma^2}{R_g}, \text{ say} \tag{5.4.6}$$

where $R_g = \mu v^2 w'$.

If two treatments are in different groups, the variance of the difference of their means is

$$2\left(\frac{v-1}{vR} + \frac{1}{R_g}\right)\sigma^2. \tag{5.4.7}$$

It will be convenient to designate treatments by two suffices, the first for the group and the second for the position within that group. To find $\boldsymbol{\tau}_{ij}$ it is necessary to know not only Q_{ij} but also $P_i = \Sigma_j Q_{ij}$. Then, since $\hat{\boldsymbol{\tau}} = \mathbf{C}^-\mathbf{Q}$,

$$\hat{\boldsymbol{\tau}}_{ij} = \frac{Q_{ij} - P_i/v}{R} + \frac{P_i}{R_g}. \tag{5.4.8}$$

Hence, the treatment sum of squared deviations is

$$\mathbf{Q}'\mathbf{C}^-\mathbf{Q} = \frac{1}{R}\left(\sum_i\sum_j Q_{ij}^2 - \sum_i P_i^2/v\right) + \frac{1}{R_g}\sum_i P_i^2. \tag{5.4.9}$$

It will be seen in both (5.4.8) and (5.4.9) that a treatment is first considered in its group and the group is then considered relative to the experiment as a whole. (Note that $\Sigma_i P_i = 0$.)

It appears from (5.4.4) that a lot depends upon the relative values of w and w'. If $w > w'$, elementary contrasts will be better estimated when both treatments belong to the same group; if $w < w'$ such contrasts will be found with less precision, not more.

Usually group-divisible designs are equireplicate. Whether that is so or not, treatments within a group are equivalent, i.e. permuting them will have no effect on **C**. Hence, they are concordant and can be merged without affecting the precision of contrasts not involved. If each group is so merged, the resulting design is itself in total balance, i.e., the groups also are equivalent. Where a group-divisible design is used in a factorial context with one factor corresponding to the groups, it is necessarily in general balance relative to the two main effects and the interaction.

To derive a group-divisible design it is often helpful first to decide the allocation of the groups to the plots. Calling ABC the group X and DEF the group Y, then, in the example, Blocks I, II, and III are made up of XXXY and Blocks IV, V, and VI of XYYY. Since there are three blocks of each kind, the symmetric allocation of treatments to the plots of a group causes little difficulty. If, however, the contrast of the factor with two levels, i.e., ABC *versus* DEF or X *versus* Y, had been regarded as of special value, the decision might have been to keep it orthogonal and hence to make each block contain XXYY. It is quite easy to allocate three pairs of treatments (i.e., of A, B and C) to the two plots of X in each block but it will be found impossible to keep w' constant unless there are nine blocks so that all three pairs of A, B and C can concur with all three of D, E and F. That is a general difficulty; group-divisible designs need rather a lot of blocks. Nevertheless, those that are available are very useful.

5.5 EQUIPARTITE DESIGNS

To continue with the argument of the last section, if there are six treatments, 3×2 factorial, to be disposed in blocks of four plots to form a group-divisible design, with the second factor estimated with full efficiency, nine blocks would be needed. What can be done if only six are available? Basically there are two reasonable possibilities. The problem has been reduced to the association of the pairs, BC, AC, AB with the pairs EF, DF, DE. One solution is to associate each pair with its counterpart, thus giving BCEF, ACDF, and ABDE, which only requires duplication to become Design α of Example 3G. The other is to associate each pair with those in the other set that are not its counterpart, which gives BCDF, BCDE, ACEF, ACDE, ABEF, and ABDF, i.e. Design β of the same example. Either way $\mathbf{W}$ may be written as

$$\mathbf{W} = \left[\begin{array}{ccc|ccc} 0 & w & w & w' & w'' & w'' \\ w & 0 & w & w'' & w' & w'' \\ w & w & 0 & w'' & w'' & w' \\ \hline w' & w'' & w'' & 0 & w & w \\ w'' & w' & w'' & w & 0 & w \\ w'' & w'' & w' & w & w & 0 \end{array}\right]. \tag{5.5.1}$$

It may be said to possess 'factorial balance'. Such designs were studied by Kurkijan and Zelen (1963) and Mukerjee (1981) in papers that repay study. In general such designs have μ groups, each of v treatments, those within each group being ordered. Any treatment makes w weighted concurrences with another of the same group, w' with the corresponding treatments in other groups and w'' with any other. Corresponding to (5.4.1) it will be seen that

$$q = (v-1)w + (\mu-1)w' + (\mu-1)(v-1)w''. \tag{5.5.2}$$

The design would not be altered in principle if the two factors were interchanged to make $\mu = 3$, $v = 2$. In that event

$$\mathbf{W} = \left[\begin{array}{cc|cc|cc} 0 & w' & w & w'' & w & w'' \\ w' & 0 & w'' & w & w'' & w \\ \hline w & w'' & 0 & w' & w & w'' \\ w'' & w & w' & 0 & w'' & w \\ \hline w & w'' & w & w'' & 0 & w' \\ w'' & w & w'' & w & w' & 0 \end{array}\right].$$

To continue with the matrix in (5.5.1) it will be found that the pattern of $\mathbf{W}$

transfers to other matrices. Thus, in $\mathbf{C}$ the analogues of 0, w, w' and w'' are respectively q, $-w$, $-w'$ and $-w''$, while in $\boldsymbol{\Xi}^{-1}$ those values are all increased by q/v, the same pattern being found also in $\boldsymbol{\Xi}$. However, as before, it will be more convenient to work with $\boldsymbol{\Xi}+\theta\mathbf{1}_v\mathbf{1}_v'$, θ being chosen in this instance so that the analogue of w'' shall equal zero. The corresponding analogues of w and w' will then be respectively

$$\frac{w-w''}{\mu\delta[w'+(v-1)w'']}=\zeta_1,\text{ say} \tag{5.5.3a}$$

and

$$\frac{w'-w''}{v\delta[w+(\mu-1)w'']}=\zeta_2, \tag{5.5.3b}$$

where $\delta = vw+\mu w'+(\mu v-\mu-v)w''$. The analogue of 0 will then equal

$$1/\delta+\zeta_1+\zeta_2=\xi,\text{ say.} \tag{5.5.3c}$$

To take Design β of Example 3G by way of illustration, $\mu=2$, $v=3$, $w=\frac{1}{2}=w'$, $w''=\frac{3}{4}$, so $\delta=\frac{13}{4}$, $\zeta_1=-\frac{1}{52}$, $\zeta_2=-\frac{4}{195}$, $\xi=\frac{209}{780}$ and the following should be the generalized inverse sought:

$$\mathbf{C}^-=\tfrac{1}{780}\begin{bmatrix} 209 & -15 & -15 & -16 & 0 & 0\\ -15 & 209 & -15 & 0 & -16 & 0\\ -15 & -15 & 209 & 0 & 0 & -16\\ -16 & 0 & 0 & 209 & -15 & -15\\ 0 & -16 & 0 & -15 & 209 & -15\\ 0 & 0 & -16 & -15 & -15 & 209 \end{bmatrix}.$$

In fact, each element differs from the corresponding value in $\boldsymbol{\Omega}$ by $\frac{4}{585}$, so validity is confirmed.

Since $\hat{\boldsymbol{\tau}}=\mathbf{C}^-\mathbf{Q}$, using double suffices to represent levels of the two factors,

$$\begin{aligned}\hat{\tau}_{ij}&=\xi Q_{ij}+\zeta_1(P_i-Q_{ij})+\zeta_2(S_j-Q_{ij})\\ &=(\xi-\zeta_1-\zeta_2)Q_{ij}+\zeta_1P_i+\zeta_2S_j\end{aligned} \tag{5.5.4}$$

where $P_i=\Sigma_jQ_{ij}$ and $S_j=\Sigma_iQ_{ij}$. It will be recalled that $\xi-\zeta_1-\zeta_2=1/\delta$. Accordingly the treatment sum of squared deviations equals

$$\begin{aligned}&(\xi-\zeta_1-\zeta_2)\sum_i\sum_j Q_{ij}^2+\zeta_1\sum_i P_i+\zeta_2\sum_j S_j^2\\ &=\left(\sum_i\sum_j Q_{ij}^2-\sum_i P_i^2/v-\sum S_i^2/\mu\right)/\delta\\ &\quad+\sum_i P_i^2/\{\mu v[w'+(v-1)w'']\}\\ &\quad+\sum_j S_j^2/\{\mu v[w+(\mu-1)w'']\}.\end{aligned} \tag{5.5.5}$$

The expression shows how the partition into two main effects and an interaction can readily be made. It also shows the design to be in general balance, the effective replications for the intra-block analysis being as follows:

$$R_1 = v[w' + (v-1)w'']$$
$$R_2 = \mu[w + (\mu-1)w''] \qquad (5.5.6)$$
$$R_{12} = \delta.$$

In the illustration $R_1 = 4$, $R_2 = \frac{15}{4}$ and $R_{12} = \frac{13}{4}$. Remembering that in Example 3G the three sets of contrasts, $\mathbf{c}_1$ and $\mathbf{c}_2$, $\mathbf{c}_3$, $\mathbf{c}_4$ and $\mathbf{c}_5$, correspond respectively to contrasts within the second factor, the first factor and the interaction, it will be seen that the two methods of evaluating effective replications are equivalent.

The general class of equipartite designs may be defined by the relationship

$$\Xi^{-1} = \begin{bmatrix} \mathbf{S}_{11} & \mathbf{S}_{12} & \mathbf{S}_{1\mu} \\ \mathbf{S}_{12} & \mathbf{S}_{22} & \mathbf{S}_{2\mu} \\ \mathbf{S}_{1\mu} & \mathbf{S}_{2\mu} & \mathbf{S}_{\mu\mu} \end{bmatrix},$$

where the sub-matrices, $\mathbf{S}_{ij}$, are all $v \times v$ in size, and have all diagonal elements equal to a_{ij} and off-diagonal elements of b_{ij}. Let $a_{ij} - b_{ij} = \kappa_{ij}$. Permutation of the levels of either factor will have no effect on Ξ^{-1} and consequently none on Ξ so Ξ must have the same pattern as Ξ^{-1} and may be represented by writing A_{ij} instead of a_{ij} and B_{ij} instead of b_{ij}. Now in any group multiply out the row for the first treatment in Ξ^{-1} by the corresponding row on Ξ and by the second also. It will appear that

$$\sum_j \mathsf{K}_{ij}\kappa_{ij} = 1$$

where $\mathsf{K}_{ij} = A_{ij} - B_{ij}$. Where, however, the multiplication takes place with lines from other groups

$$\sum_j \mathsf{K}_{ij}\kappa_{ij} = 0.$$

Writing the κ_{ij} and K_{ij} as matrices it appears that

$$\mathsf{K}\boldsymbol{\kappa} = \mathbf{I}_n$$

so

$$\mathsf{K} = \boldsymbol{\kappa}^{-1}$$

A similar approach shows that

$$\boldsymbol{\Theta} = \boldsymbol{\theta}^{-1}$$

where $\boldsymbol{\theta}_{ij} = a_{ij} + (v-1)b_{ij}$ and $\boldsymbol{\Theta}_{ij} = A_{ij} + (v-1)B_{ij}$. Hence, it is possible to find A_{ij} and B_{ij} for each sub-matrix.

Although general equi-partite designs are unusual, they are quite feasible. The following will serve as an example:

Block I	A	B	C	D	Block IV	B	C	E	F
II	A	B	C	E	V	A	C	D	F
III	A	B	C	F	VI	A	B	D	E

It was obtained by taking the first part of the now familiar group-divisible design, and adding a single component of Design α in Example 3G. Its chief value would be in estimating differences in the group A B C, but it would also show up a marked interaction, assuming that D, E, and F, related respectively to A, B, and C, represented a subsidiary group of treatments in which something was done rather differently. For the above design

$$\boldsymbol{\Omega}^{-1} = \tfrac{1}{24}\begin{bmatrix} 115 & 1 & 1 & -3 & 3 & 3 \\ 1 & 115 & 1 & 3 & -3 & 3 \\ 1 & 1 & 115 & 3 & 3 & -3 \\ -3 & 3 & 3 & 63 & 3 & 3 \\ 3 & -3 & 3 & 3 & 63 & 3 \\ 3 & 3 & -3 & 3 & 3 & 63 \end{bmatrix}.$$

Here $\boldsymbol{\kappa} = \tfrac{1}{24}\begin{bmatrix} 114 & -6 \\ -6 & 60 \end{bmatrix}$, $\mathbf{K} = \tfrac{4}{189}\begin{bmatrix} 10 & 1 \\ 1 & 19 \end{bmatrix}$,

$$\boldsymbol{\theta} = \tfrac{1}{24}\begin{bmatrix} 117 & 3 \\ 3 & 69 \end{bmatrix}, \quad \boldsymbol{\Theta} = \tfrac{1}{112}\begin{bmatrix} 23 & -1 \\ -1 & 39 \end{bmatrix}.$$

So $A_{11} - B_{11} = \frac{40}{189}$ and $A_{11} + 2B_{11} = \frac{23}{112}$ so $A_{11} = \frac{1901}{9072}$ and $B_{11} = \frac{-19}{9072}$. Proceeding similarly with the other sub-matrices it appears that

$$\boldsymbol{\Omega} = \tfrac{1}{9072}\begin{bmatrix} 1901 & -19 & -19 & 101 & -91 & -91 \\ -19 & 1901 & -19 & -91 & 101 & -91 \\ -19 & -19 & 1901 & -91 & -91 & 101 \\ 101 & -91 & -91 & 3485 & -163 & -163 \\ -91 & 101 & -91 & -163 & 3485 & -163 \\ -91 & -91 & 101 & -163 & -163 & 3485 \end{bmatrix}.$$

The method is a useful one, applicable to a number of designs.

5.6. MULTIPARTITE DESIGNS

In passing from group-divisible to equipartite designs the equality of group sizes was retained, but it was no longer required that all treatments of a group should make the same number of weighted concurrences, w', with those outside it. A different generalization is afforded by 'multipartite' designs, in which group sizes no longer have to be the same.

Such designs are chiefly used when the treatments fall into groups such that the main comparisons are within groups and only subsidiarily between them. For example, in a project on intensive systems of fruit growing an experiment could study various ways of pruning in the summer together with other ways that involve pruning during the dormant season instead. The choice between seasons depends upon considerations like the better fruit colour obtained by removing leaves during a sunny period and greater dwarfing, which have to be set against the cost of pruning when labour is scarce; consequently it will be made for reasons that lie outside the function of the experiment, which would rather be to compare methods within each season. Nevertheless, a broad comparison between the two groups of treatments could well be desired also.

Suppose that there are three treatments in one group, namely, A, B, and C, and two in the other, namely, X and Y. For an experiment in ten blocks each of three plots, it is reasonable to allocate six plots to each treatment, taking care to keep the treatments of a group together as far as possible:

Block	I	A	B	C	Block	VI	B	C	Y
	II	A	B	C		VII	A	C	Y
	III	B	C	X		VIII	A	B	Y
	IV	A	C	X		IX	X	X	Y
	V	A	B	X		X	X	Y	Y

In general, let there be μ groups with v_i treatments in the ith group. If two treatments are both in that group, let them make w_{ii} weighted concurrences. If one is in the ith group and the other in the jth, let there be w_{ij} such concurrences. Then,

$$\boldsymbol{\Xi}^{-1} = \begin{bmatrix} \mathbf{S}_{11} & \mathbf{S}_{12} & \dots & \mathbf{S}_{1\mu} \\ \mathbf{S}_{12} & \mathbf{S}_{22} & \dots & \mathbf{S}_{2\mu} \\ \vdots & \vdots & & \vdots \\ \mathbf{S}_{1\mu} & \mathbf{S}_{2\mu} & \dots & \mathbf{S}_{\mu\mu} \end{bmatrix}, \tag{5.6.1}$$

where a sub-matrix like $\mathbf{S}_{ii}$ has all diagonal elements equal to a_i and all others equal to b_{ii}, while one like $\mathbf{S}_{ij} (i \neq j)$ has all elements equal to b_{ij}. It will be seen that in the ith group

$$q_i = (v_i - 1)w_{ii} + \sum_{j \neq i} (v_j w_{ij}) \tag{5.6.2a}$$

Also

$$u = \sum_i [v_i(v_i - 1)w_{ii}] + 2 \sum_{i \neq j} \sum (v_i v_j w_{ij}) \tag{5.6.2b}$$

$$a_i = q_i + q_i^2/u \tag{5.6.3a}$$

$$b_{ij} = -w_{ij} + q_i q_j/u \tag{5.6.3b}$$

the last including the case where i equals j.

In the above design $\mu = 2$, $v_1 = 3$, $v_2 = 2$, $w_{11} = \frac{4}{3} = w_{22}$, $w_{12} = \frac{2}{3}$, $q_1 = 4$, $q_2 = \frac{10}{3}$ and $u = \frac{56}{3}$. Hence

$$\boldsymbol{\Xi}^{-1} = \frac{1}{42} \begin{bmatrix} 204 & -20 & -20 & 2 & 2 \\ -20 & 204 & -20 & 2 & 2 \\ -20 & -20 & 204 & 2 & 2 \\ 2 & 2 & 2 & 165 & -31 \\ 2 & 2 & 2 & -31 & 165 \end{bmatrix}$$

It appears that permutation of treatments within a group will have no effect on Ξ^{-1} and consequently can have none on Ξ. That means in this instance that the two matrices must have the same pattern, e.g. since the first three diagonal elements can be interchanged by a permutation, not only are they the same in Ξ^{-1} but they must be so in Ξ. Accordingly Ξ may be written replacing a_i, b_{ii} and b_{ij} in Ξ^{-1} respectively by A_i, B_{ii} and B_{ij}. Now consider the first treatment of the ith group and multiply out its row in Ξ^{-1} by its column in Ξ. It appears that

$$a_i A_i + (v_i - 1) b_{ii} B_{ii} + \sum_{l=2}^{\mu} (v_l b_{il} B_{il}) = 1 \qquad (5.6.4a).$$

If, however, the same row in Ξ^{-1} is multiplied out by a column in Ξ corresponding to some other treatment from the same group

$$a_i B_{ii} + b_{ii} A_i + (v_i - 2) b_{ii} B_{ii} + \sum_{l=2}^{\mu} (v_l b_{il} B_{il}) = 0 \qquad (5.6.4b)$$

Subtraction of (5.6.4b) from (5.6.4a) leads to a result familiar from other designs, namely

$$A_i - B_{ii} = \frac{1}{a_i b_{ii}} = \frac{1}{q_i + w_{ii}} \qquad (5.6.5)$$

Further, multiplying (5.6.4b) by $(v_i - 1)$ and then adding (5.6.4a) gives

$$\sum_l (\theta_{il} \Theta_{il}) = 1 \qquad (5.6.6.)$$

where θ_{il} is the sum of the elements of $\mathbf{S}_{il}$ in Ξ^{-1} (5.6.1) and Θ_{il} is the mean of the elements of its analogue in Ξ.

If now the row in Ξ^{-1} for the first treatment of the ith group is multiplied by the columns in Ξ for the first and second treatments of the jth group $(i \neq j)$, two more equations emerge, which can be similarly combined to give

$$\sum_l (\theta_{il} \Theta_{jl}) = 0 \qquad (5.6.7)$$

Writing the θ_{ij} and Θ_{ij} as matrices $\boldsymbol{\theta}$ and $\boldsymbol{\Theta}$, it follows from (5.6.6) and (5.6.7) that

$$\boldsymbol{\Theta} = \boldsymbol{\theta}^{-1} \qquad (5.6.8)$$

Since the elements of $\boldsymbol{\theta}$ are readily found, there should be no difficulty in finding $\boldsymbol{\Theta}$, especially as $\boldsymbol{\theta}$ has dimensions, $\mu \times \mu$, and is therefore probably small. Once its elements are known, taking them in conjunction with (5.6.5), it is easy to find Ξ.

To continue with the example, from (5.6.5)

$$A_1 - B_{11} = \tfrac{3}{16} \qquad A_2 - B_{22} = \tfrac{3}{14}$$

Also

$$\boldsymbol{\theta} = \tfrac{1}{21}\begin{bmatrix} 246 & 6 \\ 6 & 134 \end{bmatrix} \qquad \boldsymbol{\Theta} = \tfrac{1}{784}\begin{bmatrix} 67 & -3 \\ -3 & 123 \end{bmatrix}$$

so

$$\tfrac{1}{3}A_1 + \tfrac{2}{3}B_{11} = \tfrac{67}{784}$$

$$B_{12} = -\tfrac{3}{784}$$

$$\tfrac{1}{2}A_2 + \tfrac{1}{2}B_{22} = \tfrac{123}{784}$$

Hence

$$\Xi = \frac{1}{784}\begin{bmatrix} 165 & 18 & 18 & -3 & -3 \\ 18 & 165 & 18 & -3 & -3 \\ 18 & 18 & 165 & -3 & -3 \\ -3 & -3 & -3 & 207 & 39 \\ -3 & -3 & -3 & 39 & 207 \end{bmatrix}$$

It will be seen that the variance of the difference of means of two treatments in the first group is $\frac{3}{8}\sigma^2$, an efficiency of $\frac{8}{9}$. In the second group the variance is $\frac{3}{7}\sigma^2$, an efficiency of $\frac{7}{9}$. That is not as good and reflects the fact that plots of the second group diverted to establishing relationships with the first, provide no intra-group information. The design would in fact be improved by additional concurrences between Treatments X and Y. Finally, a contrast between two treatments in different groups has a variance of $\frac{27}{56}\sigma^2$ with an efficiency of only $\frac{56}{81}$. The variance of the difference between groups as a whole, i.e.

$$(\tfrac{1}{3}, \tfrac{1}{3}, \tfrac{1}{3}, -\tfrac{1}{2}, -\tfrac{1}{2})'$$

is $\frac{1}{4}\sigma^2$. The efficiency, $\frac{5}{9}$, is even lower, but the variance is quite good for a general comparison between the groups, being in fact better than that for any comparison within them.

Often there is no need to find Ξ or any other generalized inverse in full. If the chief interest lies, as usually it does, in contrasts within groups, (5.6.5) gives all that is required. It is unusual for the variance between groups as a whole to be inadequate.

When $m = 2$ the design is said to be 'bipartite'. It will be found that the variance of the contrast between the two groups is $\sigma^2/(v_1 v_2 w_{12})$ as in the example.

The problem mentioned at the end of Section 5.4, i.e. that of double supplementation, is usually met by a design that is, in fact, multipartite. Thus, the solution there proposed gives three groups, S, O and A–E. An instructive example has been presented earlier (Pearce, 1963). Two spraying trials were being planned on the control of *Botrytis* in strawberries. Eight blocks, each of four plots, were available but not more and that caused difficulties. One treatment, S, the standard preparation applied at the usual time, was to be common to both experiments. In one trial it was to be compared with the same preparation applied at four other times (Treatments A, B, C, and D). It was proposed initially to assign it five blocks and to use totally balanced design. In the other trial it was proposed to compare S with the application of two new substances at the usual time (Treatments P and Q), making use of the other three blocks for another totally balanced design. In both experiments there would have been no additional replication of Treatment S, though it was really called for, because there were not enough plots. Also, neither had a surfeit of error degrees of freedom. In each, the variance of the contrast between S and one other treatment would have been $\frac{8}{15}\sigma^2$. It was then realized that the two experiments could easily be merged so that each could have access to the plots of S assigned to the other, i.e. by using

Block I	S	B	C	D		Block V	S	A	P	Q
II	S	A	C	D		VI	S	B	P	Q
III	S	A	B	D		VII	S	C	P	Q
IV	S	A	B	C		VIII	S	D	P	Q

The variance of a contrast of S with A etc. is now $\frac{29}{70}\sigma^2$ and of S with P or Q it is $\frac{17}{40}\sigma^2$. This illustration is included to emphasize the great flexibility of multipartite designs. There is no need to fear combinatorial complexities. They do exist, but those who make the attempt, once they have obtained confidence, find that groups can often be associated in a satisfactory way.

5.7 CYCLIC DESIGNS

Cyclic designs have been known for a long time but their intensive study is a fairly recent development, so this is not a good time at which to assess their usefulness. However, they have some attractive properties and should therefore be considered seriously for those practical purposes that they appear to fit. In essence, the treatments are thought of as arranged in a cycle, e.g.

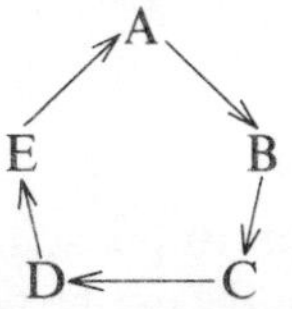

A first block is chosen and the others are derived from it by cyclic permutation. Thus if Block I contained A B D, the others would contain B C E, C D A = A C D, B D E, B D A = A B D respectively. Proceeding in that way the number of blocks necessarily equals the number of treatments or is some multiple of it. Designs are both proper and equi-replicate; also, all treatments have the same quasi-replications. It will be seen that $\mathbf{C}$ and hence $\mathbf{\Omega}^{-1}$ and $\mathbf{\Xi}^{-1}$ and hence $\mathbf{\Omega}$ and $\mathbf{\Xi}$ must all be circulant matrices, each defined by its first row, just as $\mathbf{N}'$ is so defined. It is not, however, essential to permute in steps of one; if v is a composite number one of its factors could be considered (John *et al.*, 1972).

An obvious use arises when treatments form a graduated series, e.g., of the time of application of fertilizer or a herbicide, and it is desired to find the optimal point. In short, a comparison is required between each treatment and its neighbours and a cyclic design has obvious advantages. Thus, the following could be used:

Block I	A	B	C		Block IV	A	D	E
II	B	C	D		V	A	B	E
III	C	D	E					

On general grounds many would prefer to replace the contents of Blocks IV and V by D E E and A A B. The result will not be cyclic, but in these days of plentiful computational aids there is no need to object. Another possibility is

Block I	A	B	C	Block	IV	C	D	E
II	A	B	C		V	C	D	E
III	B	C	D					

Again, it is not cyclic but it has advantages of its own.

Larger block sizes are, of course, possible. For example, to return to a familiar context, if there were six treatments to be studied in blocks each of four plots, two initial blocks are readily available, i.e. A B C D and A B B C. Comparison of the two designs is of some interest; they yield respectively the following first rows of $\mathbf{\Omega}$:

$$\tfrac{1}{2340}\,(643,\ 16,\ -29,\ -32,\ -29,\ 16)$$

and

$$\tfrac{1}{1260}\,(457,\ 64,\ -71,\ -128,\ -71,\ 64)$$

It might be expected that the second design would be worse for comparing treatments well separated in the cycle but better for the comparison of neighbours. In fact, it is worse on both counts. Lest it should seem that any advantage lies in dispersion round the cycle, a further initial block should be considered, namely, A B C E. If it is adopted, the first row of $\mathbf{\Omega}$ is

$$\tfrac{1}{360}(99,\ -5,\ 3\ -5,\ 3,\ -5)$$

For the comparison of neighbours it is not as good as the design based on A B C D, but it is better for the comparison of Treatments A and C; for A and D there is little to choose.

John (1973) has drawn attention to the suitability of some cyclic designs for factorial situations. He shows that if $\mathbf{\Omega}$ can be partitioned into $\mu \times \mu$ sub-matrices of size $v \times v$, i.e.

$$\mathbf{\Omega} = \begin{bmatrix} \mathbf{\Omega}_1 & \mathbf{\Omega}_2 & \cdots & \mathbf{\Omega}_\mu \\ \mathbf{\Omega}_\mu & \mathbf{\Omega}_1 & \cdots & \mathbf{\Omega}_{\mu-1} \\ \vdots & \vdots & & \vdots \\ \mathbf{\Omega}_2 & \mathbf{\Omega}_3 & \cdots & \mathbf{\Omega}_1 \end{bmatrix}$$

such that all the $\mathbf{\Omega}_j (j = 1, \ldots, \mu)$ are symmetric and circulant, the design permits the analysis of a $\mu \times v$ factorial set of treatments with the two main effects and the interaction estimated independently. The converse also holds. The last design considered, i.e. the one with ABCE in the initial block, exhibits that property with $\mu = 3$, $v = 2$. It is in fact an old friend, namely, the example first introduced in Section 3.4 and used several times thereafter. It is not at first recognizable, but it takes its familiar form if treatments are permuted thus:

$$\text{A} \to \text{A},\quad \text{B} \to \text{D},\quad \text{C} \to \text{B},\quad \text{D} \to \text{E},\quad \text{E} \to \text{C},\quad \text{F} \to \text{F}.$$

The fact, incidentally, that the same design can be described with complete accuracy as being group-divisible, as having factorial balance and as being cyclic, points to a difficulty in the whole subject of non-orthogonality, namely, that a design—and this applies especially to straightforward ones—can belong to several classes. Further, if its treatments are permuted, it can sometimes be shown to belong to several more. It is all very confusing, especially as affinities are often hard to detect if they appear only after permutation. To return, however, to cyclic designs, their facility of generating schemes suitable for factorial sets of treatments, especially group-divisible designs, has been considered in some detail e.g. Freeman (1976b), Patterson and Williams (1976), Jarrett and Hall (1978).

5.8 APPROXIMATIONS TO TOTAL BALANCE

Ever since total balance was regarded as desirable, there have been efforts to achieve an approximation in cases where an exact solution is not possible. Some care is needed at this point. If an approximation to total balance is so close that the slight differences between variances can be ignored, it does not matter which contrasts are estimated slightly better or slightly worse. Thus, any of the classes so far considered can be regarded as providing, at least potentially, an approximation if members can be found in which the different variances are all about the same. If, however, there are appreciable differences, it is important that they shall fit the treatment structure.

An early attempt to achieve some sort of an approximation was that of Bose and Nair (1939) with the concept of 'partial balance'. It may be summed up thus:

1. No treatment occurs more than once in any block.
2. Each treatment occurs in r blocks.
3. One treatment is an ith associate of another if they concur (i.e. come together) in λ_i blocks, the numbers $\lambda_1, \lambda_2, \ldots$, etc. being the same whichever treatment is chosen initially.
4. If two treatments are ith associates and if p^i_jk is the number of jth associates of one that are also kth associates of the other, then p^i_jk is constant irrespective of the initial choice of two ith associates.

It will be seen that these requirements are scarcely those of a practical man. The object is to make the variances of estimation of the elementary contrasts as similar as possible. That is quite admirable if anyone is studying elementary contrasts and regards them all as of equal importance, but that is not always the problem. It is true that some very useful designs are partially balanced, but others are not. The same goes for completely useless designs as well. To be fair, that reaction is to some extent emotional. Partial balance has often been advanced with too little thought and in too exclusive a spirit. Books start out hopefully to explore the intricacies of non-orthogonal designs and end having dealt effectively only with total and partial balance. People stand up at conferences and announce, as if it settled the matter under discussion, that there is in fact 'a design' to cover the case, meaning that there is one in partial balance with the desired values of v, r, b and k, all of which produces a reaction that can be as unthinking as the original action.

Partially balanced designs with two associate classes have been catalogued and classified by Bose *et al.* (1954) in a publication that contains much of value, as does the book of Ragavarao (1971). The catalogue has been amplified by Clatworthy (1973).

Nevertheless, the question remains. What are partially balanced designs intended to do? If the answer is that they are intended to give effectively uniform precision for all contrasts it is fair to comment that some do not succeed very well. Thus, the design that has been used in this text for purposes of illustration, beginning with Section 3.4, happens to be partially balanced. It is in fact No. R2 in the catalogue of Bose *et al.* and No. R94 in that of Clatworthy, but no sensible experimenter would use it if uniformity of precision were required. Nevertheless, as appeared in Example 3F, its association pattern fits very well a 3×2 factorial set of treatments. If, on the other hand, the question is answered by conceding that partially balanced designs do have marked association patterns, it is fair to concede in return that those patterns are sometimes very useful. For example, those of Latin square type obviously fit the case where there are two factors, the levels of which are the same. That happens with diallel crosses, where the same range of source material is used both for male and female parents. It happens also in fruit tree studies, where the same range of varieties may be used both as rootstock and scion to elucidate mechanisms. Again, as Kramer and Bradley (1957) and Bradley, Walpole and Kramer (1960) have pointed out, group-divisible designs are well adapted to experiments where the treatments have factorial structure. That, however, does not endorse all partially balanced designs, some of which are nearly useless. In this monograph no further attention will be given to the subject, which is the great irrelevancy of the subject of experimental design. If a design is useful, it will be mentioned; if it is not, it will be ignored, whether partially balanced or not.

Other attempts have been made to achieve an approximation to total balance, where that is called for. For example, there are the 'nearly balanced designs' of Nigam (1976b), which succeed very well though they do need a lot of blocks. In practice much greater use is made of the 'lattice designs' introduced by Yates (1936a, 1937). Their most obvious applications come when v, the number of treatments, is a perfect square. Suppose, for example, that v equals 16, then the treatments can be thought of as forming a square, thus:

A	B	C	D
E	F	G	H
I	J	K	L
M	N	O	P

The first four blocks will consist of treatments that come together in rows and the next four of those given by columns, i.e.

Block I	A	B	C	D	Block V	A	E	I	M
II	E	F	G	H	VI	B	F	J	N
III	I	J	K	L	VII	C	G	K	O
IV	M	N	O	P	VIII	D	H	L	P

That is called a 'simple lattice'. It has in fact the factorial balance described in Section 5.5, though that does not matter much. If a third replicate is required, a Latin square is imposed on the table of treatments and its 'directrices' (i.e. groups of plots to receive the same treatment) can be used to form four more blocks. However, if the process is not to end at three replicates, it is important to choose a Latin square that can support a complete orthogonal set, such as this one:

1	2	3	4
2	1	4	3
3	4	1	2
4	3	2	1

It gives

Block IX	A	F	K	P	Block XI	C	H	I	N
X	B	E	L	O	XII	D	G	J	M

This more complicated design, which is in partial balance, is that ordinarily given by a 'triple lattice'. Having got so far, a fourth replicate, if one were sought, could be obtained from the directrices of a Graeco-Latin square, for example

Block XIII	A	G	L	N	Block XV	C	E	J	P
XIV	B	H	K	M	XVI	D	F	I	O

As total balance is approached the design begins to become simpler, the latest being group-divisible. A further four blocks to associate those treatments not already associated would achieve totality of balance, namely:

Block XVII	A	H	J	O	Block XIX	C	F	L	M
XVIII	B	G	I	P	XX	D	E	K	N

In general, if $v = x^2$, $x(x+1)$ blocks, each with x plots, will be needed to achieve total balance but any multiple of x blocks will go some way towards the goal. (It is assumed that sets of orthogonal $x \times x$ squares are available, as they will be for values of all x other than 6.)

With a simple lattice there are two replications throughout and each treatment concurs once in blocks with $2(x-1)$ others, leaving $(x-1)^2$ with which it does not concur. The design is in factorial balance. The variance of the difference between two treatment means is $[(x+1)/x]\sigma^2$ if the treatments concur and $[(x+2)/x]\sigma^2$ if they do not. Passing to the triple lattice, each treatment concurs once with $3(x-1)$ others and not at all with $(x-1)(x-2)$; the corresponding variances are $[2(x+1)/3x]\sigma^2$ and $[2(x+3)/3x]\sigma^2$ respectively. The coefficient matrix, **C**, has a recognizable pattern that has received some attention under the name of the Latin square association scheme (Bose *et al.*, 1954). As further replications are added the design becomes more complicated, but it begins to simplify again as total balance is approached. When still two replicates short, the pattern of the triple lattice appears again. Each treatment now concurs with

$(x-1)^2$ others but not with the remaining $2(x-1)$ and the respective variances are

$$\frac{2(x+1)}{x(x-1)}\sigma^2 \quad \text{and} \quad \frac{2(x^2-x-1)}{x(x-1)(x-2)}\sigma^2.$$

At the next stage, when only one further replicate is needed for total balance, the design is group-divisible, the groups being defined by the blocks of the last replicate, and the variances are respectively

$$\frac{2(x+1)}{x^2}\sigma^2 \quad \text{and} \quad \frac{2}{x-1}\sigma^2.$$

Finally, all pairs of treatments concur in some block and all elementary contrasts have a variance of $(2/x)\sigma^2$. At all stages considered, the efficiency of estimation of a difference between concurring treatments has been $x/(x+1)$. Where treatments do not concur, the variance has in all cases been greater by a ratio of $1+\left[\frac{1}{(r-1)(x+1)}\right]$, where r is the degree of replication. (For total balance, of course, the last statement does not apply, but only for the lattices.) It appears then that equality of variances is only approximate unless r and x are both fairly large.

A difficulty with lattices lies in their inflexibility, since the desired number of treatments will not usually be a perfect square. There are in fact ways of relaxing this limitation. Rectangular lattices are possible and so are cubic lattices. However, their principal use lies in the study of new varieties where the precise number to be included is often rather indefinite. If, say, 46 or 53 varieties have been chosen for retention, usually no one will be much bothered by an adjustment to 49 for technical reasons. In any event, it is wise to include the established varieties, which the new ones are required to beat, and they may well be given increased replication for the reasons set out in Section 1.6. Also, they will be in good supply so it will not matter much if a larger number of plots is required for each. It is true that the inclusion of a standard variety twice or more can disturb the neat regularity of some of the designs, but in these days of general computer programs that should not matter. Of course, if the treatments do have a factorial structure, an ingenious experimenter can well relate it to his design.

A specialist study of designs for comparing varieties according to the regulations of the European Economic Community (Patterson *et al.*, 1978) contains some interesting surprises and provides useful guidance to those faced with that problem.

The more complicated forms of lattice have been usefully described by Federer (1955).

Examples 5

A. An experiment employs an unreduced balanced incomplete block design with v treatments, each block containing k plots. One of the treatments proves to be very disappointing and all plots that have received it are eliminated from the experiment. As a result the variances of adjusted treatment means are θ times what they would otherwise have been. Evaluate θ. (University of Kent at Canterbury, Diploma/M.Sc., 1969)

Answer: In any unreduced design

$$\lambda = \frac{(v-2)!}{(k-2)!\,(v-k)!}$$

because any two treatments concur once for each combination of the other $(v-2)$ treatments taken $(k-2)$ at a time. Hence from (5.2.5) $R = v\lambda/k$. Alternatively, from (5.2.3)

$$R = \frac{b(k-1)}{(v-1)} = \frac{v(v-2)!}{k(k-2)!\,(v-k)!}.$$

If now a treatment is lost, what remains consists of (a) an unreduced design of $(v-1)$ treatments in blocks of k plots and (b) another unreduced design of $(v-1)$ treatments in blocks of $(k-1)$ plots. Hence

$$R_a = \frac{(v-1)(v-3)!}{k(k-2)!\,(v-k-1)!}, \qquad R_b = \frac{(v-1)\,(v-3)!}{(k-1)\,(k-3)!\,(v-k)!}.$$

For the complete design

$$R_c = R_a + R_b = \frac{(v-1)\,(v-3)!\,(vk-v-k)}{k!\,(v-k)!},$$

so

$$\theta = R/R_c = \frac{v(v-2)\,(k-1)}{(v-1)\,(vk-v-k)}.$$

Note: $m! = (m+1)!/(m+1)$. Hence $0! = 1$. The result can be useful, especially when k or $v = 3$.

B. Design 2 is said to be the *complement* of Design 1 if their incidence matrices are such that

$$\mathbf{N}_2 + \mathbf{N}_1 = \mathbf{1}_v \mathbf{1}'_b.$$

Design 3 is said to be the *extension* of Design 1 if

$$\mathbf{N}_3 - \mathbf{N}_1 = \mathbf{1}_v \mathbf{1}'_b.$$

If Design 1 is in balanced incomplete blocks, show that its complement and its extension are both totally balanced.

Writing r_i and R_i for the actual replication and the effective replication respectively of Design i, show that

$$r_1(r_1 - R_1) = r_2(r_2 - R_2) = r_3(r_3 - R_3).$$

(University of Kent at Canterbury, Diploma/M.Sc. Notation changed to accord with that of the present text.)

Extended hint:

$$\mathbf{r}_2 = \mathbf{N}_2\mathbf{1}_b = b\mathbf{1}_v - \mathbf{r}_1 = (b - r_1)\mathbf{1}_v$$
$$\mathbf{r}_3 = \mathbf{N}_3\mathbf{1}_b = b\mathbf{1}_v + \mathbf{r}_2 = (b + r_1)\mathbf{1}_v$$
$$\mathbf{k}_2 = \mathbf{N}_2'\mathbf{1}_v = v\mathbf{1}_b - \mathbf{k}_1 = (v - k_1)\mathbf{1}_b$$
$$\mathbf{k}_3 = \mathbf{N}_3'\mathbf{1}_v = v\mathbf{1}_b + \mathbf{k}_1 = (v + k_1)\mathbf{1}_b$$

Note that $vr_i = bk_i$

$$\mathbf{N}_1\mathbf{k}_1^{-\delta}\mathbf{N}_1' = \mathbf{N}_1\mathbf{N}_1'/k_1 = [(r_1 - \lambda)\mathbf{I}_v + \lambda\mathbf{1}_v\mathbf{1}_v']/k_1, \text{ say}$$
$$\mathbf{N}_2\mathbf{k}_2^{-\delta}\mathbf{N}_2' = \mathbf{N}_2\mathbf{N}_2'/(v - k_1) = [(b - 2\mathbf{r}_1)\mathbf{1}_v\mathbf{1}_v' + \mathbf{N}_1\mathbf{N}_1']/(v - k_1)$$
$$= [k_1(r_1 - \lambda)/(v - k_1)]\mathbf{I}_v + \text{a term in } \mathbf{1}_v\mathbf{1}_v']$$
$$\mathbf{N}_3\mathbf{k}_3^{-\delta}\mathbf{N}_3' = \mathbf{N}_3\mathbf{N}_3'/(v + k_1) = [(b + 2r_1)\mathbf{1}_v\mathbf{1}_v' + \mathbf{N}_1\mathbf{N}_1']/(v + k_1)$$
$$= [k_1(r_1 - \lambda)/(v + k_1)]\mathbf{I}_v + \text{a term in } \mathbf{1}_v\mathbf{1}_v'$$
$$\mathbf{C}_1 = \left(r_1 - \frac{r_1 - \lambda}{k_1}\right)\mathbf{I}_v + \text{a term in } \mathbf{1}_v\mathbf{1}_v'$$
$$\mathbf{C}_2 = \left(r_2 - \frac{(r_1 - \lambda)}{v - k_1}\right)\mathbf{I}_v + \text{a term in } \mathbf{1}_v\mathbf{1}_v'$$
$$\mathbf{C}_3 = \left(r_3 - \frac{(r_1 - \lambda)}{v + k_1}\right)\mathbf{I}_v + \text{a term in } \mathbf{1}_v\mathbf{1}_v'$$

From (5.2.4), if all off-diagonal elements of $\mathbf{C}_i$ are equal, then Design i is in total balance. Also, its effective replication, R_i, equals the extent to which the diagonal elements of $\mathbf{C}_i$ exceed those off the diagonal, i.e. it equals the coefficient of I_v in the expression for $\mathbf{C}_i$.

C. An experiment is to be designed in six blocks, each of three plots. There are three treatments. Two of them, A and B, are intended to arrest the development of a fungus and the third, O, is an untreated control. Initially it was proposed to design the experiment in randomized blocks, i.e., with each treatment occurring once in each block, but someone has pointed out that the real task is to compare treatments A and B, comparisons with the control being of less importance. He therefore proposes the following design:

Block	I	A	B	O
	II	A	B	O
	III	A	B	O
	IV	A	B	O
	V	A	A	B
	VI	A	B	B

Assess this proposal on the basis of variances of estimation of all contrasts between pairs of treatments.
(University of Kent at Canterbury, Diploma/M.Sc. 1979)

Answer:

$$\mathbf{N} = \begin{bmatrix} 1 & 1 & 1 & 1 & 0 & 0 \\ 1 & 1 & 1 & 1 & 2 & 1 \\ 1 & 1 & 1 & 1 & 1 & 2 \end{bmatrix}, \qquad \mathbf{Nk}^{-\delta}\mathbf{N}' = \tfrac{1}{3}\begin{bmatrix} 4 & 4 & 4 \\ 4 & 9 & 8 \\ 4 & 8 & 9 \end{bmatrix}.$$

The design therefore is a supplemented balance with $v = 3, w = \frac{8}{3}$ and $w_0 = \frac{4}{3}$. Hence $R = \frac{20}{3}$ and $R_0 = \frac{10}{3}$. Accordingly the variance of the difference of two treatment parameters is

$$\begin{array}{lll} 3\sigma^2/10 & \text{for} & \text{A and B} \\ 9\sigma^2/20 & \text{for} & \text{O and A or O and B.} \end{array}$$

The optimal ratio of these two variances depends upon the relative importance of the treatments. If a large effect of sprays is expected, it could well be that even more blocks should be confined to Treatments A and B.
Corollary: To follow up that comment, the problem could be extended to cover the case of b blocks, of which a contain AAB, a contain ABB and the rest OAB. Then

$$\mathbf{Nk}^{-\delta}\mathbf{N}' = \tfrac{1}{3}\begin{bmatrix} b-2a & b-2a & b-2a \\ b-2a & b+3a & b+2a \\ b-2a & b+2a & b+3a \end{bmatrix}$$

Now $w = \frac{1}{3}(b+2a)$, $w_0 = \frac{1}{3}(b-2a)$ and $v = 3$. Therefore

$$R = \tfrac{1}{3}(3b+2a) \qquad \text{and} \qquad R_0 = \tfrac{1}{3}(3b+2a)(b-2a)/(b+2a).$$

The ratio is now, of course, $(b-2a)/(b+2a)$, which gives an indication of the value of a/b needed to achieve a desired ratio. In the main question, where $b = 6$, if a were put equal to 2 instead of 1, the ratio of variances would have been $\frac{1}{5}$ instead of $\frac{1}{2}$ as before.

D. One design mentioned in the text is the following:

Block I	O	O	A	A	B	C	Block III	O	O	A	B	C	C		(α)
II	O	O	A	B	B	C	IV	A	A	B	B	C	C		

Of the four treatments, A, B, and C represented times of applying fertilizer to blackcurrant bushes, while O represented the omission of fertilizer in that year.

When the design was presented at a meeting of the Royal Statistical Society (Pearce, 1963) there was some disagreement. The author defended his design on the grounds that he wished to compare A, B, and C with one another rather than with O. Also, he wanted to preserve symmetry between A, B, and C since, he argued, a design is judged by the worst estimate it makes of an important contrast. Other possible designs are

Block I	O	O	O	A	B	C	Block III	O	A	A	B	C	C	(β)
II	O	A	B	B	C		IV	Q	A	A	B	B	C	
Block I	O	O	O	A	B	C	Block III	A	A	B	B	C	C	(γ)
II	O	O	O	A	B	C	IV	A	A	B	B	C	C	

and

Block I	O	O	A	A	B	C	Block III	O	O	A	B	B	C		
II	O	A	B	B	C	C	IV	O	A	A	B	C	C		(δ)

What does the reader think?

Comments: This is not an answer. That is left to the reader.

For Design α

$$\mathbf{Nk}^{-\delta}\mathbf{N}' = \tfrac{1}{6}\begin{bmatrix} 12 & 8 & 8 & 8 \\ 8 & 10 & 9 & 9 \\ 8 & 9 & 10 & 9 \\ 8 & 9 & 9 & 10 \end{bmatrix}$$

The design is in supplemented balance, with $w = \frac{9}{6}$, $w_0 = \frac{8}{6}$ and $v = 4$, so $R = \frac{35}{6}$ and $R_0 = \frac{140}{27}$. For the comparison of A, B and C then the effective replication is $\frac{35}{6}$. If O is included in the comparison, the variance is not much worse.

For Design β

$$\mathbf{Nk}^{-\delta}\mathbf{N}' = \tfrac{1}{6}\begin{bmatrix} 12 & 8 & 8 & 8 \\ 8 & 10 & 9 & 9 \\ 8 & 9 & 10 & 9 \\ 8 & 9 & 9 & 10 \end{bmatrix}.$$

Apparently there is nothing to choose between α and β.

For Design γ

$$\mathbf{Nk}^{-\delta}\mathbf{N}' = \tfrac{1}{6}\begin{bmatrix} 18 & 6 & 6 & 6 \\ 6 & 10 & 10 & 10 \\ 6 & 10 & 10 & 10 \\ 6 & 10 & 10 & 10 \end{bmatrix}.$$

Now $w = \frac{10}{6}$, $w_0 = 1$ and $v = 4$, so $R = 6$ and $R_0 = \frac{18}{5}$. The proportionality of A, B and C has led to full efficiency, as would be expected.

For Design δ

$$\mathbf{Nk}^{-\delta}\mathbf{N}' = \tfrac{1}{6}\begin{bmatrix} 10 & 9 & 9 & 8 \\ 9 & 10 & 8 & 9 \\ 9 & 8 & 10 & 9 \\ 8 & 9 & 9 & 10 \end{bmatrix}.$$

The design is group-divisible, with groups O, C and A, B. Hence

$$\mathbf{\Xi}^{-1} = \tfrac{1}{12}\begin{bmatrix} 65 & -5 & -5 & -3 \\ -5 & 65 & -3 & -5 \\ -5 & -3 & 65 & -5 \\ -3 & -5 & -5 & 65 \end{bmatrix},$$

$$\mathbf{S} = \tfrac{1}{12}\begin{bmatrix} 65 & -3 \\ -3 & 65 \end{bmatrix}, \qquad \mathbf{S}_* = \tfrac{1}{12}\begin{bmatrix} -5 & -5 \\ -5 & -5 \end{bmatrix},$$

$$\mathbf{S}-\mathbf{S}_* = \tfrac{1}{12}\begin{bmatrix} 70 & 2 \\ 2 & 70 \end{bmatrix}, \qquad \mathbf{M} = (\mathbf{S}-\mathbf{S}_*)^{-1} = \tfrac{1}{204}\begin{bmatrix} 35 & -1 \\ -1 & 35 \end{bmatrix}.$$

The value of **M** could have been obtained direct from (5.4.4). A generalized inverse may therefore be written as

$$\frac{1}{204}\begin{bmatrix} 35 & 0 & 0 & -1 \\ 0 & 35 & -1 & 0 \\ 0 & -1 & 35 & 0 \\ -1 & 0 & 0 & 35 \end{bmatrix}.$$

That comes from (5.4.3). Accordingly the contrast of parameters between A and B has the variance $\frac{72}{204}\sigma^2 = \frac{6}{17}\sigma^2$. For the contrast of C with A or B it is $\frac{70}{204}\sigma^2 = \frac{35}{102}\sigma^2$.

The author readily admits that Design γ is an improvement on his own. He is not so sure about Design δ. The contrast between A and B is rather worse than with Design α and the others virtually the same. For the sake of symmetry he likes the designs in supplemented balance and would give them up only if variances were much worse.

Note: A biometrician could be forgiven for not noticing that Design δ is group-divisible. (It is made so only by permuting the treatments, though that was done only by implication in the working.) Anyone who thought it was in factorial balance would be quite right. He would supposedly proceed thus:

$$w = \tfrac{9}{6}, \quad w' = \tfrac{9}{6}, \quad w'' = \tfrac{8}{6}, \quad \mu = 2, \quad v = 2.$$

Hence

$$\delta = 2w + 2w' + (4 - 2 - 2)w'' = 6,$$

$$\zeta_1 = \frac{w - w''}{\mu\delta[w' + (v-1)w'']} = \tfrac{1}{204}$$

$$\zeta_2 = \frac{w' - w''}{v\delta[w + (\mu-1)w'']} = \tfrac{1}{204}$$

$$\xi = \tfrac{1}{6} + \tfrac{2}{204} = \tfrac{3}{17}.$$

So

$$\mathbf{C}^- = \frac{1}{204}\begin{bmatrix} 36 & 1 & 1 & 0 \\ 1 & 36 & 0 & 1 \\ 1 & 0 & 36 & 1 \\ 0 & 1 & 1 & 36 \end{bmatrix}.$$

The conclusions are as before.

E. An experiment is to be designed in randomized blocks with replicate blocks, i.e. there are to be b blocks each with v plots, one plot for each treatment. The treatments are to be

O	Untreated
S	Use of the standard treatment
A, B, C, etc.	Use of some new treatments

In view of the relative unimportance of O and the fact that the main comparisons are those of S with A, B, C, etc. someone suggests that in *a* blocks the plot assigned O should receive S instead, i.e. that there should be a blocks like this:

S S A B C, etc.

and $(b-a)$ like this:

O S A B C, etc.

Assuming the suggestion to be adopted, what are the variances of the following differences of adjusted treatment means:

(i) O and S
(ii) O and A
(iii) S and A?

(University of Kent at Canterbury. Diploma/M.Sc. 1975)

Answer:

$$\mathbf{C} = \frac{1}{v}\begin{bmatrix} (v-1)(b-a) & -(b-a) & \ldots & -(b-a) & \\ -(b-a) & (v-1)b+(v-3)a & \ldots & -(b+a) & \\ -(b-a) & -(b+a) & \ldots & (v-1)b & \text{etc.} \\ \vdots & \vdots & & \vdots & \\ -(b-a) & -(b+a) & \ldots & -b & \end{bmatrix}.$$

The columns for Treatments B, C etc. can be derived from that of A. The first element will always equal $-(b-a)$ and the second $-(b+a)$. All the rest will equal $-b$ except for the one on the diagonal and that will be $(v-1)b$.

The design is perhaps most easily regarded as multipartite with group sizes 1, 1 and $(v-2)$. In a sense that leads to a simplification, because w_{11} and w_{22} do not exist and some of the sub-matrices, i.e. $\mathbf{S}_{11}$, $\mathbf{S}_{12}$ and $\mathbf{S}_{22}$, reduce to scalars. Note that $u = (v-1)b - 2a/v$. In an arithmetical context that could well be the best way of dealing with it. Algebraically, however, there are better ways. As far as the contrast between O and S is concerned, it might be better to use(4.3.1) using $\mathbf{C}$ for $\mathbf{E}$. It then becomes apparent from (4.3.1) and (4.3.3) that Treatments S and O are concordant, so (4.3.4) holds and the variance of the difference between their parameters is

$$\frac{2vb\sigma^2}{(b-a)[vb+(v-2)a]}.$$

The other variances are more difficult but not impossibly so. Converting $\mathbf{C}$ to $\mathbf{\Omega}^{-1}$ and writing it in the form (4.3.1) it appears that $\mathbf{Y}$ is a null-matrix and $\mathbf{Z}$ is a diagonal matrix with its non-zero values equal to b, while

$$\mathbf{X} = \frac{b-a}{bv}\begin{bmatrix} vb-a & a \\ a & \dfrac{vb^2+(v-1)ab+a^2}{b-a} \end{bmatrix},$$

$$\mathbf{X}^{-1} = \frac{1}{b(b-a)[vb+(v-2)a]}\begin{bmatrix} vb^2+(v-1)ab+a^2 & -a(b-a) \\ -a(b-a) & (vb-a)(b-a) \end{bmatrix}.$$

Hence, writing

$$\boldsymbol{\Omega} = \begin{bmatrix} \mathbf{X}^{-1} & \mathbf{0}' \\ \mathbf{0} & \mathbf{I}_{v-2}/b \end{bmatrix},$$

the variance of the other differences of means are

S, A
$$\left\{\frac{vb^2 + (v-1)ab + a^2}{b(b-a)[vb + (v-2)a]} + \frac{1}{b}\right\}\sigma^2$$

$$= \left\{\frac{2vb^2 + (v-3)ab - (v-3)a^2}{b(b-a)[vb + (v-2)a]}\right\}\sigma^2$$

O, A
$$\left\{\frac{(vb - a)}{b[vb + (v-2)a]} + \frac{1}{b}\right\}\sigma^2$$

$$= \left\{\frac{2vb - (v+3)ab - (v-3)a}{b(b-a)[vb + (v-2)a]}\right\}\sigma^2$$

Since $\mathbf{X}^{-1}$ has been evaluated, there is another way of obtaining the variance of the contrast between Treatments S and O.

Chapter 6

Other blocking systems

6.1 BLOCK STRUCTURE

So far the only blocking system seriously considered has been that of a simple division of the experimental area into blocks, each again divided into plots. Following the approach of Nelder (1965a) such designs may be said to have the 'block structure' represented by

$$\text{Plots} \rightarrow \text{Blocks} \rightarrow \text{Experimental Area.}$$

It has already been used explicitly in Section 4.8. It may be expressed by saying that the plots are 'nested' within the blocks, which are nested within the experimental area. Also, since there is only one experimental area, no use can be made of it for purposes of comparing treatments so it will ordinarily be disregarded, though the situation could well be different for a co-operative experiment conducted at a number of sites.

However, other block structures are possible. For example, where several factors are required it can happen that ideally each needs its own size of plot. Thus, it is not practicable to cultivate small areas of land in any realistic way, so methods of soil preparation call perforce for large plots. On the other hand, much smaller ones must be used for varieties. If someone got the idea that economies in land preparation could be effected by choosing varieties easily satisfied in that respect, he would be committed to a factorial set of treatments. The plot size required for the cultivation treatments would be extravagant for a comparison of varieties; nevertheless, he could well design an experiment on methods of land preparation using plots of a size that suited it and divide each of those 'main plots' into 'sub-plots' for the varieties. The block structure would then be

$$\text{Sub-plots} \rightarrow \text{Main plots} \rightarrow \text{Blocks} \rightarrow \text{Experimental Area.}$$

It will be convenient here and later to refer to the smallest area used (in this instance, the sub-plot) as the 'unit plot'. That is what has to be recorded to provide the data.

The design matrix for blocks, $\mathbf{D}$, now shows how the unit plots relate to the blocks. It needs to be supplemented by a further matrix, $\mathbf{D}_m$, which shows how they relate to the main plots. It should be noted that values for block size, treatment replications and n, the total number of plots, now refer to the number of unit plots. Writing

$$\mathbf{D}_m \mathbf{1}_n = \mathbf{k}_m, \quad \mathbf{D}_m \mathbf{D}'_m = \mathbf{k}^{\delta}_m, \quad \mathbf{D}_m \mathbf{y} = \mathbf{M}, \quad \text{etc.}$$

(following 4.8.6), $\mathbf{y}'\mathbf{y}$, the sum of data squared, can be partitioned among four strata, thus:

$\mathbf{y}'\mathbf{y} - \mathbf{M}'\mathbf{k}_m{}^{-\delta}\mathbf{M}$	Sub-plots → Main plots
$\mathbf{M}'\mathbf{k}_m^{-\delta}\mathbf{M} - \mathbf{B}'\mathbf{k}^{-\delta}\mathbf{B}$	Main plots → Blocks
$\mathbf{B}'\mathbf{k}^{-\delta}\mathbf{B} - G^2/n$	Blocks → Experimental area
G^2/n	Experimental area.

Writing

$$\begin{aligned} \boldsymbol{\phi}_1 &= (\mathbf{I}_n - \mathbf{D}'_m \mathbf{k}_m^{-\delta} \mathbf{D}_m) \\ \boldsymbol{\phi}_2 &= (\mathbf{D}'_m \mathbf{k}_m^{-\delta} \mathbf{D}_m - \mathbf{D}'\mathbf{k}^{-\delta}\mathbf{D}) \\ \boldsymbol{\phi}_3 &= (\mathbf{D}'\mathbf{k}^{-\delta}\mathbf{D} - \mathbf{1}_n \mathbf{1}'_n/n) \\ \boldsymbol{\phi}_4 &= \mathbf{I}_n \mathbf{1}'_n/n \end{aligned} \qquad (6.1.1)$$

(4.8.4) and (4.8.5) have analogues

$$\mathbf{y}'\mathbf{y} = \sum_i \mathbf{y}'\boldsymbol{\phi}_i\mathbf{y} = \sum_i \mathbf{y}'_i\mathbf{y}_i, \qquad (6.1.2)$$

where $\mathbf{y}_i = \boldsymbol{\phi}_i\mathbf{y}$ ($i = 1, 4$). Also

$$\mathbf{I}_n = \sum_i \boldsymbol{\phi}_i \qquad (6.1.3)$$

$$\mathbf{y} = \sum_i \mathbf{y}_i \qquad (6.1.4)$$

$$\boldsymbol{\phi}_i = \boldsymbol{\phi}'_i, \qquad \boldsymbol{\phi}_i\boldsymbol{\phi}_i = \boldsymbol{\phi}_i \qquad \text{and} \qquad \boldsymbol{\phi}_i\boldsymbol{\phi}_j = \mathbf{0} \quad (i \neq j) \qquad (6.1.5)$$

as before. Note also that $\mathbf{y}'_i\mathbf{y}_j = 0$ ($i \neq j$)

There is no reason in principle why the sub-plots should not be further divided into sub^2-plots, which can then be divided into sub^3-plots and so on. The extension to further strata causes no great statistical difficulty and there are occasions when further splits are required by the nature of the material, but complicated systems should be introduced only for sound practical reasons and never as a display of virtuosity.

A different approach comes from the 'crossing' of block systems, as in a row-and-column design. Here a rectangular area is divided one way into rows and at right angles into columns, the plots (= unit plots) being formed by the intersections. Following Nelder (1965a) again, that can be written

(Rows × Columns) → Experimental area.

There are four strata

(1) Plots within rows and columns.
(2) Rows within the experimental area.

(3) Columns within the experimental area.
(4) The experimental area.

The partition of $\mathbf{y}'\mathbf{y}$ appropriate to the first stratum need cause no difficulty because of the analogy to an orthogonal block design under the null-hypothesis. Writing rows for blocks and columns for treatments, (4.1.6) becomes

$$\mathbf{y}'\boldsymbol{\phi}_1\mathbf{y} = \mathbf{y}'\mathbf{y} - \mathbf{B}'\mathbf{B}/c - \mathbf{C}'\mathbf{C}/b + G^2/n,$$

where $\mathbf{B}$ is the vector of row totals, $\mathbf{C}$ the vector of column totals and b and c are respectively the numbers of rows and columns, n being equal to bc. Hence

$$\boldsymbol{\phi}_1 = \mathbf{I}_n - \mathbf{D}_r'\mathbf{D}_r/c - \mathbf{D}_c'\mathbf{D}_c/b + \mathbf{1}_n\mathbf{1}_n'/n, \qquad (6.1.6)$$

where $\mathbf{D}_r$ and $\mathbf{D}_c$ are the design matrices that allocate unit plots to rows and columns respectively. It will also readily be seen that

$$\boldsymbol{\phi}_2 = \mathbf{D}_r'\mathbf{D}_r/c - \mathbf{1}_n\mathbf{1}_n'/n$$
$$\boldsymbol{\phi}_3 = \mathbf{D}_c'\mathbf{D}_c/b - \mathbf{1}_n\mathbf{1}_n'/n$$
$$\boldsymbol{\phi}_4 = \mathbf{1}_n\mathbf{1}_n'/n$$

and (6.1.1), (6.1.2), (6.1.3), (6.1.4) and (6.1.5) apply as before.

Other systems can be envisaged. For example, it is possible to split the plots of a row-and-column design, i.e.

Sub-plots → (Rows × Column) → Experimental area

or to take a design in blocks and to form a system of rows and columns on each plot, i.e.

(Rows × Columns) → Blocks → Experimental area.

The above corresponds to the case in which the main plots have to be so large that each can contain several replicates of the sub-treatments and some control of local variation within the main plots is called for. More usual, however, would be a second set of blocks, here called 'sub-blocks' to correspond to the sub-treatments, thus giving the structure

Sub-plots → Sub-blocks → Main plots → Main blocks → Experimental area.

It is not, however, necessary to enumerate all the possibilities, which are very numerous. The important thing is to devise sets of strata, based on nesting and crossing, each with its matrix, $\boldsymbol{\phi}_i$, such that equations (6.1.1) to (6.1.5), which are but generalizations of (4.8.1) to (4.8.6), hold. That is called 'simple block structure' and is the basis of almost all effective design. If, further, treatments can be added in such a way as to achieve the general balance described in Section 4.8, certain simplifications result. That is not essential, though it is often desirable.

Although many ingenious schemes can be devised, for practical purposes it is necessary in any system to have the same number of plots in each block. Where two blocking systems cross, e.g. rows and columns, it is difficult to envisage anything else, but there are no problems about having, say, five sub-plots in some main plots and four in others. At this point there is some inconsistency in statistical practice. If anyone were to propose such a scheme, he would be confronted by a demand to know how he proposed to carry out his main plot analysis when some totals would be subject to a different variance from the rest. Only a very unpractical person would attempt to talk himself out of the difficulty by suggesting that a suitable weighting of data would solve the problem. That is to assume either that main plots are formed by a random sampling of sub-plots or that there is knowledge *a priori* of strata variances, and each is absurd. If he did attempt such a line of defence he would be dismissed as uncomprehending. However, when it comes to inter-block and intra-block analyses (see Section 3.8) no such reservations appear to apply. In this text it will be taken for granted that where one blocking system is nested in another or is crossed with it, if an analysis is needed for all strata, then all blocks of a system must contain the same number of unit plots. Thus, an improper block experiment, i.e. one with blocks of varying size, is permissible only as long as no inter-block analysis is required.

It will be recognized that the $\boldsymbol{\phi}_i$ used here are residuals matrices like those introduced in (2.1.5), (3.1.6), etc. for simpler designs. Corresponding to them are similar residuals matrices, $\boldsymbol{\psi}_i$, in which the effects of treatments have been allowed for. Here also

$$\boldsymbol{\psi}_i = \boldsymbol{\phi}_i - \boldsymbol{\phi}_i \boldsymbol{\Delta}' \mathbf{C}_i^- \boldsymbol{\Delta} \boldsymbol{\phi}_i \qquad (6.1.7)$$

where $\mathbf{C}_i = \boldsymbol{\Delta}\boldsymbol{\phi}_i\boldsymbol{\Delta}'$. The analogy with (2.1.4) and (3.3.7) is plain.

6.2 THE LEAST SQUARES SOLUTION GIVEN SIMPLE BLOCK STRUCTURES

Given several blocking systems, *a*, *b*, etc., each will have a design matrix, $\mathbf{D}_a$, $\mathbf{D}_b$, etc., to show how the unit plots are assigned to its members. Further, it will be assumed for reasons just given that

$$\mathbf{D}_i \mathbf{1}_n = k_i \mathbf{1}_{b_i}, \qquad (6.2.1)$$

where b_i is the number of blocks in that particular system. Further, as is the way of design matrices,

$$\mathbf{D}_i' \mathbf{1}_{b_i} = \mathbf{1}_n, \qquad \mathbf{D}_i \mathbf{D}_i' = k_i \mathbf{I}_{b_i}. \qquad (6.2.2)$$

It will also be convenient to write

$$\boldsymbol{\Delta}\mathbf{D}_i' = \mathbf{N}_i. \qquad (6.2.3)$$

Additionally, if two systems, *i* and *j*, are crossed, then all elements of $\mathbf{D}_i\mathbf{D}_j'$ will equal one, i.e.

$$\mathbf{D}_i \mathbf{D}_j' = \mathbf{1}_{b_i} \mathbf{1}_{b_j}' \qquad (6.2.4)$$

if *j* is nested within *i*,

$$\mathbf{D}_i' \mathbf{D}_i \mathbf{D}_j' = k_i \mathbf{D}_j', \qquad \mathbf{D}_j' \mathbf{D}_j \mathbf{D}_i' = k_j \mathbf{D}_i'. \qquad (6.2.5)$$

As to parameters, when systems are crossed they produce plots from their intersections, which require no additional parameters if they are themselves used as blocks, because those for the crossing systems are taken to be additive. With nesting, parameters are needed for the smallest blocks, those for larger ones being subsumed within them. A full calculus has been given by Nelder (1965b). The above relationships, though not exhaustive, will cover the more usual cases.

For two crossed blocking systems, the model may be written

$$\mathbf{y} = \mathbf{D}_a'\boldsymbol{\beta}_a + \mathbf{D}_b'\boldsymbol{\beta}_b + \boldsymbol{\Delta}'\boldsymbol{\tau} + \boldsymbol{\eta}. \tag{6.2.6}$$

Minimization of $\boldsymbol{\eta}'\boldsymbol{\eta}$ leads to the normal equations

$$\begin{aligned} \mathbf{D}_a\mathbf{y} &= k_a\hat{\boldsymbol{\beta}}_a + \mathbf{1}\mathbf{1}'\hat{\boldsymbol{\beta}}_b + \mathbf{N}_a'\hat{\boldsymbol{\tau}} \\ \mathbf{D}_b\mathbf{y} &= \mathbf{1}\mathbf{1}'\hat{\boldsymbol{\beta}}_a + k_b\hat{\boldsymbol{\beta}}_b + \mathbf{N}_b'\hat{\boldsymbol{\tau}} \\ \boldsymbol{\Delta}\mathbf{y} &= \mathbf{N}_a\hat{\boldsymbol{\beta}}_a + \mathbf{N}_b\hat{\boldsymbol{\beta}}_b + \mathbf{r}^\delta\hat{\boldsymbol{\tau}}. \end{aligned}$$

(The number of elements in each unit vector is readily seen.) Adding equations in any of the above sets gives

$$\mathbf{1}_n'\mathbf{y} = k_a\mathbf{1}'\hat{\boldsymbol{\beta}}_a + k_b\mathbf{1}'\hat{\boldsymbol{\beta}}_b + \mathbf{r}'\hat{\boldsymbol{\tau}},$$

whence it follows that $\boldsymbol{\Delta\phi}\mathbf{y} = \mathbf{C}\hat{\boldsymbol{\tau}}$, or

$$\begin{aligned} &[\boldsymbol{\Delta} - \mathbf{N}_a\mathbf{D}_a/k_a - \mathbf{N}_b\mathbf{D}_b/k_b + \mathbf{r}\mathbf{1}_m'/m]\mathbf{y} \\ &= (\mathbf{r}^\delta - \mathbf{N}_a\mathbf{N}_a'/k_a - \mathbf{N}_b\mathbf{N}_b'/k_b + \mathbf{1}_m\mathbf{1}_m'/m)\hat{\boldsymbol{\tau}}, \end{aligned} \tag{6.2.7}$$

where $m = k_a k_b$. It will be noted that m is not necessarily the same as n because there could be other strata. In these general cases, an analysis of variance is worked in terms of unit plots, which accounts for the difference. All means will be expressed in those terms and variances will apply to means so based.

Nested systems are similar. Let the blocks of b be nested within those of a, then the analysis for the stratum of b need raise no problems, being that for a block design as set out in Section 3.3, apart from the proviso just made about distinguishing between m and n. For stratum a, however, things are rather different. First of all, its data are not $\mathbf{y}$ but $\mathbf{D}_b\mathbf{y}$, so the analogue of (3.1.1) is

$$\mathbf{D}_b\mathbf{y} = \mathbf{D}_b\mathbf{D}_a'\boldsymbol{\beta} + \mathbf{D}_b\boldsymbol{\Delta}'\boldsymbol{\tau} + \boldsymbol{\eta}. \tag{6.2.8}$$

Minimization of $\boldsymbol{\eta}'\boldsymbol{\eta}$ leads to the normal equations

$$\begin{aligned} k_b\mathbf{N}_a\mathbf{D}_a\mathbf{y} &= k_ak_b\mathbf{N}_a\hat{\boldsymbol{\beta}} + k_b\mathbf{N}_a\mathbf{N}_a'\hat{\boldsymbol{\tau}}, \\ \mathbf{N}_b'\mathbf{D}_b\mathbf{y} &= k_b\mathbf{N}_a\hat{\boldsymbol{\beta}} + \mathbf{N}_b\mathbf{N}_b'\hat{\boldsymbol{\tau}}, \end{aligned}$$

whence it follows that

$$(\mathbf{N}_b'\mathbf{D}_b/k_b - \mathbf{N}_a'\mathbf{D}_a/k_a)\mathbf{y} = (\mathbf{N}_b'\mathbf{N}_b/k_b - \mathbf{N}_a'\mathbf{N}_a/k_a)\hat{\boldsymbol{\tau}}. \tag{6.2.9}$$

By this point it is becoming apparent that both (6.2.7) and (6.2.9) can, because of (6.1.1) and (6.1.6), be written

$$\boldsymbol{\Delta\phi}_i\boldsymbol{\Delta}\hat{\boldsymbol{\tau}}_i = \boldsymbol{\Delta\phi}_i\mathbf{y}, \tag{6.2.10}$$

where $\boldsymbol{\phi}_i$ is the residuals matrix appropriate to the stratum in question. A further step for writing $\mathbf{C}_i = \boldsymbol{\Delta\phi}_i\boldsymbol{\Delta}'$ analogous to (3.1.6), and $\mathbf{Q}_i = \boldsymbol{\Delta\phi}_i\mathbf{y} = \boldsymbol{\Delta}\mathbf{y}_i$ analogous to (3.1.8), leads to the relationship

$$\mathbf{C}_i\hat{\boldsymbol{\tau}}_i = \mathbf{Q}_i \tag{6.2.11}$$

analogous to (3.1.4). It will be seen that $\hat{\boldsymbol{\tau}}_i$ has now been given a suffix as a reminder that the estimate implied by the circumflex is based on the ith stratum. A difficulty, which will be faced in Section 6.4, is that of combining several estimates if the need arises. For the moment it suffices to note that

$$\hat{\boldsymbol{\tau}}_i = \mathbf{C}_i^- \mathbf{Q}_i \tag{6.2.12}$$

as in (3.3.1). Usually the $\mathbf{C}_i$ will be such that different important contrasts appear in different strata. Thus one main effect may appear in the comparison of main plots within blocks, while another may arise from the comparison of sub-plots within main plots, and so on, but more of that later. Such, indeed, was the idea of general balance, set out in Section 4.8.

The above exposition has not covered all possible cases. It will be found, however, that more complicated situations can be met by obvious extensions of what has been said.

Passing to the variances of contrasts for both nested and crossed systems,

$$\boldsymbol{\phi}_i\mathbf{D}_b' = \mathbf{0}, \tag{6.2.13}$$

analogous to (3.1.7). For the nested case, the result follows simply from (6.2.5), because

$$(\mathbf{D}_b'\mathbf{D}_b/k_b - \mathbf{D}_a'\mathbf{D}_a/k_a)\mathbf{D}_b' = \mathbf{0}$$

and for the crossed, from (6.2.1) and (6.2.4)

$$(\mathbf{I} - \mathbf{D}_a'\mathbf{D}_a/k_a - \mathbf{D}_b'\mathbf{D}_b/k_b + \mathbf{1}_m\mathbf{1}_m'/m)\mathbf{D}_b' = \mathbf{0}.$$

Hence, in both cases,

$$\boldsymbol{\Delta}\boldsymbol{\phi}_i\mathbf{y} = \boldsymbol{\Delta}\boldsymbol{\phi}_i\boldsymbol{\Delta}'\boldsymbol{\tau}_i + \boldsymbol{\Delta}\boldsymbol{\phi}_i\boldsymbol{\eta}$$

which may be written

$$\mathbf{Q}_i = \mathbf{C}_i\boldsymbol{\tau}_i + \boldsymbol{\Delta}\boldsymbol{\phi}_i\boldsymbol{\eta}$$

Also

$$\hat{\boldsymbol{\tau}}_i = \mathbf{C}_i^- \mathbf{C}_i\boldsymbol{\tau}_i + \mathbf{C}_i^- \boldsymbol{\Delta}\boldsymbol{\phi}_i\boldsymbol{\eta} \tag{6.2.14}$$

as in (3.3.2). Whence it follows that it is possible in the ith stratum to estimate contrasts, $\mathbf{c}$, provided that $\mathbf{c}$ is composed solely of eigenvectors of $\mathbf{C}_i$ with non-zero eigenvalues. When that is so, as in (3.3.4),

$$\mathbf{c}'\hat{\boldsymbol{\tau}}_i = \mathbf{c}'\mathbf{C}_i^- \mathbf{C}_i\boldsymbol{\tau}_i + \mathbf{c}'\mathbf{C}_i^- \boldsymbol{\Delta}\boldsymbol{\phi}\boldsymbol{\eta} = \mathbf{c}'\boldsymbol{\tau}_i + \mathbf{c}'\mathbf{C}_i^- \boldsymbol{\Delta}\boldsymbol{\phi}_i\boldsymbol{\eta}$$

with variance $\mathbf{c}'\mathbf{C}_i^- \mathbf{c}\sigma_i^2$, analogous to (3.3.5), σ_i^2 being the error mean squared deviation derived from the analysis of variance for the ith stratum.

It may be noted that information from different strata is independent. Thus, the covariance of $\mathbf{c}'\hat{\boldsymbol{\tau}}_i$ and $\mathbf{c}'\hat{\boldsymbol{\tau}}_j$ $(i \neq j)$ is

$$\mathbf{c}'\mathbf{C}_i^- \boldsymbol{\Delta}\boldsymbol{\phi}_i\boldsymbol{\phi}_j\boldsymbol{\Delta}'\mathbf{C}_j^- \mathbf{c}\sigma^2 = 0$$

from (6.1.5).

Worthington (1975) has considered the use of the Kuiper–Corsten iteration in these more general cases. The following exposition is based on his work, which has been adapted to assimilate the argument to what has gone before. Writing

$$\mathbf{F}_i = \mathbf{r}^{-\frac{1}{2}\delta}\mathbf{C}_i\mathbf{r}^{-\frac{1}{2}\delta} = \mathbf{r}^{-\frac{1}{2}\delta}\boldsymbol{\Delta}\boldsymbol{\phi}_i\boldsymbol{\Delta}'\mathbf{r}^{-\frac{1}{2}\delta}, \tag{6.2.15}$$

then from Lemma 3.5.A the eigenvalues of $\mathbf{F}_i$ lie between 0 and 1, because

$$\mathbf{F}_i = (\mathbf{r}^{-\frac{1}{2}\delta}\boldsymbol{\Delta}\boldsymbol{\phi}_i)(\mathbf{r}^{-\frac{1}{2}\delta}\boldsymbol{\Delta}\boldsymbol{\phi}_i)'$$

$$\mathbf{I}_v - \mathbf{F}_i = [\mathbf{r}^{-\frac{1}{2}\delta}\boldsymbol{\Delta}(\mathbf{I}_n - \boldsymbol{\phi}_i)][\mathbf{r}^{-\frac{1}{2}\delta}\boldsymbol{\Delta}(\mathbf{I}_n - \boldsymbol{\phi}_i)]'.$$

Hence, as in (3.5.6),

$$\mathbf{r}^{-\frac{1}{2}\delta}\sum_{l=0}^{\infty}(\mathbf{I}_v - \mathbf{F}_i)^l\mathbf{r}^{-\frac{1}{2}\delta} = \mathbf{r}^{-\frac{1}{2}\delta}\mathbf{F}_i^+\mathbf{r}^{-\frac{1}{2}\delta} = \boldsymbol{\Upsilon}_i, \tag{6.2.16}$$

and $\boldsymbol{\Upsilon}_i$ is a generalized inverse of $\mathbf{C}_i$. Accordingly, it may be used in (6.2.12) and similar expressions.

The problem is rather to find the form of the iteration. Following Worthington,

$$\boldsymbol{\phi}_i = \sum_{j=0}^{p+1} a_{ij}\mathbf{H}_j \tag{6.2.17}$$

where

$$\mathbf{H}_0 = \mathbf{I}_n$$

$$\mathbf{H}_j = \mathbf{D}_j'\mathbf{k}_j^{-\delta}\mathbf{D}_j \qquad (1 \leqslant j \leqslant p)$$

$$\mathbf{H}_{p+1} = \mathbf{1}_n\mathbf{1}_n'/n$$

and $a_{ij} = +1$, 0 or -1 according to the block structure. (Here $j, 1 \leqslant j \leqslant p$, indicates the jth blocking system.) It will be noted that $\mathbf{C}_i = \boldsymbol{\Delta}\boldsymbol{\phi}_i\boldsymbol{\Delta}'$ can be written similarly with $\mathbf{r}^{\delta}$ for $\mathbf{H}_0$, $\mathbf{N}_j\mathbf{k}_j^{-\delta}\mathbf{N}_j'$ for $\mathbf{H}_j$ and $\mathbf{r}\mathbf{r}'/n$ for $\mathbf{H}_{p+1}$. Then, from (6.2.15), using values of $\mathbf{H}_j$, etc., appropriate to $\boldsymbol{\phi}_i$,

$$\mathbf{I}_v - \mathbf{F}_i = (1 - a_{i0})\mathbf{I}_v - \sum_{j=1}^{p+1}(a_{ij}\mathbf{r}^{-\frac{1}{2}\delta}\boldsymbol{\Delta}\mathbf{H}_j\boldsymbol{\Delta}'\mathbf{r}^{-\frac{1}{2}\delta}) \tag{6.2.18}$$

$$\mathbf{Q}_i = \boldsymbol{\Delta}\boldsymbol{\phi}_i\mathbf{y} = a_{i0}\mathbf{T} - \sum_{j=1}^{p+1}(a_{ij}\mathbf{N}_j\mathbf{k}_j^{-\delta}\mathbf{B}_j) \tag{6.2.19}$$

and $\mathbf{r}^{-\delta}\mathbf{Q}_i$ equals the vector of treatment means, $\mathbf{r}^{-\delta}\mathbf{T}$, multiplied by the appropriate coefficient, which could be zero, less the dual-projections of the means of the various block systems $(\mathbf{k}_j^{-\delta}\mathbf{B}_j)$ of the experiment as a whole, $(\mathbf{k}_{p+1}^{-\delta}\mathbf{B}_{p+1} = G/n)$, each multiplied by its own value of a_{ij}. Taking the starting vector, $\mathbf{v}_1$, to equal $\mathbf{r}^{-\delta}\mathbf{Q}_i$ and using the relationship

$$\mathbf{v}_{l+1} = \mathbf{r}^{-\frac{1}{2}\delta}(\mathbf{I}_v - \mathbf{F}_i)\mathbf{r}^{\frac{1}{2}\delta}\mathbf{v}_l, \tag{6.2.20}$$

it appears that

$$\sum_{l=0}^{\infty}\mathbf{v}_l = \mathbf{r}^{-\frac{1}{2}\delta}\sum_{l=0}^{\infty}(\mathbf{I}_v - \mathbf{F}_i)^l\mathbf{r}^{-\frac{1}{2}\delta}\mathbf{Q}_i$$

$$= \boldsymbol{\Upsilon}_i\mathbf{Q}_i = \hat{\boldsymbol{\tau}}_i \tag{6.2.21}$$

from (6.2.16). It is assumed that there is no confounding in the stratum. If there is, then as in (3.5.9) Υ_i can be obtained column by column by applying the iteration twice to each column of $\mathbf{r}^{-\delta}\mathbf{C}_i\mathbf{r}^{-\delta}$.

The enquiry, however, concerned the form of the iteration. It is found by substituting (6.2.18) into (6.2.20), which gives

$$\mathbf{v}_{l+1} = (1 - a_{i0})\mathbf{I}_v - \sum_{j=1}^{p+1} a_{ij}\mathbf{r}^{-\delta}\mathbf{\Delta}\mathbf{H}_j\mathbf{\Delta}'$$

$$= [(1 - a_{i0})\mathbf{I}_v - \sum_{j=1}^{p} a_{ij}\mathbf{r}^{-\delta}\mathbf{N}_j\mathbf{k}_j^{-\delta}\mathbf{N}_j' - a_{i,p+1}\mathbf{r}\mathbf{r}'/n]\mathbf{v}_l. \qquad (6.2.22)$$

In short, the form of the iteration can be derived directly from the values of a_{i0}, a_{ij} and $a_{i,p+1}$. Thus, for a block design, in which there is only one blocking system, so $p = 1$, the coefficients, a_{i0}, a_{i1}, and a_{i2}, are respectively 1, -1 and 0 for the intra-block analysis (3.3.1) and 0, 1, -1 for the inter-block (3.8.4), whence the expressions already given in (3.5.3) and (3.8.6a) are readily derived. More complicated cases can be dealt with similarly. Given the constraint that $\mathbf{r}'\mathbf{v}_1$ shall equal zero, the last term of (6.2.22) has no effect.

6.3 THE ANALYSIS OF VARIANCE WITH COMPLICATED BLOCK STRUCTURES

In principle the methods of analysis already described can be used in any stratum of a simple block structure, i.e., one formed by nesting and crossing. A general method is set out in Example 6A. It is there illustrated for the intra-block and inter-block strata of a block design. Most of the results have already appeared in Sections 3.3 and 3.8. It can, however, be used for any other stratum provided the appropriate $\boldsymbol{\phi}_i$ can be found. Given $\boldsymbol{\phi}_i$, $\mathbf{\Delta}$, and $\mathbf{y}$ everything else follows.

Nevertheless, in complicated situations difficulties arise because too many contrasts are either confounded or to be merged with error. If a contrast is of full efficiency in one stratum, it must be confounded in all the rest, which leads to an additional zero eigenvalue in each of the other $\mathbf{C}_i$. Also, if there are many factors and if error degrees of freedom are short in some strata, it is wise to regard some higher-order interactions as necessarily negligible and to include their sums of squared deviations in error.

Neither of these difficulties is insuperable. A generalized inverse can always be found however many zero eigenvalues there may be. Also, there is no great difficulty about finding the sum of squared deviations for a contrast based on eigenvectors of either $\mathbf{C}_i$ or $\mathbf{F}_i$. As (3.6.1) shows, a natural contrast contributes $(\mathbf{u}_i'\mathbf{Q})^2/\lambda_i$. Similarly (3.6.12) shows that a basic contrast contributes $(\mathbf{s}_i'\mathbf{Q})^2/\varepsilon_i$. Nevertheless the calculations can become very complicated. It will be noted that $\lambda_i = R_i$, the effective replication and ε_i is the efficiency factor.

It is accordingly often easier to start with the total sum of squared deviations for the stratum and to take away the effects of those contrasts that are to be estimated in it. What is left constitutes error. The objection lies in the difficulty of checking the arithmetic, but with computers it hardly applies. Once programmed properly they can usually be relied upon.

Further, if there is general balance a fairly simple routine is available, known as 'sweeping'. It consists of continually adjusting the data to remove the effects of successive groups of parameters. Something like it has been met with in forming $\mathbf{y}_i = \boldsymbol{\phi}_i \mathbf{y}$ from $\mathbf{y}$ (6.1.2). Thus for a completely randomized design, $\boldsymbol{\phi}_i = \mathbf{I}_n - \mathbf{1}_n \mathbf{1}_n'/n$. In practice that implies evaluating $\mathbf{1}_n' \mathbf{y}/n = G/n$, the general mean of data, and subtracting it from each element of $\mathbf{y}$. Alternatively, for a block design it requires the evaluation of the block means, $\mathbf{k}^{-\delta}\mathbf{D}\mathbf{y}$, and subtracting the appropriate mean from each element of $\mathbf{y}$ to give $(\mathbf{I}_n - \mathbf{D}'\mathbf{k}^{-\delta}\mathbf{D})\mathbf{y}$. Where there are rows and columns the operation is more complicated but not too much so. The need is to subtract from each element of $\mathbf{y}$ both the appropriate row mean and column mean and to add the general mean. Programming a computer to perform these tasks causes little difficulty. In the terminology of Wilkinson (1970) the effect is to sweep the data by the blocking systems.

He went on to show how the data could be swept further by treatment contrasts. He was concerned only with natural contrasts and worked within the context of general balance as explained in Section 4.8. Let $\mathbf{u}_{j1}$ be the first contrast in the jth group. Since it is an eigenvector of $\mathbf{C}_i$,

$$\mathbf{C}_i \mathbf{u}_{j1} = R_{ij} \mathbf{u}_{j1}. \tag{6.3.1}$$

To sweep by it, totals of $\mathbf{y}_i$ are first formed ($\boldsymbol{\Delta}\mathbf{y}_i = \mathbf{Q}_i$) and made into means ($\boldsymbol{\Delta}\mathbf{y}_i/R_{ij} = \mathbf{C}_i^+ \mathbf{Q}_i = \hat{\mathbf{t}}_i$), which are then used to evaluate the contrast ($\mathbf{u}_{j1}' \boldsymbol{\Delta}\mathbf{y}_i/R_{ij}$). The value so obtained is allocated to treatments, weighting by the appropriate element of $\mathbf{u}_{j1}$, thus giving $\mathbf{u}_{j1}\mathbf{u}_{j1}' \boldsymbol{\Delta}\mathbf{y}_i/R_{ij}$. The treatment values are then assigned to plots by a premultiplication by $\boldsymbol{\Delta}'$; there is a further sweep by blocks to remove any effects introduced by non-orthogonality and finally the result is subtracted from what was there before. The outcome is

$$\begin{aligned} \mathbf{y}_{ij1} &= \mathbf{y}_i - \boldsymbol{\phi}_i \boldsymbol{\Delta}' \mathbf{u}_{ij1} \mathbf{u}_{j1}' \boldsymbol{\Delta}\mathbf{y}_i/R_{ij} \\ &= \boldsymbol{\phi}_i (\mathbf{I}_n - \mathbf{S}_{ij1}) \boldsymbol{\phi}_i \mathbf{y}, \end{aligned} \tag{6.3.2}$$

where $\mathbf{S}_{ij1} = \boldsymbol{\Delta}' \mathbf{u}_{j1} \mathbf{u}_{j1}' \boldsymbol{\Delta}/R_{ij}$ defines the sweeping operation. Some useful conclusions follow. Thus

$$\mathbf{S}_{ijp} \boldsymbol{\phi}_i \mathbf{S}_{ijp} = \mathbf{S}_{ijp}, \quad \mathbf{S}_{ijp} \boldsymbol{\phi}_i \mathbf{S}_{ijq} = \mathbf{0}, \tag{6.3.3}$$

where $p \neq q$, because

$$\begin{aligned} \mathbf{u}_{jp}' \boldsymbol{\Delta} \boldsymbol{\phi}_i \boldsymbol{\Delta}' \mathbf{u}_{jp} &= \mathbf{u}_{jp}' \mathbf{C}_i \mathbf{u}_{jp} \\ &= R_{ij} \mathbf{u}_{jp}' \mathbf{u}_{jp} = R_{ij} \end{aligned}$$

from (6.3.1) and similarly

$$\mathbf{u}_{jp}' \boldsymbol{\Delta} \boldsymbol{\phi}_i \boldsymbol{\Delta}' \mathbf{u}_{jq} = R_{ij} \mathbf{u}_{jp}' \mathbf{u}_{jq} = 0.$$

If then, $\mathbf{y}_{ij1}$ is swept by the second contrast in the jth group

$$\begin{aligned} \mathbf{y}_{ij12} &= \boldsymbol{\phi}_i (\mathbf{I}_n - \mathbf{S}_{ij2}) \boldsymbol{\phi}_i (\mathbf{I}_n - \mathbf{S}_{ij1}) \boldsymbol{\phi}_i \mathbf{y} \\ &= \boldsymbol{\phi}_i (\mathbf{I}_n - \mathbf{S}_{ij1} - \mathbf{S}_{ij2}) \boldsymbol{\phi}_i \mathbf{y} \qquad \text{from (6.3.3)} \end{aligned}$$

and, to complete the jth group,

$$\mathbf{y}_{ij} = \prod_p [\boldsymbol{\phi}_i(\mathbf{I} - \mathbf{S}_{ijp})]\boldsymbol{\phi}_i \mathbf{y}$$

$$= \boldsymbol{\phi}_i(\mathbf{I}_n - \sum_p \mathbf{S}_{ijp})\boldsymbol{\phi}_i \mathbf{y} \tag{6.3.4}$$

from (6.3.3). The reduction now brought about is from $\mathbf{y}'\boldsymbol{\phi}_i\mathbf{y}$ to

$$\mathbf{y}'_{ij}\mathbf{y}_{ij} = \mathbf{y}'\boldsymbol{\phi}_i(\mathbf{I}_n - \sum_p \mathbf{S}_{ijp})\boldsymbol{\phi}_i(\mathbf{I}_n - \sum_p \mathbf{S}_{ijp})\mathbf{y}.$$

A further application of (6.3.3) leads to a new sum of squared deviations of

$$\mathbf{y}'\boldsymbol{\phi}_i(\mathbf{I}_n - \sum_p \mathbf{S}_{ijp})\boldsymbol{\phi}_i\mathbf{y} = \mathbf{y}'\boldsymbol{\phi}_i\mathbf{y} - \sum_p (\mathbf{y}'\boldsymbol{\phi}_i\mathbf{S}_{ijp}\boldsymbol{\phi}_i\mathbf{y}). \tag{6.3.5}$$

The difference arises from sweeping by the contrasts of the jth group; it therefore corresponds to a sum of squared deviations of

$$\sum_p (\mathbf{y}'\boldsymbol{\phi}_i\boldsymbol{\Delta}\mathbf{u}_{jp}\mathbf{u}'_{jp}\boldsymbol{\Delta}'\boldsymbol{\phi}_i\mathbf{y}/R_{ij})$$

$$= \mathbf{Q}'_i\sum_p (\mathbf{u}_{jp}\mathbf{u}'_{jp})\mathbf{Q}_i/R_{ij} = \mathbf{Q}'_i\mathbf{H}_j\mathbf{Q}_i/R_{ij} \tag{6.3.6}$$

from the definition of $\mathbf{H}_j$ in Section 4.8. Just as it does not matter in what order contrasts are swept within a group, so by an extension of (6.3.3) it does not matter in what order the groups are swept. Computationally everything is very simple, as will appear in the Examples 6C and 6E at the end of the chapter. From the literature it might be supposed that the evaluation of the various R_{ij} presented some difficulty, but it need not do so, because the $\mathbf{C}_i$ and $\mathbf{H}_j$ are necessarily known and (4.8.8a) may be written as

$$\mathbf{H}_j\mathbf{C}_i\mathbf{H}_j = R_{ij}\mathbf{H}_j,$$

that indeed being the definition of general balance.

So far sweeping has been carried out using eigenvectors of $\mathbf{C}_i$, corresponding to natural contrasts, but it is possible to use basic contrasts instead (Pearce *et al.*, 1974). The sweeping matrix, analogous to (6.3.2) is now

$$\mathbf{S}_{ijp} = \boldsymbol{\Delta}'\mathbf{s}_{jp}\mathbf{s}'_{jp}\boldsymbol{\Delta}/\varepsilon_{ij}. \tag{6.3.7}$$

Thus

$$\mathbf{y}_{ij1} = (\mathbf{I}_n - \mathbf{S}_{ij1})\boldsymbol{\phi}_i\mathbf{y}$$

The basic contrasts are here defined by (3.6.11), namely

$$\mathbf{r}^{-\delta}\mathbf{C}_i\mathbf{s}_{jp} = \varepsilon_{ij}\mathbf{s}_{jp}. \tag{6.3.8}$$

On account of (3.6.10)

$$\mathbf{S}_{ijp}\mathbf{S}_{ijp} = \mathbf{S}_{ijp}, \qquad \mathbf{S}_{ijp}\mathbf{S}_{ijq} = 0 \quad \text{if } p \neq q. \tag{6.3.9}$$

Accordingly, corresponding to (6.3.4),

$$\mathbf{y}_{ij} = \prod_p (\mathbf{I}_n - \mathbf{S}_{ijp})\boldsymbol{\phi}_i \mathbf{y}$$

$$= (\mathbf{I}_n - \sum_p \mathbf{S}_{ijp})\boldsymbol{\phi}_i \mathbf{y}. \qquad (6.3.10)$$

Before summing the squares of the elements of $\mathbf{y}_{ij}$ to help in the partition of the treatment sum of squared deviations, it is necessary to consider whether a further sweep of blocks, i.e. premultiplication by $\boldsymbol{\phi}_i$, is not required. Clearly $\boldsymbol{\phi}_i \mathbf{y}_{ij}$ is going to equal $\mathbf{y}_{ij}$ if $\boldsymbol{\phi}_i \boldsymbol{\Delta}' \mathbf{s}_{jp} = \mathbf{0}$ for all p and often that is so. However, exceptions do arise. When programming a computer it is usually easier to include a sweep by blocks at this point, even if it will mostly have no effect, rather than test if it is going to be needed.

If the design is equi-replicate, which with general balance is very likely, (6.3.8) and (6.3.11) are respectively equivalent to (6.3.2) and (6.3.6).

Example 6D shows how the calculations are performed.

6.4 COMBINING INFORMATION FROM VARIOUS STRATA

A difficulty arose in Section 3.8 that has yet to be resolved, namely the combination of information from different strata. There it concerned only intra- and inter-block analyses but the problem is now seen to be more general.

Where two or more strata afford information about the same contrasts, their various estimates will be independent. Generalizing (3.3.5) it appears that the variance of a contrast, $\mathbf{c}$, in the ith stratum is $\mathbf{c}'\mathbf{C}_i^- \mathbf{c}\sigma_i^2$ because $\mathbf{C}_i^- \boldsymbol{\Delta}\boldsymbol{\phi}_i\boldsymbol{\phi}_i\boldsymbol{\Delta}'\mathbf{C}_i^-$ equals $\mathbf{C}^-\mathbf{C}\mathbf{C}^-$. If, however, a covariance is sought, $\mathbf{C}_i^- \boldsymbol{\Delta}\boldsymbol{\phi}_i\boldsymbol{\phi}_j\boldsymbol{\Delta}'\mathbf{C}_j^-$ is needed instead and from (6.1.5) that is a null-matrix. Hence, in a multi-stratum analysis the various estimates are independent and there are advantages in combining them, weighting each by the reciprocal of its own variance.

There are, however, difficulties. For example, a stratum may give an estimate with a variance that is unknown because there are no degrees of freedom for error. That happened with the inter-block estimate in the example in Section 3.8. (Even if there had been a few error degrees of freedom, many people would hesitate to use the resulting variance for fear of introducing a component into the weighted sum of squared deviations that was so poorly determined as to weaken the whole.)

The matter has been studied in depth by Nelder (1968) for the case of general balance. First he evolved a vector, here called $\mathbf{Y}$, which was intended to combine information from the various strata and so construct an estimate of $\mathbf{y}$, the actual data. Where there is no information about the jth group of contrasts in the ith stratum the weight, w_{ij} must be kept equal to zero despite any subsequent formula that would give it some other value. Otherwise w_{ij} needs to be estimated to best advantage. The constraint

$$\sum_i w_{ij} = 1 \qquad (6.4.1)$$

is needed. Once $\mathbf{Y}$ has been determined, it is intended to use $(\mathbf{y}-\mathbf{Y})'\boldsymbol{\phi}_p(\mathbf{y}-\mathbf{Y})$ as the error sum of squared deviations for the pth stratum, a value greater than that found previously, where $\mathbf{Y}$ was estimated entirely from within the stratum in such a way as to minimize the estimate of error. Here other strata are being used as well. (It will be assumed that the mean of $\mathbf{Y}$ is the same as that of $\mathbf{y}$. If it is not, a change of origin is easily made, but in practice the assumption causes little difficulty.)

The contrasts of the kth group, i.e., $\mathbf{u}_{k1}, \mathbf{u}_{k2}, \ldots$, etc., will be summed up by the matrix

$$\mathbf{H}_k = \sum_l (\mathbf{u}_{kl}\mathbf{u}'_{kl}). \tag{6.4.2}$$

Note that

$$\mathbf{H}_k\mathbf{H}_k = \mathbf{H}_k, \qquad \mathbf{H}_k\mathbf{H}_p = \mathbf{0} \quad (k \neq p). \tag{6.4.3}$$

With general balance, $\mathbf{C}_i\mathbf{H}_k = R_{ik}\mathbf{H}_k$, so $\mathbf{C}_i^+\mathbf{H}_k = \mathbf{H}_k/R_{ik}$, where R_{ik} is the constant effective replication. Hence, in the ith stratum the contrasts of the kth group give quantities like

$$\mathbf{h}_{ik} = \mathbf{H}_k\mathbf{C}_i^+\mathbf{Q}_i = \mathbf{H}_k\mathbf{Q}_i/R_{ik} = \mathbf{H}_k\hat{\mathbf{t}}_i, \tag{6.4.4}$$

where

$$\mathbf{Y} = \boldsymbol{\Delta}'\sum_i\sum_k w_{ik}\mathbf{h}_{ik}.$$

In fact, the values of $\mathbf{h}_{ik}$ are mostly going to arise in products, so it will be convenient to write

$$U_{ij.k} = \mathbf{h}'_{ij}\mathbf{h}_{jk}. \tag{6.4.5}$$

To consider now the error sum of squared deviations for the pth stratum, it equals

$$(\mathbf{y}-\mathbf{Y})'\boldsymbol{\phi}_p(\mathbf{y}-\mathbf{Y}) = \mathbf{y}'\boldsymbol{\phi}_p\mathbf{y} - 2\sum_i\sum_k(w_{ik}u_{ip.k}) + \sum_i\sum_j\sum_k(w_{ik}w_{jk}u_{ij.k}). \tag{6.4.6}$$

If $\mathbf{Y}$ had been estimated solely from the pth stratum, (6.4.6) would have read

$$\mathbf{y}'\boldsymbol{\phi}_p\mathbf{y} - \sum_k U_{pp.k}.$$

Hence, the inflation is

$$\sum_k U_{pp.k} - 2\sum_i\sum_k(w_{ik}U_{ip.k}) + \sum_i\sum_j\sum_k(w_{ik}w_{jk}U_{ij.k}). \tag{6.4.7}$$

On account of the constraints in (6.4.1), it is permissible to write (6.4.7) as

$$\sum_i\sum_j\sum_k[R_{pk}w_{ik}w_{jk}(U_{pp.k} - U_{ip.k} - U_{jp.k} + U_{ij.k})].$$

From (6.4.3), (6.4.5) and (6.4.4) that is the same as

$$\sum_i\sum_j\sum_k[R_{pk}w_{ik}w_{jk}(\hat{\mathbf{t}}_p - \hat{\mathbf{t}}_i)'\mathbf{H}_k(\hat{\mathbf{t}}_p - \hat{\mathbf{t}}_j)]. \tag{6.4.8}$$

The inflation of the sum of squared deviations having been found, the degrees of freedom next need consideration. Nelder's solution is to take the degrees of freedom for each group of contrasts, i.e. $d_k = \text{tr}(H_k)$ and to apportion according to the weights, i.e. the pth stratum receives an addition of $(1 - w_{pk})d_k$ to its error degrees of freedom. That gives a new σ_i^2 for each stratum and weights should be assigned to that, subject to (6.4.1),

$$w_{pk} \propto R_{pk}/\sigma_p^2. \tag{6.4.9}$$

To do that will require iteration. It will be noticed that the summation over i and j in (6.4.8) can omit the values $i, j = p$. Consequently, when, as is often the case, there are only two relevant strata, summation takes place only over k. If, further, $k = 1$, the weights can be determined directly. Let those for the first and second strata be w and $(1 - w)$ respectively. The calculations for a simple case are set out in Example 6F.

6.5 SPLIT-PLOTS

Split-plots were introduced in Section 6.1 on grounds of practicality. One factor may require large plots and another small; one can be applied only to long, narrow plots and another only to compact areas and so on. If each kind of treatment is to be applied to best advantage, it will not be possible to have only one kind of plot.

There is, however, another reason for using split-plots, namely the placing of emphasis, for it is widely believed that the error variance from the sub-plot stratum will necessarily be smaller than that from main plots. Often the belief is justified but not always, nor is there any reason why it should be. If, for example, there is a fertility trend across the blocks and main plots are formed so that each contains its share of good land and of bad land, those main plots are likely to prove quite uniform. If now sub-plots are formed across them, they will fall on different kinds of land and they will be diverse. There is no mystery about that. The sum of squared deviations from the two strata together (sub-plots → main plots and main plots → blocks) will be the same however the main plots are formed. The more successful the experimenter is at reducing the variance in one stratum, the greater the variance in the other. (It is the same in forming the blocks themselves. The more effective they are the greater the precision of the intra-block analysis and the less that of the inter-block.) Nevertheless, it is usually true that the sub-plot error variance will be the lower and, if that can be assumed, a factor that needs to be studied with special precision would do better on the sub-plots. Also, there is sometimes a lack of error degrees of freedom in the main plot stratum and that too can lead to imprecise estimation.

Usually the sub-plot treatments are applied orthogonally to the main plots, i.e. each occurs equally often on each main plot, but that is not essential. Occasionally there is some constraint on the number of sub-plots in a main plot just as there can be on the number of plots in a block, so it is wise to bear in mind that non-orthogonality is permissible. The following example may clarify what is involved.

Suppose that there are four treatments, A, B, C, and D, being levels of a factor that should be applied to main plots, of which ideally there should be three to a block. That suggests something like

Block I	B	C	D		Block III	A	B	D
II	A	C	D		IV	A	B	C

Suppose also that there are three levels a, b and c of a factor best applied to sub-plots, each being half a main plot. Possible combinations are bc, ac and ab. Fortunately there are three main plots of each of A, B, C and D and three in each block, so the following is a feasible design:

Block I	B(ab)	C(bc)	D(ac)
II	A(bc)	C(ac)	D(ab)
III	A(ab)	B(bc)	D(bc)
IV	A(ac)	B(bc)	C(ab)

It will be instructive to study it in some detail. For one thing, it represents the sort of situation that does arise in practice when people become interested in non-orthogonal possibilities. For another, regular as it may seem, it is full of traps and to be avoided. Study of the design will both exemplify the method and reveal the traps. (Of course, the above is a systematic representation to show the scheme. In practice the main plots would be randomized within each block and the sub-plots within each main plot.)

There are twelve treatment combinations, i.e.,

Aa Ab Ac Ba Bb Bc Ca Cb Cc Da Db Dc.

Taking the first group of contrasts to be the main effect of the treatments on the main plots, it has three components:

$$\mathbf{u}'_{11} = (1\ \ 1\ \ 1\ -1\ -1\ -1\ \ 0\ \ 0\ \ 0\ \ 0\ \ 0\ \ 0)/\sqrt{6}$$
$$\mathbf{u}'_{12} = (1\ \ 1\ \ 1\ \ 1\ \ 1\ \ 1\ -2\ -2\ -2\ \ 0\ \ 0\ \ 0)/\sqrt{18}$$
$$\mathbf{u}'_{13} = (1\ \ 1\ \ 1\ \ 1\ \ 1\ \ 1\ \ 1\ \ 1\ \ 1\ -3\ -3\ -3)/\sqrt{36}$$

These contrasts have to be related to three coefficient matrices, $\mathbf{C}_1$, $\mathbf{C}_2$ and $\mathbf{C}_3$, corresponding respectively to sub-plots within main plots, main plots within blocks and blocks within the experimental area. Since $\mathbf{C}_i = \mathbf{\Delta}\boldsymbol{\phi}_i\mathbf{\Delta}$, it appears from (6.1.1) that

$$\mathbf{C}_1 = \mathbf{r}^{\delta} - \mathbf{N}_m\mathbf{N}'_m/k_m$$
$$\mathbf{C}_2 = \mathbf{N}_m\mathbf{N}'_m/k_m - \mathbf{N}\mathbf{N}'/k$$
$$\mathbf{C}_3 = \mathbf{N}\mathbf{N}'/k - \mathbf{r}\mathbf{r}'/n,$$

where $\mathbf{N}_m (= \mathbf{\Delta}\mathbf{D}'_m)$ is the incidence matrix of treatments and main plots and k_m, the number of unit plots in a main plot, is two, while $\mathbf{N} (= \mathbf{\Delta}\mathbf{D}')$ is a similar incidence matrix for treatments and blocks, k being six. It will be found that $\mathbf{r}^{\delta}$, $\mathbf{N}_m\mathbf{N}'_m/k_m$ and $\mathbf{r}\mathbf{r}'/n$ are all highly patterned and regular, but $\mathbf{N}\mathbf{N}'/k$ is relatively arbitrary. However, for all j

$$\begin{aligned} \mathbf{u}'_{1j}\mathbf{C}_1\mathbf{u}_{1j} &= 0 = R_{11} \\ \mathbf{u}'_{1j}\mathbf{C}_2\mathbf{u}_{1j} &= \tfrac{16}{9} = R_{21} \\ \mathbf{u}'_{1j}\mathbf{C}_3\mathbf{u}_{1j} &= \tfrac{2}{9} = R_{31}. \end{aligned}$$

From (4.9.8b) those are indeed the effective replications. It will be seen that they sum to two, that being the actual replication. Taking now the treatments on the sub-plots, their effect requires two contrasts (Group 2):

$$\begin{aligned} \mathbf{u}'_{21} &= (1\ \ -1\ \ \ 0\ \ 1\ \ -1\ \ \ 0\ \ 1\ \ -1\ \ \ 0\ \ 1\ \ -1\ \ \ 0)/\sqrt{8} \\ \mathbf{u}'_{22} &= (1\ \ \ 1\ \ -2\ \ 1\ \ \ 1\ \ -2\ \ 1\ \ \ 1\ \ -2\ \ 1\ \ \ 1\ \ -2)/\sqrt{24} \end{aligned}$$

from which it appears that $R_{12} = 1\frac{1}{2}$, $R_{22} = \frac{1}{2}$ and $R_{32} = 0$. The contrasts of the third group represent the interaction. There are six of them and they are found in the usual way by multiplying out all pairs of vectors, one from each main effect, element by element, i.e.

$$\begin{aligned} \mathbf{u}'_{31} &= (1, -1, 0, -1, 1, 0, 0, 0, 0, 0, 0, 0)/\sqrt{4} \\ \mathbf{u}'_{32} &= (1, 1, -2, -1, -1, 2, 0, 0, 0, 0, 0, 0)/\sqrt{12} \\ \mathbf{u}'_{33} &= (1, -1, 0, 1, -1, 0, -2, 2, 0, 0, 0, 0)/\sqrt{12} \\ \mathbf{u}'_{34} &= (1, 1, -2, 1, 1, -2, -2, -2, 4, 0, 0, 0)/\sqrt{36} \\ \mathbf{u}'_{35} &= (1, -1, 0, 1, -1, 0, 1, -1, 0, -3, 3, 0)/\sqrt{24} \\ \mathbf{u}'_{36} &= (1, 1, -2, 1, 1, -2, 1, 1, -2, -3, -3, 6)/\sqrt{72} \end{aligned}$$

At this point general balance has broken down. For $\mathbf{u}_{31}$, $\mathbf{u}_{33}$, and $\mathbf{u}_{34}$ the effective replications in the three strata are respectively $1\frac{1}{2}, \frac{1}{4}, \frac{1}{4}$; for $\mathbf{u}_{35}$ and $\mathbf{u}_{36}$ they are $1\frac{1}{2}, \frac{5}{18}, \frac{4}{18}$, and for $\mathbf{u}_{32}$ they are $1\frac{1}{2}, \frac{11}{36}, \frac{7}{36}$. Examination of the design shows that the arbitrariness of $\mathbf{N}_m\mathbf{N}'_m$ is not just a matter of the ordering of blocks and treatments but corresponds to genuine anomalies. For example, Aa and Bc concur twice in blocks, whereas Ba and Ac do not concur at all and several similar irregularities arise. It is not, of course, being suggested that sub-plot treatments should never be applied except orthogonally, only that apparently regular schemes can be deceptive. At least, it is necessary to work out a skeleton analysis before starting and not leave everything till data actually appear. What has occurred here is an unbalanced allocation of treatment combinations to blocks even though they are balanced with regard to both main plots and main plot treatments. That can happen very easily.

The example shows also the difficulties that can arise with degrees of freedom. In strata that relate to the larger areas, e.g. main plots, blocks, etc., there are usually few degrees of freedom anyway, yet non-orthogonality will often push contrasts into them. Thus, in the third stratum here there are only three degrees of freedom in all, yet it is supposed to provide information about a main plot effect with three degrees of freedom and an interaction with six. Of course, the situation is eased with more plots. Here there are only two replicates of the twelve treatment combinations. Given, say, six replications, a cyclic permutation of a, b and c would have helped balance and there would have been more degrees of freedom. Nevertheless, the example shows the dangers and points the need for circumspection when complicated split-plot designs are proposed.

However, everything is much simpler if the sub-plot treatments can be assigned orthogonally to the main plots. Thus, if there had been three sub-plots to a main plot instead of two, it would have been possible to have had a, b and c on each main plot. As a consequence their main effect and the interaction would have lain entirely in the first stratum and the other main effect would have been divided between the second and third just as it would have been divided between the intra-block and inter-block strata in an ordinary block design.

In general, if the main plot treatments would have given rise to a given matrix, $\mathbf{M}$, supposing there to have been no sub-plots, and if there are s sub-plots in each main plot, the sub-treatments being applied orthogonally, then $\mathbf{N}_m = s\mathbf{M}$ and $\mathbf{N}$ is formed by replacing each element m_{ij} by a vector $m_{ij}\mathbf{1}_s$. It is now easy to establish the foregoing rule as of general application, i.e. when a factor is applied orthogonally to plots that are nested within larger ones, its main effect and all interactions with factors applied to the larger plots lie solely in the one stratum formed by the nesting.

Thus, for the example with three sub-plots to a main plot instead of two and with a, b and c represented in each, effective replications are

$$\begin{array}{lll} R_{11} = 0 & R_{12} = 3 & R_{13} = 3 \\ R_{21} = \frac{8}{3} & R_{22} = 0 & R_{23} = 0 \\ R_{31} = \frac{1}{3} & R_{32} = 0 & R_{33} = 0. \end{array}$$

As to sums of squared deviations, values in the first two parts of the analysis will be one-third those that would have been obtained had there been no splitting. The reason is simply that the unit plots are now sub-plots instead of main plots. Since, however, means will be worked out on the basis of this smaller area and since replications will be three times as high, the final results will be equivalent. It may also be noted that the last part of the analysis could have been obtained by thinking of four experiments in randomized blocks, one on each of the three plots assigned to A, B, C and D, with treatments a, b, and c. Each such experiment would have had two degrees of freedom for treatments and four for error. Adding degrees of freedom and sums of squared deviations for treatments over all four experiments will give the sum of the main effect and the interaction in the third analysis; adding similarly the four errors will give the error line. This ability to divide an experiment in split-plots between the main treatments can be useful if something goes wrong because it enables the effect of a mistake to be confined.

When splitting plots care is needed about the interpretation of interactions, which can be thought of in more than one way. It will here be convenient to return to the case in which there are several levels of soil preparation in conjunction with several varieties. (It will not affect the argument if there are only two soil methods, A and B, and only two varieties, a and b, the combinations being taken in the order

Aa, Ab, Ba and Bb.) The contrast studied on the main plots will be

$$(1,\ 1,\ -1,\ -1)/2$$

and on the sub-plots

$$(1,\ -1,\ 1,\ -1)/2 \quad \text{and} \quad (1,\ -1,\ -1,\ 1)/2.$$

However, once an interaction has been established it becomes necessary to interpret it. By adding and subtracting the sub-plot contrasts estimates are obtained of

$$(1,\ -1,\ 0,\ 0) \quad \text{and} \quad (0,\ 0,\ 1,\ -1)$$

which show the effect of variety for each method of soil management, but that is not the question. The enquiry concerns the method of soil management to be preferred for each variety, i.e.

$$(1,\ 0,\ -1,\ 0) \quad \text{and} \quad (0,\ 1,\ 0,\ -1).$$

It is supposed, subject to confirmation, that for one variety the method of soil management may be of little importance, whereas for the other it may matter a lot. Consequently the contrasts of interest involve both a main plot and a sub-plot comparison and that can give rise to a certain awkwardness. The interaction as such belongs to the sub-plot analysis above; its interpretation involves the main plots as well and some of the relevant contrasts may be badly determined. For combining contrasts from two levels the method of Satterthwaite (1946) can be useful. Sometimes the method of Section 6.4 is available.

In general, whenever one kind of block is nested within another, randomization is needed in the same way that it was needed for plots nested within blocks (Section 3.3). Sometimes, however, that is not feasible. For example, when the same plots are recorded on several occasions, as with crops from a perennial species, the occasions may be regarded as a sub-plot treatment, though necessarily they cannot be applied at random. (A similar situation arises when the same plants are sampled at successive heights and questions arise about differences due to heights.) Suppose that the main treatments are applied orthogonally and randomly within the blocks and the sub-treatments orthogonally within the main plots, a partition of the sub-plot error will put everything right. If there are b blocks, each divided into m main plots for as many main treatments and each is further divided into s sub-plots for that number of sub-treatments, the second error will have $m(b-1)(s-1)$ degrees of freedom, being composed of the interaction of sub-treatments and blocks with $(b-1)(s-1)$ degrees of freedom and that of main treatments, sub-treatments and blocks with $(b-1)(m-1)(s-1)$. If the sub-treatments are applied systematically, as would necessarily be the case if they represented seasons or heights, their main effect is vitiated, as is its interaction with blocks, but the interaction of main treatments and sub-treatments and its further interaction with blocks is all right. Since, supposedly, no one is studying seasons, etc. but only their interaction with the main plot treatments, no harm will have been done. All that is needed is a partition of the sub-plot error.

Sometimes a factor is introduced only to provide an interaction, its own main effect being so obvious or so well explored as to need no study. For example, an experimenter may wish to investigate different fertilizer treatments both under irrigation and with only natural rainfall. In that case it would be convenient to put the irrigation factor on main plots. Further, a few large plots may suffice, the resulting lack of degrees of freedom for error being of no importance in a stratum that gives information only about an effect not under investigation. Also, if the main plot analysis is not to be used, there is no real need to randomize. It would be unwise, though perhaps convenient, to put all irrigated main plots at one end of the area and all unirrigated at the other because then the effect of irrigation could be confounded with some other factor, such as depth of soil. Nevertheless, a systematic arrangement like A B B A might well be adopted. In that case the lack of randomization would vitiate the main plot analysis, but that would not matter since no use is to be made of it anyway.

6.6 ROW-AND-COLUMN DESIGNS

It is an old device to use two orthogonal blocking systems at right angles, the plots being formed from their intersections. Designs that are formally similar can however arise in other ways. For example, an experiment on long-lived plants may be designed initially in randomized blocks or with some other orthogonal design. When the first set of treatments has been studied for long enough, it may be discontinued and a fresh set applied that is not expected to interact with the first. The design problem is the same as with rows and columns, namely, how to apply a third non-interacting classification to two existing orthogonal classifications and to do so to best advantage. The classical design of that sort is the Latin square, but it is rather restrictive since the numbers of rows, columns and treatments all have to be the same. However, numerous other possibilities exist. In general, (3.1.1) has to be widened to read

$$\mathbf{y} = \mathbf{D}_r'\boldsymbol{\beta}_r + \mathbf{D}_c'\boldsymbol{\beta}_c + \boldsymbol{\Delta}'\boldsymbol{\tau} + \boldsymbol{\eta}, \tag{6.6.1}$$

where the suffices r and c indicate respectively rows and columns instead of blocks. Also $\mathbf{D}_r\mathbf{D}_c' = \mathbf{1}_b\mathbf{1}_c'$, i.e. all rows intersect all columns in one plot and one plot only. The number of rows is taken to be b and of columns, c.

Although general balance may be sought in row-and-column designs as in block designs, there are often so few degrees of freedom for the stratum of rows within the experimental area and the corresponding stratum for columns that the question scarcely arises, the only stratum of any importance being that which results from the crossing of rows and columns. For it, from (6.1.6),

$$\boldsymbol{\phi} = \mathbf{I}_n - \mathbf{D}_r'\mathbf{D}_r/c - \mathbf{D}_c'\mathbf{D}_c/b + \mathbf{1}_n\mathbf{1}_n'/n \tag{6.6.2}$$

Hence,

$$\mathbf{C} = \mathbf{r}^\delta - \mathbf{N}_r\mathbf{N}_r'/c - \mathbf{N}_c\mathbf{N}_c'/b + \mathbf{r}\mathbf{r}'/n \tag{6.6.3}$$

$$= (\mathbf{r}^\delta - \mathbf{N}_r\mathbf{N}_r'/c) + (\mathbf{r}^\delta - \mathbf{N}_c\mathbf{N}_c'/b) - (\mathbf{r}^\delta - \mathbf{r}\mathbf{r}'/n)$$

$$= \mathbf{C}_r + \mathbf{C}_c - \mathbf{C}_o, \tag{6.6.4a}$$

where $\mathbf{C}_r$ and $\mathbf{C}_c$ are respectively coefficient matrices for the block designs using only rows as blocks and only columns, while $\mathbf{C}_0$ is the matrix for an orthogonal design with the same replications. The two block designs are known respectively as the 'row component' and the 'column component'. Taken together they form a design with $2\mathbf{r}$ as the vector of replications and $2n$ as the number of plots, known as the 'amalgamated design' with $\mathbf{C}_a$ as its coefficient matrix. An alternative form of (6.6.2) is therefore

$$\mathbf{C} = \mathbf{C}_a - \mathbf{C}_o$$

It is equally possible to write

$$\boldsymbol{\Omega}^{-1} = \boldsymbol{\Omega}_r^{-1} + \boldsymbol{\Omega}_c^{-1} - \mathbf{r}^{\delta} \qquad (6.6.4b)$$

or, from (3.5.3)

$$\mathbf{F} = \mathbf{F}_r + \mathbf{F}_c - \mathbf{F}_o. \qquad (6.6.4c)$$

To take an example, one of the designs used to illustrate Section 3.5 was the row component of the following (Pearce, 1963):

O	B	O	C	C	A
A	O	A	B	O	C
C	C	B	A	A	B
B	A	C	O	B	O

Here the row component is in supplemented balance and the column component is orthogonal, i.e.

$$\mathbf{C}_r = \frac{1}{6}\begin{bmatrix} 26 & -9 & -9 & -8 \\ -9 & 26 & -9 & -8 \\ -9 & -9 & 26 & -8 \\ -8 & -8 & -8 & 24 \end{bmatrix}, \qquad \mathbf{C}_c = \frac{1}{2}\begin{bmatrix} 9 & -3 & -3 & -3 \\ -3 & 9 & -3 & -3 \\ -3 & -3 & 9 & -3 \\ -3 & -3 & -3 & 9 \end{bmatrix},$$

$$\mathbf{C}_a = \frac{1}{6}\begin{bmatrix} 53 & -18 & -18 & -17 \\ -18 & 53 & -18 & -17 \\ -18 & -18 & 53 & -17 \\ -17 & -17 & -17 & 51 \end{bmatrix}.$$

The design has several important features. For one thing, the column component being orthogonal, $\mathbf{C}_c = \mathbf{C}_0$, $\mathbf{C} = \mathbf{C}_r$, i.e., the variances of contrasts for the whole design derive solely from the row component. In a Latin square, where both components are orthogonal, $\mathbf{C} = \mathbf{C}_0$ and the row-and-column design also estimates all contrasts with full efficiency.

To examine the position further let O represent orthogonality, T total balance, S supplemented balance, G group-divisibility, F factorial balance and so on, the list being extended as far as may be needed. Then the above row-and-column design may be described as being of Type O:OS. The letter before the colon shows how the columns are disposed relative to the rows. For the designs here considered it is always O. The two letters after the colon show respectively how the treatments are disposed relative to rows and columns. The design could also be described as of Type O: (S), the letter within the parentheses referring to the amalgamated design and also to the row-and-column design as a whole. Consideration will show that any design of Type O : OX or O : XO, whatever X may mean, will also be of Type O :(X). The same will be true of designs of Types O : TX and O : XT, because the totally balanced component will add nothing to the pattern introduced by the other, though **C** will no longer equal $\mathbf{C}_r$ or $\mathbf{C}_c$. An interesting study is that of Type O : GG with a factorial set of treatments. If the groups in each component are based on the same factor the result would be of Type O : (G); if not of Type O : (F).

The situation can be illustrated by examining a 3 × 3 factorial set of treatments, i.e. A1, A2, A3, B1, B2, B3, C1, C2 and C3, with six rows and six columns. The first component to be considered (α) is the one in which the groups are formed on the basis of the lettered factor, i.e.

Block I	A_1 A_2 A_3 B_1 B_2 B_3	Block IV	A_1 A_2 A_3 B_1 B_2 B_3
II	A_1 A_2 A_3 C_1 C_2 C_3	V	A_1 A_2 A_3 C_1 C_2 C_3
III	B_1 B_2 B_3 C_1 C_2 C_3	VI	B_1 B_2 B_3 C_1 C_2 C_3

It has a counterpart (β) in which groups are formed from the numbered factor, i.e.

Block I	A_1 A_2 B_1 B_2 C_1 C_2	Block IV	A_1 A_2 B_1 B_2 C_1 C_2
II	A_1 A_3 B_1 B_3 C_1 C_3	V	A_1 A_3 B_1 B_2 C_1 C_2
III	A_2 A_3 B_2 B_3 C_2 C_3	VI	A_2 A_3 B_2 B_3 C_2 C_3

A third possibility (γ) is to disperse the two factors relative to one another, e.g.

Block I	A_1 A_2 B_2 B_3 C_1 C_3	Block IV	A_1 A_2 B_1 B_3 C_2 C_3
II	A_1 A_3 B_1 B_2 C_2 C_3	V	A_1 A_3 B_2 B_3 C_1 C_2
III	A_2 A_3 B_1 B_3 C_1 C_2	VI	A_2 A_3 B_1 B_2 C_1 C_3

Taking these components as simple block designs and using l, n and i to indicate respectively the lettered and numbered factors and their interaction, effective replications are

$$\begin{array}{llll} \alpha & R_l = 3 & R_n = 4 & R_i = 4 \\ \beta & R_l = 4 & R_n = 3 & R_i = 4 \\ \gamma & R_l = 4 & R_n = 4 & R_i = 3\tfrac{1}{2} \end{array}$$

Bearing in mind that the interaction has twice as many degrees of freedom as either main effect, the mean effective replication overall is $3\frac{3}{4}$ for each proposed component.

Various row-and-column designs can be formed, for example, by using α for the row component and α, β and γ in turn for the column component:

($\alpha\alpha$)	A_1	A_2	B_2	B_1	A_3	B_3
	C_1	C_2	C_3	B_3	B_1	B_2
	A_3	C_3	C_2	C_1	A_1	A_2
	A_2	A_3	B_1	B_2	B_3	A_1
	C_3	C_1	B_3	C_2	B_2	B_1
	C_2	A_1	C_1	C_3	A_2	A_3

$(\alpha\beta)$	B_2	A_3	A_1	B_1	B_3	A_2	$(\alpha\gamma)$	A_2	B_2	B_1	A_1	B_3	A_3
	C_2	A_2	C_3	C_1	A_3	A_1		B_1	A_3	A_2	B_3	A_1	B_2
	B_1	B_2	C_1	B_3	C_3	C_2		C_1	B_1	B_3	C_3	B_2	C_2
	A_1	B_3	B_1	A_3	A_2	B_2		A_3	C_3	C_2	A_2	C_1	A_1
	A_2	C_3	A_3	A_1	C_2	C_1		B_3	C_1	C_3	B_2	C_2	B_1
	C_1	C_2	B_3	C_3	B_2	B_1		C_2	A_2	A_1	C_1	A_3	C_3

These designs will repay some attention. Thus, $\alpha\alpha$ is of Type O:(G) and has the characteristics of α intensified, $\alpha\beta$ is of Type O:(F) because the main effects have been combined by crossing them, while $\alpha\gamma$ is again of Type O:(F) but with different elements in **C**.

Effective replications are:

$\alpha\alpha$	$R_l = 2$	$R_n = 4$	$R_i = 4$
$\alpha\beta$	$R_l = 3$	$R_n = 3$	$R_i = 4$
$\alpha\gamma$	$R_l = 3$	$R_n = 4$	$R_i = 3\frac{1}{2}$

The overall mean is $3\frac{1}{2}$. A certain regularity now becomes apparent. It arises because both components have the same natural contrasts. Hence from Section 3.6 and (6.6.4a)

$$R = R_r + R_c - R_o \tag{6.6.5}$$

where R, R_r, R_c and R_o are respectively the effective replications of a contrast in the row-and-column design, the row component, the column component and an orthogonal design with the same replications. (In the example $r = 4$ for all contrasts.) In short, it is not possible for one component to compensate for another. By using two complementary components it is possible to make the effective replications more uniform, as in this design, which is Type O:GG or O:(T) (Pearce, 1963):

B	D	A	C	F	E
H	E	F	D	J	G
F	J	G	A	D	C
G	C	J	H	B	A
D	G	B	E	A	H
C	B	E	J	H	F

However, that is not the same. Further, though it happens only in artificial examples, if $R_r + R_c = R_o$ the contrast concerned will be confounded in the row-and-column design though it can be estimated by either component alone. Also, if $R_r = 0$, i.e. a contrast is confounded in one component, $R = R_c - R_o$. That is to say, $R = 0$ because $R_o \geqslant R_c$, so the contrast is necessarily confounded in the design as a whole. Connectedness from the other component does not help matters.

For a row-and-column design to be orthogonal, i.e. of full efficiency for all contrasts, each component must be orthogonal also, which requires that (4.1.3) shall hold for each.

One important form of row-and-column design is the 'lattice square' (Yates, 1937) in which the various groupings of a lattice design (Section 5.8) are associated, some with rows and some with columns. A generalization has been afforded by Singh and Dey (1979). Other special designs have been suggested by Freeman (1957a, 1957b, 1975).

Another important row-and-column design is the Youden square (Youden, 1935) in which one component is in randomized complete blocks and the other is in balanced incomplete blocks. Such a design is of Type O:OT or O:TO, e.g.

A	B	C	D	E	F	G
B	E	F	G	D	A	C
C	F	D	B	A	G	E

As with block designs some attention has been given to efficiency balance as opposed to variance balance (Nigam, 1976a; Singh and Dey, 1978; Singh, Dey and Nigam, 1979).

Finding a generalized inverse of $\mathbf{C}$ is not difficult. There is no convenient analogue of Ξ because, unlike their role in a block design, the replications play an essential part, even if only because the two components must fit one another. However, $\boldsymbol{\Omega}$ is readily obtained as in (3.4.1). Also, if the natural contrasts are known, $\mathbf{C}^+$ can be obtained as in (3.4.5). The Kuiper–Corsten iteration can be made available by the simple expedient of using the incidence matrix of the amalgamated design, i.e. $(\mathbf{N}_r/c \,\vdots\, \mathbf{N}_c/b)$, in conjunction with an initial vector,

$$\mathbf{v}_1 = \mathbf{r}^{-\delta}\mathbf{Q} = \mathbf{r}^{-\delta}(\mathbf{T} - \mathbf{N}_r\mathbf{B}_r/c - \mathbf{N}_c\mathbf{B}_c/r + \mathbf{1}_v G/n), \tag{6.6.6}$$

where $\mathbf{B}_r$ and $\mathbf{B}_c$ are respectively totals for the two blocking systems, i.e., row totals and column totals. Accordingly (3.5.1) becomes

$$\mathbf{u}_j = \begin{bmatrix} \mathbf{N}_r'/c \\ \mathbf{N}_c'/b \end{bmatrix} \mathbf{v}_j,$$

$$\mathbf{v}_{j+1} = \mathbf{r}^{-\delta}(\mathbf{N}_r \,\vdots\, \mathbf{N}_c)\mathbf{u}_j$$

which makes

$$\begin{aligned} \mathbf{v}_{j+1} &= \mathbf{r}^{-\delta}(\mathbf{N}_r\mathbf{N}_r'/c + \mathbf{N}_c\mathbf{N}_c'/b)\mathbf{v}_j \\ &= \mathbf{r}^{-\delta}(\mathbf{r}^{\delta} - \mathbf{C} - \mathbf{r}\mathbf{r}'/n)\mathbf{v}_j \\ &= \mathbf{r}^{-\frac{1}{2}\delta}(\mathbf{I}_v - \mathbf{F})\mathbf{r}^{\frac{1}{2}\delta}\mathbf{v}_j \end{aligned} \tag{6.6.7}$$

because $\mathbf{r}'\mathbf{v}_j = 0$ as before. However, (6.6.7) is the same as (3.5.3), albeit with certain revised definitions, and the rest of the argument as far as (3.5.7) applies in the present context.

It follows that estimated residuals may be found as before at (3.3.7), the expression

$$\hat{\boldsymbol{\eta}} = (\boldsymbol{\phi} - \boldsymbol{\phi}\boldsymbol{\Delta}'\mathbf{C}^{-}\boldsymbol{\Delta}\boldsymbol{\phi})\mathbf{y} = \boldsymbol{\psi}\mathbf{y}, \quad \text{say} \tag{6.6.8}$$

With a row-and-column design it is sufficient to write down a configuration with the desired properties and to permute both rows and columns at random. Let

$\theta\sigma^2$ be the expectation of $\hat{\eta}_i^2$, where i represents any plot of the design, then on account of the proposed randomization, which could allocate any design plot to any field plot with equal probability, all diagonal elements of $\boldsymbol{\eta}\boldsymbol{\eta}'$ have the same expectation. Taking two design plots in the same row, $\hat{\eta}_i\hat{\eta}_j$ has an expectation of $-\theta\sigma^2/(c-1)$, because $\boldsymbol{\phi}\mathbf{D}_r' = \mathbf{0}$ and the sum of estimated residuals in a row therefore sum to zero. Similarly if i and j represent plots in the same column, the expectation of $\hat{\eta}_i\hat{\eta}_j$ is $-\theta\sigma^2/(b-1)$ and, finally if i and j have neither row nor column in common the expectation is $\theta\sigma^2/[(b-1)(c-1)]$. In short,

$$\text{ex}\,(\boldsymbol{\eta}\boldsymbol{\eta}') = \frac{bc\theta}{(b-1)(c-1)}\boldsymbol{\phi}\sigma^2. \tag{6.6.9}$$

Section 4.9 now applies as before apart from a constant multiplier that does not affect the argument, so the lack of bias is established as before. It is necessary to note only that tr $(\boldsymbol{\phi}) = (b-1)(c-1)$, the total degrees of freedom for the stratum.

Some of the above points are illustrated in Examples 6G, 6H and 6I.

If a row-and-column design is orthogonal, like a block design it can readily be analysed by the use of summation terms (see Section 4.1). Five are needed, total, row, columns, treatments and correction, which can be written respectively as S, S_r, S_c, S_t and S_o. Then the sums of squared deviations for rows, columns and treatments are respectively $(S_r - S_o)$, $(S_c - S_o)$ and $(S_t - S_o)$ with respectively $(b-1)$, $(c-1)$ and $(v-1)$ degrees of freedom. The error line has a sum of squared deviations of $(S - S_r - S_c - S_t + 2S_o)$ with $(bc - b - c - v + 2)$ degrees of freedom giving a total of $(S - S_o)$ with $(bt - 1)$.

6.7 STRIP-PLOT (OR CRISIS-CROSS) DESIGNS

Sometimes there are two factors in an experiment and both would be difficult to apply to small plots. If that is so it is sometimes a good plan to apply them at right angles in a series of rectangles that can be regarded as blocks. Thus given one factor with three levels, A, B, C and another with four, a, b, c, d, a block might look like this, the two factors both having been randomized, one onto rows and the other onto columns:

Bb	Ba	Bd	Bc
Cb	Ca	Cd	Cc
Ab	Aa	Ad	Ac.

Since everything is orthogonal, there need be no difficulties with the analysis of variance, there being three strata with useful information. In such a design there is usually higher precision for the estimation of the interaction than for either of the main effects, nor is that necessarily a bad thing. (In fact, the trouble with many designs is that they show up main effects well while missing interactions that would be quite important if they could be established, so the opposite tendency can be quite welcome.) Of course, that assumes that row and column effects are truly additive. If they are not, the discrepancy will appear as error in the third stratum, i.e., the crossing of rows and columns, which involves the interaction. It will be seen that each effect has its own interaction with blocks as error. If one factor is difficult to apply at random, a systematic application will vitiate its own analysis but not others.

The validation of the analysis for the first stratum by randomization of rows and columns follows from the same argument as before, noting only that

$$\boldsymbol{\phi}_1 = \mathbf{I}_n - \mathbf{D}_r'\mathbf{D}_r/x - \mathbf{D}_c'\mathbf{D}_c/w + \mathbf{D}'\mathbf{D}/(wx) \tag{6.7.1}$$

where $\mathbf{D}_r$, $\mathbf{D}_c$ and $\mathbf{D}$ show respectively the allocation of the $n\ (= bwx)$ plots to the bw rows, the bx columns and the b blocks. The interaction sum of squares is $\mathbf{y}'(\boldsymbol{\phi} - \boldsymbol{\psi})\mathbf{y}$, where

$$\boldsymbol{\phi} - \boldsymbol{\psi} = \boldsymbol{\Delta}'[\mathbf{I}_{wx} - \mathbf{W}'\mathbf{W}/x - \mathbf{X}'\mathbf{X}/w + \mathbf{1}_{wx}\mathbf{1}_{wx}'/(\mathbf{wx})]\boldsymbol{\Delta}/b. \tag{6.7.2}$$

Here $\mathbf{W}$ and $\mathbf{X}$ are design matrices that allocate the wx treatment combinations to the two main effects. Adapting the argument already presented in Sections 2.1, 3.3 and 6.6, there is no difficulty in seeing that the expectation of $\hat{\boldsymbol{\eta}}\hat{\boldsymbol{\eta}}'$ is given by $\theta\boldsymbol{\phi}_1$ and that the expectation of $\mathbf{y}'\boldsymbol{\phi}_1\mathbf{y}$, the total sum of squared deviations for the stratum, is $\theta\,\mathrm{tr}\,(\boldsymbol{\phi}) = b(w-1)(x-1)$. The problem is rather that of seeking the expectation of the interaction sum of squared deviations, which equals $\mathrm{tr}\,[(\boldsymbol{\phi} - \boldsymbol{\psi})\boldsymbol{\phi}]\theta$. It will be easier if the plots are arranged first in order of blocks, then of rows within blocks and then of plots within rows according to column. It then appears that $\boldsymbol{\phi}_1$ may be written as $b \times b$ sub-matrices such that those on the diagonal equal

$$\frac{1}{wx}\begin{bmatrix} (w-1)\mathbf{K} & -\mathbf{K} & -\mathbf{K} \\ -\mathbf{K} & (w-1)\mathbf{K} & -\mathbf{K} \\ -\mathbf{K} & -\mathbf{K} & (w-1)\mathbf{K} \end{bmatrix},$$

where $\mathbf{K}$ is an $x \times x$ matrix with diagonal elements equal to $(x-1)$ and off-diagonal equal to -1. Off-diagonal sub-matrices of $\boldsymbol{\phi}_1$ are null. Turning now to $\boldsymbol{\phi} - \boldsymbol{\psi}$, the same partition applies but now the off-diagonal sub-matrices are as before apart from being divided by b, while the diagonals are the same as the others, i.e.

$$\frac{1}{bwx}\begin{bmatrix} -\mathbf{K} & -\mathbf{K} & -\mathbf{K} \\ -\mathbf{K} & -\mathbf{K} & -\mathbf{K} \\ -\mathbf{K} & -\mathbf{K} & -\mathbf{K} \end{bmatrix}.$$

It will be apparent from previous arguments that the expectations of the sum of squared deviations for interaction and for error will be in the ratio of $\mathrm{tr}\,[(\boldsymbol{\phi} - \boldsymbol{\psi})\boldsymbol{\phi}]$ and $\mathrm{tr}\,(\boldsymbol{\phi}\boldsymbol{\phi})$, i.e. of $(w-1)(x-1)$ and $(b-1)(w-1)(x-1)$. Since that is also the ratio of their degrees of freedom, the F-test is unbiassed and Section 4.9 applies. [*Note on nomenclature:* The author first heard about the above design in 1950 when it was still unpublished. He was delighted with its practical usefulness and included it in a publication (Pearce, 1953), calling it 'the criss-cross design of Cochran and Cox' without realizing that its inventors had given it the name of 'strip-plot design' (Cochran and Cox, 1950). Credit for the idea belongs solely to Cochran and Cox; discredit for confusing the nomenclature is entirely the author's.]

Examples 6

A. Examine block designs in the light of Sections 6.1 and 6.2.
Summary: The block structure is

Plots → Blocks → Experimental area.

There are then two strata, namely, (1) plots within blocks and (2) blocks within the experimental area. (They correspond respectively to the intra-block and the inter-block cases of Chapter 3.) A third stratum, i.e. that of the experimental area, will be ignored because it has nothing to contribute. Corresponding to the two strata

$$\boldsymbol{\phi}_1 = \mathbf{I}_n - \mathbf{D}'\mathbf{k}^{-\delta}\mathbf{D} \qquad = \mathbf{H}_0 - \mathbf{H}_1 \tag{3.1.6}$$

$$\boldsymbol{\phi}_2 = \mathbf{D}'\mathbf{k}^{-\delta}\mathbf{D} - \mathbf{1}_n\mathbf{1}_n'/\mathbf{n} = \mathbf{H}_1 - \mathbf{H}_2. \tag{3.8.4}$$

For each stratum, there is a coefficients matrix, $\mathbf{C}$:

$$\mathbf{C}_1 = \boldsymbol{\Delta}\boldsymbol{\phi}_1\boldsymbol{\Delta}' = \mathbf{r}^{\delta} - \mathbf{N}\mathbf{k}^{-\delta}\mathbf{N}' \tag{3.1.7}$$

$$\mathbf{C}_2 = \boldsymbol{\Delta}\boldsymbol{\phi}_2\boldsymbol{\Delta}' = \mathbf{N}\mathbf{k}^{-\delta}\mathbf{N}' - \mathbf{r}\mathbf{r}'/n. \tag{3.8.6a}$$

Data are summed by the appropriate vector, $\mathbf{Q}$:

$$\mathbf{Q}_1 = \boldsymbol{\Delta}\boldsymbol{\phi}_1\mathbf{y} = \mathbf{T} - \mathbf{N}\mathbf{k}^{-\delta}\mathbf{B} \tag{3.1.8}$$

$$\mathbf{Q}_2 = \boldsymbol{\Delta}\boldsymbol{\phi}_2\mathbf{y} = \mathbf{N}\mathbf{k}^{-\delta}\mathbf{B} - (G/n)\mathbf{r}. \tag{3.8.6b}$$

Treatment parameters can be estimated thus:

$$\hat{\boldsymbol{\tau}}_1 = \mathbf{C}_1^-\mathbf{Q}_1 \tag{3.3.1}$$

$$\hat{\boldsymbol{\tau}}_2 = \mathbf{C}_2^-\mathbf{Q}_2. \tag{3.8.5}$$

A convenient generalized inverse is commonly found in $\boldsymbol{\Omega}_i = (\mathbf{C}_i + \mathbf{r}\mathbf{r}'/n)^{-1}$. Thus

$$\boldsymbol{\Omega}_1^{-1} = \mathbf{r}^{\delta} - \mathbf{N}\mathbf{k}^{-\delta}\mathbf{N}' + \mathbf{r}\mathbf{r}'/n \tag{3.4.1}$$

$$\boldsymbol{\Omega}_2^{-1} = \mathbf{N}\mathbf{k}^{-\delta}\mathbf{N}'. \tag{3.8.7}$$

For purposes of finding the analysis of variance, the total sums of squared deviations are respectively

$$\mathbf{y}'\boldsymbol{\phi}_1\mathbf{y} = \mathbf{y}'\mathbf{y} - \mathbf{B}'\mathbf{k}^{-\delta}\mathbf{B} \tag{3.3.10}$$

$$\mathbf{y}'\boldsymbol{\phi}_2\mathbf{y} = \mathbf{B}'\mathbf{k}^{-\delta}\mathbf{B} - G^2/n. \qquad \text{(Above 3.8.4)}$$

The treatment sums of squared deviations are respectively

$$\mathbf{Q}_1'\mathbf{C}_1^-\mathbf{Q}_1 = \mathbf{Q}_1'\hat{\boldsymbol{\tau}}_1 \tag{3.3.10}$$

$$\mathbf{Q}_2'\mathbf{C}_2^-\mathbf{Q}_2 = \mathbf{Q}_2'\hat{\boldsymbol{\tau}}_2. \tag{3.8.11}$$

The error sums of squared deviations are respectively

$$\mathbf{y}'\boldsymbol{\psi}_1\mathbf{y} = \mathbf{y}'(\boldsymbol{\phi}_1 - \boldsymbol{\phi}_1\boldsymbol{\Delta}'\mathbf{C}_1^-\boldsymbol{\Delta}\boldsymbol{\phi}_1)\mathbf{y} \tag{3.3.9}$$

$$\mathbf{y}'\boldsymbol{\psi}_2\mathbf{y} = \mathbf{y}'(\boldsymbol{\phi}_2 - \boldsymbol{\phi}_2\boldsymbol{\Delta}'\mathbf{C}_2^-\boldsymbol{\Delta}\boldsymbol{\phi}_2)\mathbf{y}. \tag{3.8.11}$$

The total degrees of freedom for each stratum are

$$\text{tr}\,(\boldsymbol{\phi}_1) = n - b \qquad (3.3.13)$$

$$\text{tr}\,(\boldsymbol{\phi}_2) = b - 1. \qquad (\text{Above } 3.8.14)$$

The degrees of freedom for treatments in each stratum are

$$\text{tr}\,(\mathbf{C}_1\mathbf{C}_1^-) = h_1, \quad \text{say} \qquad (3.3.12)$$

$$\text{tr}\,(\mathbf{C}_2\mathbf{C}_2^-) = h_2, \quad \text{say.} \qquad (3.8.15)$$

The integers, h_1 and h_2, give the number of independent contrasts estimated, either wholly or partly, in each stratum. If information about a contrast is provided by several strata, it should be counted in each. The degrees of freedom for error are therefore respectively $n-b-h_1$ and $b-1-h_2$.

B. Find the form of the Kuiper–Corsten iteration for the two strata of a block design.
Answer: In the first stratum, from (6.2.20)

$$\begin{aligned}\mathbf{v}_{j+1} &= (\mathbf{I}_v - \mathbf{r}^{-\delta}\mathbf{C}_1)\mathbf{v}_j \\ &= (\mathbf{I}_v - \mathbf{I}_v + \mathbf{r}^{-\delta}\mathbf{N}\mathbf{k}^{-\delta}\mathbf{N}')\mathbf{v}_j \\ &= \mathbf{r}^{-\delta}\mathbf{N}\mathbf{k}^{-\delta}\mathbf{N}'\mathbf{v}_j \qquad \text{Cf. } (3.5.2).\end{aligned}$$

In the second

$$\begin{aligned}\mathbf{v}_{j+1} &= (\mathbf{I}_v - \mathbf{r}^{-\delta}\mathbf{C}_2)\mathbf{v}_j \\ &= (\mathbf{I}_v - \mathbf{r}^{-\delta}\mathbf{N}\mathbf{k}^{-\delta}\mathbf{N}' + \mathbf{1}_v\mathbf{r}'/n)\mathbf{v}_j \\ &= (\mathbf{I}_v - \mathbf{r}^{-\delta}\mathbf{N}\mathbf{k}^{-\delta}\mathbf{N}')\mathbf{v}_j \qquad \text{Cf } (3.8.12).\end{aligned}$$

Since $\mathbf{r}'\mathbf{v}_1 = 0$, so also $\mathbf{r}'\mathbf{v}_j$ for all j.

C. Analyse the data of Example 3A using the method of sweeping by natural contrasts, as described in Section 6.3.
Answer: It appears at once that the natural contrasts are not those that were in mind when the experiment was designed. Those intended were

$$(1,\ 0,\ 0,\ 0,\ -1)'/\sqrt{2},$$
$$(0,\ 1,\ 0,\ 0,\ -1)'/\sqrt{2},$$
$$(0,\ 0,\ 1,\ 0,\ -1)'/\sqrt{2},$$
$$(0,\ 0,\ 0,\ 1,\ -1)'/\sqrt{2}.$$

The natural contrasts can be written

$$\begin{aligned}\mathbf{u}_1 &= (1,\ \ 1,\ \ 1,\ \ 1,\ -4)'/\sqrt{20}, && R_1 = \tfrac{50}{7} \\ \mathbf{u}_2 &= (1,\ -1,\ \ 0,\ \ 0,\ \ 0)'/\sqrt{2} && \\ \mathbf{u}_3 &= (1,\ \ 1,\ -2,\ \ 0,\ \ 0)'/\sqrt{6}, && R_2 = R_3 = R_4 = \tfrac{34}{7} \\ \mathbf{u}_4 &= (1,\ \ 1,\ \ 1,\ -3,\ \ 0)'/\sqrt{12}. &&\end{aligned}$$

So long as there is no need to partition the treatment sum of squared deviations but only to evaluate it as a whole, that does not matter. The reader would, however, perhaps like to see how a partition is effected. That need cause no difficulty because the natural contrasts do in fact make good sense. Three of them, $\mathbf{u}_2$, $\mathbf{u}_3$ and $\mathbf{u}_4$, relate to differences between herbicides, while $\mathbf{u}_1$ relates to the contrast between the untreated control and the herbicides as a group, supposing that no differences have been found between them. Those were not the questions being asked at the time; no one thought that four chemically distinct herbicides would have much in common. Nevertheless, for purposes of illustration it will be convenient to find the treatments sum of squares in two parts, one for $\mathbf{u}_1$, with one degree of freedom, and the other for $\mathbf{u}_2$, $\mathbf{u}_3$, and $\mathbf{u}_4$, with three.

The calculations are set out in Table 6C. The first two columns set out respectively the block and the treatment for each plot and the data are in the third. The first step is to sweep by blocks and that is done by reducing each

Table 6C

The data of Example 3A analysed by the method of sweeping

(1)	(2)	(3)	(4)	(5)	(6)	(7)	(8)	(9)
I	D	107	−50.43	−41.21	−37.26	−17.00	−13.05	−11.48
I	A	166	+8.57	+17.79	+21.74	+6.77	+10.72	+12.29
I	B	133	−24.43	−15.21	−11.26	−11.47	−7.52	−5.95
I	C	166	+8.57	+17.79	+21.74	+0.86	+4.81	+6.38
I	O	177	+19.57	−17.29	−13.34	−17.29	−13.34	−11.77
I	A	163	+5.57	+14.79	+18.74	+3.77	+7.72	+9.29
I	O	190	+32.57	−4.29	−0.34	−4.29	−0.34	+1.23
II	A	136	−0.14	+9.08	+13.03	−1.94	+2.01	+1.47
II	O	146	+9.86	−27.00	−23.05	−27.00	−23.05	−23.59
II	D	104	−32.14	−22.92	−18.97	+1.29	+5.24	+4.70
II	C	152	+15.86	+25.08	+29.03	+8.15	+12.10	+11.56
II	B	119	−17.14	−7.92	−3.97	−4.18	−0.23	−0.77
II	O	164	+27.86	−9.00	−5.05	−9.00	−5.05	−5.59
II	B	132	−4.14	+5.08	+9.03	+8.82	+12.77	+12.23
III	C	118	−20.71	−11.49	−7.54	−28.42	−24.47	−22.06
III	A	117	−21.71	−12.49	−8.54	−23.51	−19.56	−17.15
III	O	176	+37.29	+0.43	+4.38	+0.43	+4.38	+6.79
III	B	132	−6.71	+2.51	+6.46	+6.25	+10.20	+12.59
III	C	139	+0.29	+9.51	+13.46	−7.42	−3.47	−1.06
III	O	186	+47.29	+10.43	+14.38	+10.43	+14.38	+16.79
III	D	103	−35.71	−26.49	−22.54	−2.28	+1.67	+4.08
IV	O	173	+38.43	+1.57	+5.52	+1.57	+5.52	+2.06
IV	D	95	−39.57	−30.35	−26.40	−6.14	−2.19	−5.65
IV	D	109	−25.57	−16.35	−12.40	+7.86	+11.81	+8.35
IV	A	130	−4.57	+4.65	+8.60	−6.37	−2.42	−5.88
IV	B	103	−31.57	−22.35	−18.40	−18.61	−14.66	−18.12
IV	O	185	+50.43	+13.57	+17.52	+13.57	+17.52	+14.06
IV	C	147	+12.43	+21.65	+25.60	+4.72	+8.67	+5.21

value in Column (3) by the mean value for the block in which it occurs, namely,

I, 157.43; II, 136.14; III, 138.71; IV, 134.57.

The reduced values appear in Column (4) and represent $\mathbf{y}_i$, the component of $\mathbf{y}$ for the stratum. The sum of their squares, which is 20 490, gives the total sum of squared deviations for the stratum, a figure that was earlier found in Example 3A.

The next step is to sweep by the first contrast, $\mathbf{u}_1$. It is first evaluated from the treatment totals in Column (4), namely

A, −12.28; B, −83.99; C, +16.44; D, −183.42; O, +263.30.

The totals, which represent $\boldsymbol{\Delta\phi}\mathbf{y} = \mathbf{Q}$, should sum to zero. They do not do so exactly but the approximation is good enough. (Rounding errors are a difficulty with sweeping. However, since the method is essentially one for computers and since they mostly carry a lot of decimal places the trouble is more apparent in demonstrations than in practice.) The figures just found represent $\boldsymbol{\Delta\phi}_i\mathbf{y}$ in (6.3.2). The next step is to multiply out by the contrast, thus:

$$\begin{aligned}\mathbf{u}_1'\boldsymbol{\Delta\phi}\mathbf{y} &= [-12.28 - 83.99 + 16.44 - 183.42 - 4(263.30)]/\sqrt{20}\\ &= -1316.45/\sqrt{20}.\end{aligned}$$

A treatment vector is then formed according to the elements of $\mathbf{u}_1$, thus

$$\mathbf{u}_1\mathbf{u}_1'\boldsymbol{\Delta\phi}_i\mathbf{y}/R_1 = \frac{-1316.45 \times 7}{20 \times 50}\begin{bmatrix}1\\1\\1\\1\\-4\end{bmatrix} = \begin{bmatrix}-9.22\\-9.22\\-9.22\\-9.22\\36.86\end{bmatrix}.$$

It follows that sweeping by $\mathbf{u}_1$ involves adding 9.22 to all values in Column (4) that relate to Treatments A, B, C, or D and subtracting 36.86 from all that relate to Treatment O. In that way Column (5) is formed.

Since it has been decided to partition the treatment sum of squared deviations, it is necessary at this point to sweep by blocks again and so produce Column (6). Squaring its values and summing gives 8359. Hence the reduction between Columns (4) and (6) is 20 490 − 8359 = 12 131, which should represent the sum of squared deviations of $\mathbf{u}_1$, which had been swept out of (6) but not out of (4). Use of (3.6.1) shows that the value should be $(\mathbf{u}_1'\mathbf{Q})^2/R_1$, which it is.

However, Column (6) was found only to effect the partition. It is not necessarily needed. The next step is to sweep by $\mathbf{u}_2$, $\mathbf{u}_3$, and $\mathbf{u}_4$. Since they have the same effective replication, $R = \frac{34}{7}$, they can be considered together.

Here,

$$\mathbf{u}_2\mathbf{u}_2' - \mathbf{u}_3\mathbf{u}_3' + \mathbf{u}_4\mathbf{u}_4' = \tfrac{1}{4}\begin{bmatrix} 3 & -1 & -1 & -1 & 0 \\ -1 & 3 & -1 & -1 & 0 \\ -1 & -1 & 3 & -1 & 0 \\ -1 & -1 & -1 & 3 & 0 \\ 0 & 0 & 0 & 0 & 0 \end{bmatrix}.$$

For the moment the sweep will be applied to Column (5), the values of which will be termed $\mathbf{y}_5$, since Column (6) does not necessarily exist. The quantities to be subtracted to form Column (7) are

$$(\mathbf{u}_2\mathbf{u}_2' + \mathbf{u}_3\mathbf{u}_3' + \mathbf{u}_4\mathbf{u}_4')\boldsymbol{\Delta}\mathbf{y}_5/R = \begin{bmatrix} +11.02 \\ -3.74 \\ +16.93 \\ -24.21 \\ 0.00 \end{bmatrix}.$$

The quantities $\boldsymbol{\Delta}\mathbf{y}_5$ are, of course, the treatment totals in Column (5), namely,

A, +33.82; B, −37.89; C, +62.54; D, −137.32; O, −31.58.

In that way Column (7) is arrived at. A final sweep by blocks gives Column (9).

As each column is formed in the computer it can overwrite its predecessor. Accordingly it would be inconvenient to have to store Column (5) when Column (6) was generated. There is in fact no need to do so. If Column (6) is swept for $\mathbf{u}_2$, $\mathbf{u}_3$, and $\mathbf{u}_4$ using the above method to give Column (8) and if that column is then swept by blocks, the same Column (9) will be found, possible rounding errors expected. It represents the data swept for blocks and for all the treatment contrasts and is therefore $\hat{\boldsymbol{\eta}}$, an estimate of the residuals, $\boldsymbol{\eta}$, first introduced in (3.1.1). The sum of elements squared is 3460, which is nearly correct, the true value for the error sum of squared deviations being 3461. However, after several sweeps rounding errors accumulate and more decimal places in the working are really required. All that is needed here is a description of the method.

The reader would perhaps like to explore the calculations a little. There is, for example, that figure of 3.95 which comes in several times. Again, what would happen if the contrasts, $\mathbf{u}_2$, $\mathbf{u}_3$, $\mathbf{u}_4$, were dealt with before $\mathbf{u}_1$?

D. Analyse the data of Example 3A using the method of sweeping by basic contrasts, which also is described in Section 6.3.

Answer: It is first necessary to find the eigenvalues and eigenvectors of $\mathbf{F}$. They can be written:

$$\begin{aligned} \mathbf{p}_1 &= (1,\ 1,\ 1,\ 1,\ -\sqrt{10})'/\sqrt{14} & \varepsilon_1 &= 1, \\ \mathbf{p}_2 &= (1,\ -1,\ 0,\ 0,\ 0)'/\sqrt{2}. & & \\ \mathbf{p}_3 &= (1,\ 1,\ -2,\ 0,\ 0)'/\sqrt{6}, & \varepsilon_2 &= \varepsilon_3 = \varepsilon_4 = \tfrac{34}{35}, \\ \mathbf{p}_4 &= (1,\ 1,\ 1,\ -3,\ 0)'/\sqrt{12}. & & \end{aligned}$$

Since $\mathbf{s}_i = \mathbf{r}^{-\frac{1}{2}\delta}\mathbf{p}_i$

$$\begin{aligned}
\mathbf{s}_1 &= (2, \quad 2, \quad 2, \quad 2, -5)'/\sqrt{280},\\
\mathbf{s}_2 &= (1, -1, \quad 0, \quad 0, \quad 0)'/\sqrt{10},\\
\mathbf{s}_3 &= (1, \quad 1, -2, \quad 0, \quad 0)'/\sqrt{30},\\
\mathbf{s}_4 &= (1, \quad 1, \quad 1, -3, \quad 0)'/\sqrt{60}.
\end{aligned}$$

It will be seen that in this instance the basic contrasts correspond to the natural, so it will be possible to relate the two methods of sweeping from a common example. Indeed, there is no need of a second table, the one developed for the last example serving for this also. The first four columns are in any case the same, representing respectively blocks, treatments, data and, in Column (4), data swept by blocks.

The next step is to find treatment totals of the values in Column (4) as in the other method, but this time $\mathbf{s}_1'\boldsymbol{\Delta\phi}y = \mathbf{s}_1'\mathbf{Q}$ is needed instead of $\mathbf{u}_1'\boldsymbol{\Delta\phi}\mathbf{y}$. It is

$$\begin{aligned}
&[2(-12.28-83.99+16.44-183.42)-5(263.30)]/\sqrt{280}\\
&= -1843.00/\sqrt{280}.
\end{aligned}$$

Hence

$$\mathbf{s}_1\mathbf{s}_1'\boldsymbol{\Delta\phi}_i\mathbf{y} = \frac{-1843.00}{280}\begin{bmatrix}2\\2\\2\\2\\-5\end{bmatrix} = \begin{bmatrix}-13.16\\-13.16\\-13.16\\-13.16\\32.91\end{bmatrix}.$$

Subtracting these quantities from those in Column (4) according to treatment gives Column (6) without the intermediary of Column (5). (Actually 13.17 would be better than 13.16, but a rounding error has intervened.) The fact that no further sweep is needed for blocks is shown by summing the values within each block and finding that all block totals are zero. With a computer the sweep may as well be completed.

To consider now sweeping by $\mathbf{s}_2$, $\mathbf{s}_3$ and $\mathbf{s}_4$, since they all have the same efficiency factor, ε, they can be taken together, just as $\mathbf{u}_2$, $\mathbf{u}_3$, and $\mathbf{u}_4$ could be taken together because they had the same effective replication, R. Actually, since in this instance

$$(\mathbf{s}_2\mathbf{s}_2' + \mathbf{s}_3\mathbf{s}_3' + \mathbf{s}_4\mathbf{s}_4')/\varepsilon = (\mathbf{u}_2\mathbf{u}_2' + \mathbf{u}_3\mathbf{u}_3' + \mathbf{u}_4\mathbf{u}_4')/R,$$

the two methods converge. Sweeping Column (6) by $\mathbf{s}_1$, $\mathbf{s}_2$ and $\mathbf{s}_3$ gives Column (8) and a further sweep by blocks gives Column (9) as before. It will be noted that Columns (4), (6) and (9), which are summed to give values for partition, are the same by either method.

E. Find the inter-block analysis for the data in Example 3A, again using the method of sweeping by contrasts.

Answer: Really this is very simple. The sweeping matrix for the blocking system is $\boldsymbol{\phi}_0 = \mathbf{D}'\mathbf{k}^{-\delta}\mathbf{D} - \mathbf{1}_n\mathbf{1}_n'/n$. That is to say, the general mean is subtracted from the appropriate block mean. In Table 6E the first three columns are the same as in Table 6C; at the fourth appears the difference between the appropriate block mean

(I, 157.43; II, 136.14; III, 138.71; IV, 134.57)

and the general mean, 141.71. The sum of squares in Column (4), i.e.

$$7(15.72^2 + 5.57^2 + 3.00^2 + 7.14^2) = 2367,$$

represents the total sum of squared deviations in the inter-block stratum. (It is actually a slight over-estimate on account of rounding errors.)

The next task is to sweep by $\mathbf{u}_1$, which is the same as using $\mathbf{s}_1$. It is estimated with full efficiency in the intra-block analysis; it is therefore

Table 6E
Calculations involved in finding the inter-block analysis of variance of the data of Example 3A using the method of sweeping

(1)	(2)	(3)	(4)	(5)	(6)
I	D	107	+15.72	+65.72	+0.00
I	A	166	+15.72	−94.30	+0.00
I	B	133	+15.72	+54.73	+0.00
I	C	166	+15.72	+36.56	+0.00
I	O	177	+15.72	+15.74	+0.00
I	A	163	+15.72	−94.30	+0.00
I	O	190	+15.72	+15.72	+0.00
II	A	136	−5.57	−115.59	+0.00
II	O	146	−5.57	−5.57	+0.00
II	D	104	−5.57	+44.43	+0.00
II	C	152	−5.57	+15.45	+0.00
II	B	119	−5.57	+33.44	+0.00
II	O	164	−5.57	−5.57	+0.00
II	B	132	−5.57	+33.44	+0.00
III	C	118	−3.00	+18.02	−0.00
III	A	117	−3.00	−113.02	−0.00
III	O	176	−3.00	−3.00	−0.00
III	B	132	−3.00	+36.01	−0.00
III	C	139	−3.00	+18.02	−0.00
III	O	186	−3.00	−3.00	−0.00
III	D	103	−3.00	+47.00	−0.00
IV	O	173	−7.14	−7.14	−0.00
IV	D	95	−7.14	+42.86	−0.00
IV	D	109	−7.14	+42.86	−0.00
IV	A	130	−7.14	−117.16	−0.00
IV	B	103	−7.14	+31.87	−0.00
IV	O	185	−7.14	−7.14	−0.00
IV	C	147	−7.14	+13.88	−0.00

confounded in other strata. However, $\mathbf{u}_2$, $\mathbf{u}_3$, and $\mathbf{u}_4$, or alternatively $\mathbf{s}_2$, $\mathbf{s}_3$, and $\mathbf{s}_4$, have an effective replication, R, of $\frac{1}{7}$ and efficiency, ε, of $\frac{1}{35}$, because

$$\mathbf{C}_0\mathbf{u}_i = (\mathbf{N}\mathbf{k}^{-\delta}\mathbf{N}' - \mathbf{r}\mathbf{r}'/n)\mathbf{u}_i = \tfrac{1}{7}\mathbf{u}_i \qquad (i = 2, 3, 4)$$
$$\mathbf{r}^{-}\boldsymbol{\delta}\mathbf{C}_0\mathbf{s}_i = (\mathbf{r}^{-\delta}\mathbf{N}\mathbf{k}^{-\delta}\mathbf{N}' - \mathbf{1}_v\mathbf{r}'/n)\mathbf{s}_i = \tfrac{1}{35}\mathbf{s}_i \qquad (i = 2, 3, 4).$$

Sweeping by the three contrasts simultaneously is quite feasible but first it is necessary to have the treatment totals from Column (4).

A, 15.73; B, −5.56; C, −2.99; D, −7.13; O, 0.02.

Already the rounding errors are beginning to show, though with a computer they would be much less serious. The totals thus found may be written $\boldsymbol{\Delta}\mathbf{y}_4 = \boldsymbol{\Delta}\boldsymbol{\phi}_0\mathbf{y} = \mathbf{Q}_0$. The values to be subtracted in the sweep may now be found either from the natural contrasts or the basic, i.e.

$$\begin{aligned}
&(\mathbf{u}_2\mathbf{u}_2' + \mathbf{u}_3\mathbf{u}_3' + \mathbf{u}_4\mathbf{u}_4')\boldsymbol{\Delta}\mathbf{y}_4/R \\
&= (\mathbf{s}_2\mathbf{s}_2' + \mathbf{s}_3\mathbf{s}_3' + \mathbf{s}_4\mathbf{s}_4')\boldsymbol{\Delta}\mathbf{y}_4/\varepsilon
\end{aligned}$$

$$= \frac{7}{4}\begin{bmatrix} 3 & -1 & -1 & -1 & 0 \\ -1 & 3 & -1 & -1 & 0 \\ -1 & -1 & 3 & -1 & 0 \\ -1 & -1 & -1 & 3 & 0 \\ 0 & 0 & 0 & 0 & 0 \end{bmatrix}\begin{bmatrix} 15.73 \\ -5.56 \\ -2.99 \\ -7.13 \\ 0.02 \end{bmatrix} = \begin{bmatrix} 110.02 \\ -39.01 \\ -21.02 \\ -50.00 \\ 0.00 \end{bmatrix}.$$

In that way Column (5) is obtained. The block means are now all equal to $0.03/7 = 0.00$, so a further sweeping by the blocking system gives Column (6), which consists entirely of values of 0.00, thus confirming that the error sum of squared deviations for the stratum is zero.

F. Combine the information gained from the analyses in Examples 6C and 6E. *Answer:* There is no problem about $\mathbf{u}_1$ since the only information about it is gained from the intra-block analysis. For $\mathbf{u}_2$, $\mathbf{u}_3$ and $\mathbf{u}_4$, however, there is an estimate from each analysis and an intra-block variance. On account of the lack of error degrees of freedom, however, no such variance can be found inter-block. It is already known that the appropriate $\mathbf{H}$ in Section 4.8, i.e.

$$\mathbf{H} = \mathbf{u}_2\mathbf{u}_2' + \mathbf{u}_3\mathbf{u}_3' + \mathbf{u}_4\mathbf{u}_4' = \frac{1}{4}\begin{bmatrix} 3 & -1 & -1 & -1 & 0 \\ -1 & 3 & -1 & -1 & 0 \\ -1 & -1 & 3 & -1 & 0 \\ -1 & -1 & -1 & 3 & 0 \\ 0 & 0 & 0 & 0 & 0 \end{bmatrix}.$$

Hence, $\mathbf{h} = \mathbf{H}\boldsymbol{\Delta}\boldsymbol{\phi}_i\mathbf{y}/R$ will give estimates of the treatment parameters that will at least suffice to estimate contrasts and it will be recalled that $\boldsymbol{\Delta}\boldsymbol{\phi}_i\mathbf{y}$ has been found both intra-block and inter-block. For the intra-block information

$$\mathbf{h} = \tfrac{7}{34}\mathbf{H}\begin{bmatrix} -12.28 \\ -83.99 \\ +16.44 \\ -183.42 \\ +263.30 \end{bmatrix} = \begin{bmatrix} +11.02 \\ -3.74 \\ +16.93 \\ -24.21 \\ 0.00 \end{bmatrix}.$$

This vector is the one used to sweep Column (5) in Table 6C to form Column (7). It would have been used if $\mathbf{u}_2$, $\mathbf{u}_3$ and $\mathbf{u}_4$ had been swept before $\mathbf{u}_1$. Similarly, for the inter-block information

$$\mathbf{h}_0 = 7\mathbf{H}\begin{bmatrix} 15.73 \\ -5.56 \\ -2.99 \\ -7.13 \\ 0.02 \end{bmatrix} = \begin{bmatrix} 110.02 \\ -39.01 \\ -21.02 \\ -50.00 \\ 0.00 \end{bmatrix}.$$

That vector is familiar from Example 6D, where in fact $\mathbf{u}_2$, $\mathbf{u}_3$ and $\mathbf{u}_4$ were the first to be swept. Treatment parameters having been evaluated from each stratum, the task is to combine the two estimates. To do that they will be given weights respectively of w and $(1-w)$, so that the final estimate is to be $w\mathbf{h} + (1-w)\mathbf{h}_0$. Although direct solutions are possible in certain simple cases, it is mostly necessary to iterate to find the weights and that approach will be followed here.

Since there are three degrees of freedom inter-block and 24 intra-block an initial guess for w might be $\frac{24}{27}$ or about 0.90. In general, however, the intra-block analysis gives a lower error variance and is therefore more informative, so w will be given a starting value of 0.95. The error degrees of freedom in the two strata will accordingly be amended

$$20 + 3(0.05) = 20.15$$
$$0 + 3(0.95) = \ \ 2.85.$$

The figures of 20 and 0 represent the error degrees of freedom for the two strata when three degrees of freedom for treatment have been removed from each. With a common estimate of the treatment effects, the degrees of freedom should be removed only once. (Here, there is only one such effect to be considered but in more complicated cases there could be several.)

Turning to (6.4.3) the next task is to find the sums of squared deviations appropriate to the additional degrees of freedom. In this instance

$$\mathbf{L} = \frac{\mathbf{Q}}{R} - \left(0.95\frac{\mathbf{Q}}{R} + 0.05\frac{\mathbf{Q}_0}{R_0}\right)$$

$$\mathbf{L}_0 = \frac{\mathbf{Q}_0}{R_0} - \left(0.95\frac{\mathbf{Q}}{R} + 0.05\frac{\mathbf{Q}_0}{R_0}\right).$$

It will be seen that $0.95\mathbf{L} = -0.05\mathbf{L}_0$. More specifically

$$\frac{\mathbf{Q}}{R} = \frac{7}{34}\begin{bmatrix} -12.28 \\ -83.99 \\ +16.44 \\ -183.42 \\ +263.30 \end{bmatrix} = \begin{bmatrix} -2.53 \\ -17.29 \\ +3.38 \\ -37.76 \\ +54.21 \end{bmatrix}, \quad \frac{\mathbf{Q}_0}{R_0} = 7\begin{bmatrix} +15.73 \\ -5.56 \\ -2.99 \\ -7.13 \\ +0.02 \end{bmatrix} = \begin{bmatrix} +110.02 \\ -39.01 \\ -21.02 \\ -50.00 \\ +0.05 \end{bmatrix}.$$

In the case of $\mathbf{Q}_0/R_0$, the elements of which should sum to zero, the origin has been changed to make that condition hold, the discrepancy clearly being due to cumulative rounding errors. Hence

$$\frac{0.95\mathbf{Q}}{R} + \frac{0.05\mathbf{Q}_0}{R_0} = \begin{bmatrix} +3.10 \\ -18.38 \\ +2.16 \\ -38.37 \\ +51.50 \end{bmatrix}, \quad \mathbf{L} = \begin{bmatrix} -5.63 \\ +1.09 \\ +1.22 \\ +0.61 \\ +2.71 \end{bmatrix}, \quad \mathbf{L}_0 = \begin{bmatrix} +106.92 \\ -20.63 \\ -23.18 \\ -11.63 \\ -51.45 \end{bmatrix}.$$

It will be seen that the vectors $\mathbf{L}$ and $\mathbf{L}_0$ measure the extent to which treatment parameters, as estimated from the separate analyses, differ from the combined estimate based on the weights provisionally adopted.

From (6.4.3) the two error sums of squared deviations should be augmented respectively by

$$0.95\mathbf{L}'\mathbf{HL} \qquad \text{and} \qquad 0.05\mathbf{L}_0'\mathbf{HL}_0$$

respectively, i.e. by 35 and 662. That makes the two variances

$$(3460+35)/20.15 = 173.4 = \sigma^2$$
$$(0+662)/2.85 = 232.3 = \sigma_0^2.$$

Weights are being sought that will make

$$\frac{w}{w_0} = \frac{R\sigma_0^2}{R_0\sigma^2} = 34\frac{\sigma_0^2}{\sigma^2} = 45.55.$$

That is to say, in the initial choice of w too much importance was ascribed to the inter-block analysis. Better estimates of w and $w_0 = (1-w)$ would have been 0.979 and 0.021. Further cycles would find the optimal values.

In this instance it is possible to find a direct solution quite readily. Thus, the amended degrees of freedom are $(23-3w)$ and $3w$ respectively. To find the augmentations of the sums of squared deviations it will be convenient to write $\mathbf{P} = \mathbf{Q}/R - \mathbf{Q}_0/R_0$, which makes $\mathbf{L} = (1-w)\mathbf{P}$ and $\mathbf{L}_0 = -w\mathbf{P}$. Hence the respective augmentations are

$$w(1-w)^2\mathbf{P}'\mathbf{HP} = 13\,150w(1-w)^2 \qquad \text{and}$$
$$w^2(1-w)\mathbf{P}'\mathbf{HP} = 13\,150w^2(1-w).$$

Accordingly

$$\sigma^2 = \frac{3460 + 13\,150w(1-w)^2}{23-3w}$$

$$\sigma_0^2 = \frac{13\,150w^2(1-w)}{3w}.$$

Writing $w/w_0 = 34\sigma_0^2/\sigma^2$ as before, it appears that

$$\frac{w}{1-w} = \frac{447\,100w^2(1-w)(23-3w)}{3w[3460+13\,150w(1-w)^2]},$$

so the desired value of w is 0.966.

Once the weights are determined, there is no difficulty about finding a combined estimate and its associated variances, because information is obtained independently from the various strata.

G. If there are six rows, six columns and four treatments and if all contrasts are of equal importance, Design α has been suggested (Pearce and Taylor, 1948). However, a computer suggests Design β (Jones, 1979). Evaluate and compare the two designs.

(α)

C	D	D	C	B	A
C	A	C	B	B	D
B	A	A	C	C	D
B	D	B	A	A	C
D	B	B	D	A	C
A	C	D	A	D	B

(β)

A	B	C	C	D	D
B	A	A	B	D	C
B	A	D	D	B	C
C	D	B	A	A	B
D	C	B	C	A	A
C	D	D	B	C	A

Answer: The two designs are for practical purposes equivalent because they yield the same amalgamated design. Most would prefer α on account of its having intelligible components, but there is nothing wrong with β.

H. Design α in the above example has in fact been used with a factorial set of treatments (Pearce, 1963). The enquiry concerned two sources of strawberry plant (O, Virus free; V, Virus infected) and two levels of fertilization (O, No added nitrogen; N, added nitrogen). The object was to find out if virus infected plants need as much nitrogen as those that are virus free, i.e. the chief interest lay in the interaction. Writing A = O, B = N, C = V, D = NV, and randomizing gave what was wanted. Later the author wondered (Pearce, 1975) why he had not used the following:

(γ)

O	O	VN	N	V	N
V	VN	N	V	VN	O
VN	V	O	VN	N	V
N	O	V	VN	O	N
V	N	N	O	VN	O
VN	N	O	V	V	VN

In that way the chief loss of information would have fallen on the main effect of virus. There is, however, a further possibility in Design δ, namely

(δ)						
	O	N	V	N	VN	VN
	O	V	N	O	VN	V
	VN	O	V	V	N	O
	N	O	O	VN	V	V
	V	N	VN	N	O	VN
	N	VN	VN	O	V	N

In each row and column there are three plots of either O or VN and three of either N or V, so the interaction is estimated with full efficiency in both components and consequently in the design as a whole. Choose between Designs α, γ and δ.

Answer: It will be helpful to work out a generalized inverse for each design.

$$\mathbf{\Omega}_\alpha = \frac{1}{450}\begin{bmatrix} 53 & -1 & -1 & -1 \\ -1 & 53 & -1 & -1 \\ -1 & -1 & 53 & -1 \\ -1 & -1 & -1 & 53 \end{bmatrix},$$

$$\mathbf{\Omega}_\gamma = \frac{1}{126}\begin{bmatrix} 15 & 1 & -1 & -1 \\ 1 & 15 & -1 & -1 \\ -1 & -1 & 15 & 1 \\ -1 & -1 & 1 & 15 \end{bmatrix},$$

and

$$\mathbf{\Omega}_\delta = \frac{1}{144}\begin{bmatrix} 17 & 0 & 0 & -1 \\ 0 & 17 & -1 & 0 \\ 0 & -1 & 17 & 0 \\ -1 & 0 & 0 & 17 \end{bmatrix}.$$

If the sole purpose is to study the interaction, i.e. $\mathbf{c} = (1, -1, -1, 1)'$, then Design δ cannot be bettered, though Design γ is as good. It will be seen that $\mathbf{c}'\mathbf{\Omega}_\alpha\mathbf{c} = \frac{12}{25}$, $\mathbf{c}'\mathbf{\Omega}_\gamma\mathbf{c} = \mathbf{c}'\mathbf{\Omega}_\delta\mathbf{c} = \frac{4}{9}$, the latter being the value for an orthogonal design. However, once an interaction has been established questions will be asked about the particular effects of nitrogen, i.e. those for the two virus statuses considered separately, namely, $(-1, 1, 0, 0)'$ and $(0, 0, -1, 1)$. For Design α $\mathbf{c}'\mathbf{\Omega}\mathbf{c}$ is now $\frac{6}{25}$, for Design γ it is $\frac{2}{9}$, and for Design δ $\frac{17}{72}$, so altogether Design γ should be preferred.

It may be asked why the particular effects of nitrogen have been considered and not those of virus. The answer lies in the objects of the experiment and the preconceptions of the era. The destruction of virus by heat treatment was then a new thing and the first virus-free strawberry strains had been greeted with enthusiasm. Early trials, however, had shown that they grew excessively and cropped rather poorly, which is what can happen when strawberry plants are given too much nitrogen. It was then natural to ask about the comparative response to nitrogen of plants of differing virus status. Despite the novelty of the clean material, no one at that stage was asking about the effects of virus infection.

I. Consider the following row-and-column design (Pearce, 1975)

C	E	A	E	D	B	A	D
A	C	E	D	B	A	D	E
E	B	D	D	A	C	E	A
B	A	E	A	C	D	D	E

Evaluate $\mathbf{\Omega}$, as far as possible from the combinatorial properties.

Answer: It should be recalled that the loss of information in a row-and-column design equals the sum of losses of information in the two components (6.6.5).

In any row or column if B occurs, so does C. The two treatments are therefore proportionate in both components, so there is no information loss in either; consequently their contrast is of full efficiency in the design as a whole. Further, D and E are proportionate in rows; any information loss will therefore derive from the columns. Here they are equivalent. The matrix $\mathbf{X}$ (4.5.2) equals

$$\begin{bmatrix} 5 & -2 \\ -2 & 5 \end{bmatrix} + \tfrac{1}{32}\begin{bmatrix} 64 & 64 \\ 64 & 64 \end{bmatrix} = \begin{bmatrix} a & b \\ b & a \end{bmatrix},$$

so the variance of $(*\mathrm{D} - *\mathrm{E})$ is $2\sigma^2/7$. Further, A occurs once in each row and twice in each column. It is therefore orthogonal to both blocking systems, so off-diagonal elements of $\mathbf{\Omega}^{-1}$ that relate to it equal zero. Finally, B and C are concordant, as are D and E. When the pairs are merged, it appears that BC and DE are both disposed orthogonally to rows. Consequently any information loss must come from columns, where the three treatments (A, BC and DE) give a design such that

$$\mathbf{\Omega}^{-1} = \begin{bmatrix} 8 & 0 & 0 \\ 0 & 6 & 2 \\ 0 & 2 & 14 \end{bmatrix}, \qquad \mathbf{\Omega} = \tfrac{1}{40}\begin{bmatrix} 5 & 0 & 0 \\ 0 & 7 & -1 \\ 0 & -1 & 3 \end{bmatrix}.$$

(That is the one matrix inversion that has been needed.) Hence, any off-diagonal terms in $\mathbf{\Omega}$ for the row-and-column design that relate to B or C on the one hand and D or E on the other must also equal $-\frac{1}{40}$ from (5.6.6).

It is now possible to construct $\mathbf{\Omega}$ for the complete design from what is known. On account of symmetry of B and C and of D and E,

$$\mathbf{\Omega} = \begin{bmatrix} a & b & b & c & c \\ b & d & e & f & f \\ b & e & d & f & f \\ c & f & f & g & h \\ c & f & f & h & g \end{bmatrix}.$$

From the orthogonality of A it follows that $b = c = 0$. Since $\mathbf{r} = 8\mathbf{1}_v$, $a = \frac{1}{8}$. Also $f = -\frac{1}{40}$ and $2(d-e) = \frac{1}{4}+\frac{1}{4} = \frac{1}{2}$ on account of the contrast of B and C being estimated with full efficiency. Again invoking the relationship $\mathbf{\Omega r} = \mathbf{1}_v$, $4d+4e+16f = 1$, so $d+e = \frac{3}{20}$. Similarly $g-h = \frac{1}{7}$ and $g+h = \frac{3}{20}$. Accordingly

$$\mathbf{\Omega} = \frac{1}{280}\begin{bmatrix} 35 & 0 & 0 & 0 & 0 \\ 0 & 84 & 14 & -7 & -7 \\ 0 & 14 & 84 & -7 & -7 \\ 0 & -7 & -7 & 41 & 1 \\ 0 & -7 & -7 & 1 & 41 \end{bmatrix},$$

which is correct.

Chapter 7

The spoilt experiment

7.1 MISHAPS, BLUNDERS AND DISASTERS

Nothing shows the extent of a writer's practical experience better than his attitude to accidents. Some statistical texts proceed suavely as if nothing ever went wrong. On the other hand, discussions with experimenters often turn to mistakes, whether actual or narrowly averted. The author knows best the conditions of experimentation with fruit trees in England. A trial is being picked and recorded. The pickers, mostly women, have no scientific training and are blindly following instructions; their children are everywhere. Mostly they are country-bred and show respect for growing crops, but the little ones play games to amuse themselves and sometimes innocently wreak havoc. (A game of 'shops' with the picked crop as merchandise can ruin an experiment.) Through it all the recording staff have to keep cool, be aware of everything, solve human problems and at the end produce a clean sheet of data, which is taken indoors and treated as if it came from the layman's idea of a scientific experiment, in which white-coated, aloof beings watch dials. In real conditions mistakes will certainly be made. The difference between a well-conducted institute and the other kind is that one is aware of the dangers and is continually improving its procedures as faults are discovered; the other goes on making the same mistakes unaware that they occur. It is the same with the application of treatments. A visitor will sometimes express wonder at the way an agronomist can take him over the experiment to a desired plot. He does not always consider how good the organization must have been that each plot should have received its correct treatment, especially if those treatments were applied before much was visible above ground. More than that, operations like recording and applying treatments do not always take place in good weather. What happens if it suddenly comes on to rain?

The best policy with regard to mistakes is to prevent them, impossible though that counsel may be. The best insurance that the non-scientific staff shall do things correctly is the presence of the scientific staff, not as supervisors but as fellow workers. Unfortunately a postgraduate qualification in biology implies no skill in the practicalities of running a field experiment so the novice has a lot to learn and should quickly be told so. Quite the worst results are obtained when the scientific staff retreat to their several laboratories and leave the field work to 'experts', who may indeed be very expert in certain matters but who were ignored when the scientific niceties of the experiment were being discussed. Also, the biometricians must know how things are done in the field. For one thing, they have to design experiments in the first place and their proposals should be feasible. Further, it is often the biometrician who knows best what faults have occurred in the past and is accordingly the best person to warn the new agronomist of what can go wrong. A few cautionary words in handing over the plan of an experiment or a sampling scheme can work wonders.

In this chapter two sorts of error will be considered. First, there are those that arise from the wrong application of treatments. They can be very vexatious and can call for a lot of labour, but to anyone who knows the general theory of experimental design they are rarely disastrous. When someone has come white-faced into his office to report such an error, the author has recalled some advice given to medical practitioners in a 15th century manuscript. On first seeing the patient, say 'His condition is indeed grave but with skilful treatment he may yet recover'. (As a matter of fact, nothing raises the biometrician more in the esteem of experimenters than an ability to cope with blunders. It should always be made clear, however, that the resulting analysis of data is only second-best, otherwise faults will recur, the impression having been given that they are of no consequence.) The other kind of error concerns faulty data. Here there are no rules. Once data are suspect, little credence can or should be attached to results based upon them. Certain standard cases will be dealt with, e.g. missing data and mixed-up data, but even they require caution.

7.2 ERRORS IN THE APPLICATION OF TREATMENTS

Errors will sometimes be made in the application of treatments and will need to be dealt with. If it is just a matter of two plots with interchanged treatments, or of one treatment having an additional plot while another has lost one, things are fairly simple. The trouble is that treatments often involve several stages and a fault at one of them may produce an entirely new treatment that has no place in the experimental scheme at all. To take an example, it may be intended to use four treatments, O, 2E, 2L, EL, in which, respectively, no fertilizer is applied, a full dose is applied early, a full dose is applied late and half is applied on each occasion. To consider now eight plots in two of the blocks, the randomization may have given

| O EL 2L 2E | 2E EL O 2L |.

If someone makes mistakes at the early application and produces

| O 2E O E | E E O O |,

there is still time to restore the position in the left-hand block, albeit with a new randomization. (If the mistake genuinely occurred at random, the new is as good as the old.) The right-hand block, however, is more difficult, but having two plots of EL and none of 2E, though awkward, is scarcely a disaster. In short, a hurried replanning is called for, because

| O 2E 2L EL | EL EL O 2L |,

which uses only the intended treatments, is better than

| O 2EL 2L E | E EL O 2L |,

the latter being the result of leaving the fertilizer schedule unaltered. However, even if the mistakes at the first stage were discovered too late for effective modification, perhaps as a result of efforts at concealment, it might still be possible to use most of the data. In the example above, the single plot of 2EL would have to be discarded, but the two plots of E could be retained. It is true that the additional treatment parameter would not enter into any contrast of interest, but the two plots would help in the estimation of the error variance.

In general, errors in the application of treatments are less serious with simple designs. As appeared in Sections 4.2 and 4.4, if a sub-set of treatments is orthogonal to blocks or proportionate, contrasts between them will be estimated with full efficiency no matter what mistakes were made in applying the other treatments. Also, if the omission of a phase in applying treatments leads to merging, then as Section 4.3 shows, the effects on the remaining contrasts of interest are possibly beneficial and never detrimental. All that, however, can be cold comfort.

Usually the problem does not come in finding the error sum of squared deviations. It lies rather in partitioning the sum of squares for treatments, because general balance, supposing that to have been introduced, is likely to have been destroyed. As a consequence interactions are no longer estimated independently of main effects, which are no longer estimated independently of one another and so on. It is the same with other partitions. The loss of precision can be serious too, but usually that is of less importance.

7.3 MISSING DATA

Of all the things that can go wrong in the conduct of an experiment it is loss of data that has received the most attention. In so far as the fault occurs fairly often, the vastness of the literature about it is justified. There is, however, need for some care in applying the various techniques that have been suggested.

First of all, the data may have been lost because the treatment itself is a poor one. Failure to control weeds, for example, may lead to little growth and little crop; it may lead to none at all on some plots, which are thereupon marked down as 'missing'. They are nothing of the kind; they are present but not yielding. Their data should therefore be recorded as zero and analysis should proceed on that basis. (If it is objected that such a body of data scarcely fulfils the requirements of a good variate, that is an argument for excluding the treatment altogether. After all, it has produced some infertile plots and therefore is not to be recommended.) There can be no justification for the evident absurdity of replacing the zeroes with positive values on the grounds of better performance elsewhere in the experimental area. (Incidentally, field recorders must be taught never to record yields with a dash. They must put either 0 for zero or M for missing.)

What then does constitute a missing plot? Statistically the term should be confined to one that has perforce been excluded for reasons that have nothing to do with the experiment. However, that definition has to be used with discretion because there are too many borderline cases. Some mishaps are clearly within it. If a vehicle runs off the road and ploughs into the experiment, the damaged plot may fairly be excluded, but what if animals break through the fence and start eating? Their preference may well depend upon the treatment. Perhaps the best procedure is to exclude the plot on the grounds that invading animals are outside the scope of the experiment. But are they? If fields are ordinarily unfenced and marauding animals are an accepted hazard, it can fairly be held against a treatment that it attracts damage. (No one questions such an approach if the marauders are insects, but elephants for some reason are considered to be different.) The author recalls a fine series of maize fertilizer trials spread over the islands of the West Indies. Several came to grief, notably three of them that suffered from parasites of various kinds. In the first, goats got in and ate a corner plot. Later the herd divided, one group working along each side but doing ever less damage as they went. In the second, monkeys settled on a plot in the middle and played riotous games as well as eating cobs. The third adjoined a cricket field and suffered damage from spectators who jumped over the fence and helped themselves along that side. (In the last experiment, fortunately, recorders had notes of the number of cobs awaiting harvest.) The writer has no perfect solution for such mishaps; he merely comments that missing plot technique as generally understood is expected to cope with a rather wide range of contingencies.

However, once it has been decided that certain plots should be excluded there is a bewildering range of techniques available. Broadly they fall into three groups:

(1) Accepting the data as they stand without reference to what was intended at the inception of the experiment.
(2) Calculating 'missing plot values' to fill the gaps left by the missing data and proceeding as if nothing had happened.
(3) Replacing the missing values with approximate values and using the analysis of covariance to make any necessary adjustments.

All have in fact much in common. However, the first is the least contentious and may therefore be regarded as the norm. If it is adopted there is not really anything to add to what has already been said. For block designs the solution in Chapter 3 is completely general and applies to the achieved design as to any other. It is nevertheless instructive to see what does happen if an arbitrary plot is excluded. The following demonstration was presented by Burns *et al.* (1978).

Let $\mathbf{d}$ and $\boldsymbol{\delta}$ be two matrices corresponding to $\mathbf{D}$ and $\boldsymbol{\Delta}$ respectively. They indicate how the missing plot stands with regard to blocks and treatments. Let

$$\mathbf{D}_0 = \mathbf{D} - \mathbf{d}, \quad \boldsymbol{\Delta}_0 = \boldsymbol{\Delta} - \boldsymbol{\delta} \tag{7.3.1}$$

Here the suffix 0 indicates the achieved design. Neither $\mathbf{D}_0$ nor $\boldsymbol{\Delta}_0$ is a true design matrix because each has a void column corresponding to the missing plot, but it is still permissible to write

$$\mathbf{k}_0^\delta = \mathbf{D}_0\mathbf{D}_0', \qquad \mathbf{r}_0^\delta = \mathbf{\Delta}_0\mathbf{\Delta}_0', \qquad \mathbf{N}_0 = \mathbf{\Delta}_0\mathbf{D}_0'$$

analogous to (3.1.2). Writing $\mathbf{V} = \boldsymbol{\delta}\mathbf{d}'$, it will be more convenient to use the forms

$$\mathbf{k}_0^\delta = \mathbf{k}^\delta - \mathbf{V}'\mathbf{V}, \qquad \mathbf{r}_0^\delta = \mathbf{r}^\delta - \mathbf{V}\mathbf{V}', \qquad \mathbf{N}_0 = \mathbf{N} - \mathbf{V}. \tag{7.3.2}$$

Further

$$\mathbf{k}_0^{-\delta} = \mathbf{k}^{-\delta} + \frac{\mathbf{V}'\mathbf{V}}{k_j(k_j-1)},$$

where k_j is the intended number of plots in Block j, i.e. the one that is defective. It follows that

$$\mathbf{C}_0 = \mathbf{C} - \frac{\mathbf{E}\mathbf{E}'}{k_j(k_j-1)}, \tag{7.3.3}$$

where

$$\mathbf{E} = (\mathbf{k}_j\mathbf{V} - \mathbf{N})\mathbf{V}'$$

because $\mathbf{V}'\mathbf{V}\mathbf{V}' = \mathbf{V}'$ and $\mathbf{V}\mathbf{V}'\mathbf{V} = \mathbf{V}$.

Examination of $\mathbf{E}$ shows that it has a row and a column for each treatment. Further it is null apart from the column that represents the treatment of the lost plot; that holds the contrast between the lost treatment and the others applied in the same block. It will be more convenient to write that contrast as $\mathbf{e}$, the divisor $k_j(k_j-1)$ serving to standardize $\mathbf{e}'\mathbf{e}$. Expansion (7.3.3) can then be written

$$\mathbf{C}_0 = \mathbf{C} - \mathbf{e}\mathbf{e}' \tag{7.3.4}$$

It follows that the effective replication of a contrast, $\mathbf{c}$, is then reduced by ρ^2, where ρ is the correlation coefficient between the elements of $\mathbf{c}$ and those of $\mathbf{e}$, i.e.

$$\rho^2 = (\mathbf{c}'\mathbf{e})^2/(\mathbf{c}'\mathbf{c}\mathbf{e}'\mathbf{e}) = (\mathbf{c}'\mathbf{e})^2.$$

To take an example, suppose that the intended design was in balanced incomplete blocks, thus:

Block I	C	D	E	F		Block V	A	B	E	F
II	B	D	F	G		VI	A	D	E	G
III	B	C	E	G		VII	A	B	C	D
IV	A	C	F	G						

Then all contrasts are estimated with the same effective replication, namely $3\frac{1}{2}$. Suppose, however, that damage occurs to the plot in Block VII with Treatment A, and it has to be excluded. For the achieved design

$$\mathbf{C}_0 = \tfrac{1}{12}\begin{bmatrix} 27 & -3 & -3 & -3 & -6 & -6 & -6 \\ -3 & 35 & -7 & -7 & -6 & -6 & -6 \\ -3 & -7 & 35 & -7 & -6 & -6 & -6 \\ -3 & -7 & -7 & 35 & -6 & -6 & -6 \\ -6 & -6 & -6 & -6 & 36 & -6 & -6 \\ -6 & -6 & -6 & -6 & -6 & 36 & -6 \\ -6 & -6 & -6 & -6 & -6 & -6 & 36 \end{bmatrix}.$$

The contrast immediately affected by the loss is

$$\mathbf{e} = (3, -1, -1, -1, 0, 0, 0)'/\sqrt{12}$$

and its effective replication, $\mathbf{e}'\mathbf{C}_0\mathbf{e}$, is only $2\frac{1}{2}$, ρ being $+1$. If, however, the contrast considered is uncorrelated with $\mathbf{e}$, for example,

$$\mathbf{c}_1 = (0, 1, -2, 1, -1, 2, -1)'/\sqrt{12}$$

the effective replication, $\mathbf{c}_1'\mathbf{C}_0\mathbf{c}_1$ remains $3\frac{1}{2}$ because $\rho_1 = \mathbf{c}_1'\mathbf{e} = 0$. To take an intermediate case, e.g.,

$$\mathbf{c}_2 = (0, 1, 1, 1, -1, -1, -1)'/\sqrt{6},$$

then $\rho_2 = \mathbf{c}_2'\mathbf{e} = -1/\sqrt{8}$ and $\mathbf{c}_2'\mathbf{C}_0\mathbf{c}_2$ equals $3\frac{3}{8}$, as would be expected. (Note, however, that $\mathbf{c}_2$ is not an eigenvector.)

It sometimes happens that lost data leads to a block containing only one plot. Where that happens the best course usually is to omit the block altogether. It will be making no contribution to $\mathbf{C}_0$, so it is at best an irrelevancy and at worst a nuisance.

The above study applies to block designs only. For row-and-column designs the non-orthogonality between rows and columns introduced by defective data is so disturbing in its effects that some other method will almost certainly be better.

With split-plot designs it often happens that the design can be thought of as an aggregated set of experiments, one on each of the main plot treatments. Thus, if there are b blocks, each with m main plots, one for each of the main plot treatments, and if each main plot has s sub-plots, one for each of the sub-treatments, then the error has $m(b-1)(s-1)$ degrees of freedom. Further examination will show that the sum of squared deviations likewise represents the summation of values for m experiments, one on each of the main treatments, having s treatments in b randomized blocks. That being so, if a sub-plot is lost, it can be fitted within its own component, the others being unaffected. The loss of a complete main plot is more awkward but not a disaster to anyone conversant with general theory.

As just mentioned, the estimation of error from missing data is often not enough. Where there is a partition of treatment effects, factorial or otherwise, it is not always easy to deal with the achieved design as it stands. In principle there is nothing against doing so; in practice the fitting of missing values or the use of covariance can be much simpler, though any loss of general balance remains.

7.4 MISSING PLOT VALUES

Quite the oldest way of dealing with incomplete data (Allen and Wishart, 1931; Yates, 1933b) is to fill the gaps with values chosen to minimize the error sum of squared deviations. The following argument has been presented by various writers in various notations, e.g. Wilkinson (1958ab). The form given here follows that of Pearce and Jeffers (1971).

Let $\mathbf{x}$ be the data for the achieved design with zeroes introduced to fill out the vector so that it relates to the intended design and let $\mathbf{z}$ be a column vector of m values to fill the gaps. Then effectively the data will be

$$\mathbf{y} = \mathbf{x} + \mathbf{\Lambda}'\mathbf{z} \tag{7.4.1}$$

where $\mathbf{\Lambda}$ is an $m \times n$ matrix. It is similar to a design matrix, all elements being zero except those that equal one and assign a particular element of $\mathbf{z}$ to a particular missing plot. As a result, the error sum of squared deviations equals

$$\mathbf{y}'\boldsymbol{\psi}\mathbf{y} = (\mathbf{x}' + \mathbf{z}'\mathbf{\Lambda})\boldsymbol{\psi}(\mathbf{x} + \mathbf{\Lambda}'\mathbf{z}), \tag{7.4.2}$$

where $\boldsymbol{\psi}$ is the matrix required to give the estimated residuals from the data, i.e.

$$\hat{\boldsymbol{\eta}} = \boldsymbol{\psi}\mathbf{y}. \tag{7.4.3}$$

Expressions for $\boldsymbol{\psi}$ have been given previously for various designs at (2.1.3), (3.3.7) and (6.6.8). Minimizing (7.4.2) with respect to $\mathbf{z}$ gives

$$\hat{\mathbf{z}} = -(\mathbf{\Lambda}\boldsymbol{\psi}\mathbf{\Lambda}')^{-1}(\mathbf{\Lambda}\boldsymbol{\psi}\mathbf{x}) = -\mathbf{V}^{-1}\mathbf{U}, \quad \text{say.} \tag{7.4.4}$$

From this point, where $\mathbf{y}$ in (7.4.1) is assigned the values given by (7.4.4) it will be written $\hat{\mathbf{y}}$.

Incidentally, if the missing plots have already been assigned approximate values, $\boldsymbol{\mu}$, (7.4.4) can still be used, because

$$-(\mathbf{\Lambda}\boldsymbol{\psi}\mathbf{\Lambda}')^{-1}\mathbf{\Lambda}\boldsymbol{\psi}(\mathbf{x} + \mathbf{\Lambda}'\boldsymbol{\mu}) = \hat{\mathbf{z}} - \boldsymbol{\mu}. \tag{7.4.5}$$

The task is however much easier if the analysis program calculates estimated residuals explicitly. That happens, for example, with the Kuiper–Corsten iteration (Section 3.5) and the method of sweeping (Section 6.3). The vector, $\mathbf{U}$, in (7.4.4) sets out the residuals that arise when gaps are assigned the value zero. One approach is that of Rubin (1972), who took $\mathbf{x}$ as a vector of genuine data, worked out the estimated residuals, $\boldsymbol{\psi}\mathbf{x}$, and selected those that related to missing plots, thus obtaining $\mathbf{U} = \mathbf{\Lambda}\boldsymbol{\psi}\mathbf{x}$. He then formed m pseudo-variates, one for each missing plot. In any pseudo-variate the element corresponding to the missing plot is set equal to one and all others to zero. (It will be seen that $\mathbf{\Lambda}'$ holds the pseudo-variates in its columns.) He then analysed the pseudo-variate to obtain estimated residuals, $\boldsymbol{\psi}\mathbf{\Lambda}'$, and selected those that related to missing plots, thus finding $\mathbf{\Lambda}\boldsymbol{\psi}\mathbf{\Lambda}'$. An inversion then gave $(\mathbf{\Lambda}\boldsymbol{\psi}\mathbf{\Lambda}')^{-1}$ and hence $\hat{\mathbf{z}}$ from (7.4.4). The method is illustrated in Example 7B.

If the missing plot values are incorrectly estimated as $\hat{\mathbf{z}} + \mathbf{d}$ instead of $\hat{\mathbf{z}}$, the error sum of squared deviations will be

$$\begin{aligned}&(\mathbf{x}' + \hat{\mathbf{z}}'\mathbf{\Lambda} + \mathbf{d}'\mathbf{\Lambda})\boldsymbol{\psi}(\mathbf{x} + \mathbf{\Lambda}'\mathbf{z} + \mathbf{\Lambda}'\mathbf{d})\\ &= [\mathbf{x}'\boldsymbol{\psi}\mathbf{x} - \mathbf{x}'\boldsymbol{\psi}\mathbf{\Lambda}'(\mathbf{\Lambda}\boldsymbol{\psi}\mathbf{\Lambda}')^{-1}\mathbf{\Delta}\boldsymbol{\psi}\mathbf{x}] + \mathbf{d}'\mathbf{\Lambda}\boldsymbol{\psi}\mathbf{\Lambda}'\mathbf{d}.\end{aligned} \tag{7.4.6}$$

The expression in square brackets represents the minimized value when the missing plot values are correctly estimated. The last term shows that errors in estimation lead to the error sum of squared deviations forming a paraboloid when plotted in terms of the discrepancies, $\mathbf{d}$. That leads to the method of finding $\hat{\mathbf{z}}$ set out in Example 7A. It is derived from an approach suggested by Hartley (1956).

There is an interesting possibility here. When chemists are trying to find the best conditions for a reaction they sometimes assume that the response surface of yield plotted against factors they can vary is approximated by a paraboloid and methods have been evolved for finding a maximum, which would apply equally well for a minimum (Myers, 1971). Here the response surface is known to be exactly a paraboloid. It looks as if an opportunity is being missed.

Iterative methods are also available. For example, Healy and Westmacott (1956) suggested starting with guessed values, γ_0, for the missing values and working out the residuals so produced i.e. $\mathbf{\Lambda}'\boldsymbol{\psi}(\mathbf{x}+\mathbf{\Lambda}\gamma_0)$. They found a better approximation, γ_1, by subtracting the residuals from the original guesses, i.e.

$$\gamma_1 = \gamma_0 - \mathbf{\Lambda}'\boldsymbol{\psi}(\mathbf{x}+\mathbf{\Lambda}\gamma_0) \tag{7.4.7}$$

and so on. Their iteration always converges though some have found it slow (Dempster *et al.*, 1977, Discussion). For that reason there has been interest in 'accelerated methods' in which (7.4.7) is replaced by

$$\gamma_{i+1} = \gamma_i - \boldsymbol{\zeta}^{-\delta}\mathbf{\Lambda}'\boldsymbol{\psi}(\mathbf{x}+\mathbf{\Lambda}\gamma_i) \tag{7.4.8}$$

where $\boldsymbol{\zeta}$ is some vector of positive constants. It is possible to approximate $\mathbf{\Lambda}\boldsymbol{\psi}\mathbf{\Lambda}'$ to $\boldsymbol{\zeta}^{\delta}$ in various ways (Pearce and Jeffers, 1971). Mostly the off-diagonal elements will be small anyway, while those in its diagonal may be approximated by

$$(1-1/k_i)(1-\Omega_{jj}) \tag{7.4.9}$$

for a block design, where k_i is the number of plots intended for the block that contains the missing plot and Ω_{jj} is the diagonal element $\mathbf{\Omega}$ corresponding to its treatment. For a row-and-column design the corresponding approximation is

$$\frac{(b-1)(c-1)}{bc}(1-\Omega_{jj}) \tag{7.4.10}$$

where there are b rows and c columns. Such values can be used to give the elements of $\boldsymbol{\zeta}$. A further approximation is to use $1/r_j$ instead of Ω_{jj}. Finally, and probably most usefully, unless blocks are of widely different size or treatments have very different replications, a satisfactory approximation is given by $\boldsymbol{\zeta}^{\delta} = (f/n)\mathbf{I}_m$ where f is the number of error degrees of freedom for the intended design and n is the number of plots intended. (For many designs, notably all equi-replicate orthogonal ones, that expression is not an approximation but exact.) With $\mathbf{\Lambda}\boldsymbol{\psi}\mathbf{\Lambda}$ thus reduced to a diagonal matrix, ζ^{δ}, it is a simple matter to offer a vector, $\mathbf{z}_j$, of approximations to the missing plot values, to form

$$\mathbf{y}_j = \mathbf{x} + \mathbf{\Lambda}'\mathbf{z}_j,$$

to find the estimated residuals that result, namely $\boldsymbol{\psi}\mathbf{y}_j$, and to select those that relate to missing plots, i.e. $\mathbf{\Lambda}\boldsymbol{\psi}\mathbf{y}_j$, to form a new vector of approximations, namely

$$\mathbf{z}_{j+1} = \mathbf{z}_j - \boldsymbol{\zeta}^{-\delta}\mathbf{\Lambda}\boldsymbol{\psi}\mathbf{y}_j. \tag{7.4.11}$$

If $\boldsymbol{\zeta}^{\delta}$ were exactly equal to $\mathbf{\Lambda}\boldsymbol{\psi}\mathbf{\Lambda}'$, the iteration would end after the first cycle, because

$$\Lambda\psi y_2 = \Lambda\psi(x + \Lambda' z_2) = \Lambda\psi(x + \Lambda' z_1 - \Lambda'\zeta^{-\delta}\Lambda\psi y_1)$$
$$= \Lambda\psi(y_1 - \Lambda'\zeta^{-\delta}\Lambda\psi y_1)$$

would be a null-vector. In fact, convergence is usually rapid and depends both upon the adequacy of ζ^{δ} and the initial guess, z_1. A good starting vector can be found by assigning to each missing plot the mean of other plots with the same treatment. The method is illustrated in Example 7C.

The trouble with accelerated methods is that they can in special cases multiply the residuals by so large a factor that the iteration will diverge instead of converge. For example, Jarrett (1978) has pointed out that for a 5 × 5 Graeco-Latin square, n/f equals $\frac{25}{8}$, which is much too large. In a valuable paper Preece (1971) has shown that no element of ζ should exceed $2/\lambda_{max}$, where λ_{max} is the largest eigenvalue of $\Lambda\psi\Lambda'$.

Equation (7.4.11) may thus be used in an iterative context, because the correct missing plot values give estimated residuals equal to zero, i.e. from (7.4.1) and (7.4.4)

$$\Lambda\psi\hat{y} = \Lambda\psi[x - \Lambda'(\Lambda\psi\Lambda')^{-1}\Lambda\psi x] = 0 \tag{7.4.12}$$

That provides a neat way of fitting missing values in a number of cases, as Example 7D shows.

Although the fitting of missing plot values is standard practice for many people, it suffers from the disadvantage of over-estimating the treatment sum of squared deviations. There should be a second minimization like that used to find (2.1.7) and (3.1.6) for comparison with (2.1.4) and (3.3.9) respectively. The values chosen for the missing plots have been those that will minimize $y'\psi y$. In general, they will not minimize $y'\phi y$ also. Hence that quantity and the treatment sum of squared deviations, namely $y'(\phi - \psi)y$, will both be over-estimated. The correct procedure is to evaluate the missing plot values again with the treatment effects merged in error and in that way to find $y'\phi y$ correctly. In fact, that is not often done, though many would have their doubts about accepting as significant a treatment mean square that only just exceeded the critical F-value. (Of course, a small excess over a critical value is not really of much importance in the interpretation of an analysis of variance, but assertions in a report about significance should be made honestly.)

Substituting (7.4.4) into (7.4.2) shows that in terms of x, the error sum of squared deviations is

$$x'\psi[I_n - \Lambda'(\Lambda\psi\Lambda')^{-1}\Lambda]\psi x$$

Writing $U = \Lambda\psi x$ and $V = \Lambda\psi\Lambda'$, that becomes

$$x'\psi x - U'V^{-1}U. \tag{7.4.13}$$

If now treatments are merged with error and a new minimization takes place, that becomes

$$x'\phi x - U_0'V_0^{-1}U_0, \tag{7.4.14}$$

where $U_0 = \Lambda\phi x$ and $V_0 = \Lambda\phi\Lambda'$. If, however, no fresh minimization takes place, the sum of squared deviations for error and treatment together is taken to be $\hat{y}'\phi\hat{y}$, where $\hat{\mathbf{y}} = \mathbf{x} + \boldsymbol{\Lambda}'\hat{\mathbf{z}}$, i.e. $\hat{\mathbf{y}} = \mathbf{x} - \boldsymbol{\Lambda}'\mathbf{V}^{-1}\mathbf{U}$, so it is found as

$$\mathbf{x}'\boldsymbol{\phi}\mathbf{x} - \mathbf{U}'\mathbf{V}^{-1}\mathbf{U}_0 - \mathbf{U}_0'\mathbf{V}^{-1}\mathbf{U} + \mathbf{U}'\mathbf{V}^{-1}\mathbf{V}_0\mathbf{V}^{-1}\mathbf{U}. \tag{7.4.15}$$

The difference between (7.4.15) and (7.4.14) is

$$(\mathbf{U}_0'\mathbf{V}_0^{-1} - \mathbf{U}'\mathbf{V}^{-1})\mathbf{V}_0(\mathbf{V}_0^{-1}\mathbf{U}_0 - \mathbf{V}^{-1}\mathbf{U}). \tag{7.4.16}$$

It will be noted that $-\mathbf{V}_0^{-1}\mathbf{U}_0 = \hat{\mathbf{z}}_0$ and $-\mathbf{V}^{-1}\mathbf{U} = \hat{\mathbf{z}}$ are the two estimates that have been made of the missing plot values. In fact, (7.4.16) can well be written as

$$(\hat{\mathbf{z}}' - \hat{\mathbf{z}}_0')\mathbf{V}_0(\hat{\mathbf{z}} - \hat{\mathbf{z}}_0). \tag{7.4.17}$$

Only if $\hat{\mathbf{z}} = \hat{\mathbf{z}}_0$ will there be no inflation of the treatment sum of squared deviations and that is most unlikely to happen.

7.5 FITTING MISSING VALUES BY A COVARIANCE ADJUSTMENT

This is in fact a very old method (Bartlett, 1937), which appears to have gained in popularity over recent years (Smith, 1981) possibly on account of the proliferation of computer programs that make covariance adjustments without difficulty. First, the gaps in the data are filled in any convenient way, so that (7.4.1) again applies, except that in the present context $\mathbf{z}$ is an arbitrary vector, corresponding to the μ in (7.4.5). Then, pseudo-variates are formed, one for each missing plot, each with n elements. In any pseudo-variate all elements are zero except that for the missing plot corresponding to that pseudo-variate; for that the element equals one. It will be seen that $\boldsymbol{\Lambda}'$ holds those pseudo-variates in its columns. Reading Section 2.3 in the light of Section 3.1, a reasonable model is seen to be

$$\mathbf{y} = \mathbf{D}'\boldsymbol{\beta} + \boldsymbol{\Delta}'\boldsymbol{\tau} + \boldsymbol{\Lambda}'\boldsymbol{\theta} + \boldsymbol{\eta}, \tag{7.5.1}$$

whence it appears that the least squares estimator of $\hat{\boldsymbol{\theta}}$ is

$$(\boldsymbol{\Lambda}\boldsymbol{\psi}\boldsymbol{\Lambda}')^{-1}\boldsymbol{\Lambda}\boldsymbol{\psi}\mathbf{y} = (\boldsymbol{\Lambda}\boldsymbol{\psi}\boldsymbol{\Lambda}')^{-1}\boldsymbol{\Lambda}\boldsymbol{\psi}(\mathbf{x} + \boldsymbol{\Lambda}'\mathbf{z}) = -\hat{\mathbf{z}} + \mathbf{z},$$

where $\hat{\mathbf{z}}$ is the estimate of the missing plot values given in (7.4.4), as modified in (7.4.5). Consequently

$$\mathbf{y} - \boldsymbol{\Lambda}\hat{\boldsymbol{\theta}} = \mathbf{x} + \boldsymbol{\Lambda}\hat{\mathbf{z}} = \hat{\mathbf{y}}, \tag{7.5.2}$$

the same variate that results from fitting missing plot values. The two approaches are therefore essentially the same. John and Prescott (1975) have argued that such methods are in general quicker than iterative ones, though a lot depends on circumstances and the possibility of acceleration.

It will have been noticed that $\hat{\mathbf{y}}$ was formed in (7.5.2) by taking the standard values of all pseudo-variates to be zero, not the mean, which is $1/n$ for each. That need not matter much since only contrasts between adjusted treatment means need to be estimated, though it is more satisfactory if the correct adjustment can be made. Example 7E sets out the calculations.

The advantages of using a covariance approach are, first, that the second minimization, which should really be insisted upon, will with most programs be carried out as a standard procedure. Secondly, that the variances of contrasts will be calculated correctly since the program will allow for variation in concomitant variables. Choice of approach must, however, be determined partly by what is available. If the standard computer program makes provision for only a few concomitant variables but it does calculate residuals explicitly, that could be a good reason for modifying it to obtain missing plot values, either by iteration or by a matrix inversion, whichever may be more convenient. On the other hand, given a program that will make adjustment, if need be, by a sufficient number of independent variables, it needs little modification before it can be used with missing data.

Although the method has been described in a context of block designs, it may be used for any other sort provided only that appropriate values are given to ψ and ϕ.

7.6 VALIDATION PROCEDURES

One use of missing plot values is to test if data that raise doubts are in fact very different from the rest. Here, a note of warning is needed. In any body of data there are bound to be some outliers, just as in a crowd of people some are on the outside. Consequently the removal of outliers from data is always possible and the operation always leaves more to be removed at the next stage if anyone is so minded. There are occasions, admittedly, when a value looks so absurd that no one really believes in it. If moving the decimal point one place brings it into line, most people would be content to make the change, knowing that such a mistake can readily be made, but caution is needed. If the plot was as different as the data make it out to have been, surely someone noticed it and can recall the extraordinary phenomenon. That applies especially to a high value for crop. For the contrary case inspection of the site may show that the plot was stony and the low value probably genuine. At one time 'validation' had a great vogue. Before being analysed data were submitted to a program that eliminated extreme values. That is all very well if there are sound prior reasons for believing that the data must follow some particular distribution, but usually there are not. It is also useful if the validation can be done quickly while there is still time to check questionable values. Here there can be surprises. The writer recalls the only validation program he ever wrote, which had one notable success. It indicated nine data out of about 70 as being different from the rest. A hurried investigation revealed that they were the only ones that were correct, having been recorded first before the weighing apparatus went wrong. With an automated validation procedure they would have been the ones excluded. In any case, to return to the point about prior distributions, extreme values can be genuine and, if so, are of especial interest. In fact, nothing distinguishes a mathematician from a biologist as much as his reaction to an unexpected result. If, for example, an unsprayed plot nevertheless shows no signs of disease, the pathologist will want to look at it to find an explanation or a hint for further study. Is it sheltered? or exposed? on a dry patch? or a wet one? The mathematician regards it only as a nuisance and 'validates' it out of existence. (What would happen, one may wonder, if a study were made of the migration of insects and gave rise to nightly figures of captures on the wing like

1 0 2 1 9714 2 0 1?

A good validation program would establish beyond doubt that the species in fact never migrated!)

All that is an extreme reaction to an extreme position. Sensible people use validation programs not to reject data but to point out that some are unusual and need consideration, which can be valuable, whether it does lead to rejection or provokes effective thinking. One point needs to be noted. Novices are sometimes told to look through the data from an experiment and to check extreme values, but that is a mistake. One treatment may give generally high values, the highest but not the lowest being called into question and possibly amended. It is the same with another treatment that gives low values, the lowest stand the risk of being rejected for one reason or another, but not the others. In fact, the method results in treatment means, which may genuinely be different, being collapsed towards the general mean. (That, of course, only illustrates the tendency of a validation program to produce data that conform to the preconceptions of the person who wrote it.) The remedy is to look at residuals instead of data. It is quite possible that some quite undistinguished data in the middle of the range are far from their expected values.

The following is a possible approach. If one datum is to be called into question, so must all the others if there is no prior reason for distinguishing them and all are equally open to disturbing factors. Hence, for each in turn it is necessary to take a vector, $\boldsymbol{\lambda}$, like the rows of $\boldsymbol{\Lambda}$ in (7.4.1), i.e., the plot in question—call it the jth—has an element of one, all others being zero. Then its observed value is $\boldsymbol{\lambda}'\mathbf{y}$ with a residual $\rho_j = \boldsymbol{\lambda}'\boldsymbol{\psi}\mathbf{y}$. If, however, the plot had been missing, the vector $\mathbf{x}$ in (7.4.1) would have been $(\mathbf{I}_n - \boldsymbol{\lambda}\boldsymbol{\lambda}')\mathbf{y}$, which from (7.4.4) gives a missing plot value of

$$-(\boldsymbol{\lambda}'\boldsymbol{\psi}\boldsymbol{\lambda})^{-1}\boldsymbol{\lambda}\boldsymbol{\psi}\mathbf{x} = -\rho_j/\psi_{jj} + \boldsymbol{\lambda}'\mathbf{y} \tag{7.6.1}$$

because $\boldsymbol{\lambda}'\boldsymbol{\lambda} = 1$. In short, any datum, j, differs from its missing plot value by ρ_j/ψ_{jj}, where ρ_j is its residual. Since the variance of the residuals is estimated by the error mean square, $\hat{\sigma}^2$, the standard error of that deviation is $\hat{\sigma}/\psi_{jj}$. Between (7.4.9) and (7.4.11) some attention is given to approximations to ψ_{jj}. For many designs all diagonal elements of $\boldsymbol{\psi}$ are in fact f/n, where f is the degrees of freedom for error in the design. There are other cases where the approximation is reasonably good. If not, other expressions can be used. In difficult cases it is always possible to use Rubin's device, and use $\boldsymbol{\lambda}$ as a variate to produce an error sum of squared deviations of $\boldsymbol{\lambda}'\boldsymbol{\psi}\boldsymbol{\lambda} = \psi_{jj}$.

To take the matter further, accepting all data as they stand, the error sum of squared deviations is $f\hat{\sigma}^2$ with f degrees of freedom. If, however, plot j is given its missing plot value, then in (7.4.6) $\mathbf{d}$ is a scalar, ρ_j/ψ_{jj}, and $\boldsymbol{\Lambda}\boldsymbol{\psi}\boldsymbol{\Lambda}' = \psi_{jj}$, so the error sum of squared deviations becomes

$$f\hat{\sigma}^2 - \rho_j^2/\psi_{jj} \qquad \text{with } (f-1) \text{ degrees of freedom.}$$

The significance of the change can therefore be judged by

$$F = \frac{(f-1)\rho_j^2}{\psi_{jj}(f\hat{\sigma}^2 - \rho_j^2)}. \tag{7.6.2}$$

Given a computer program that has analysed the complete data so that $\hat{\sigma}^2$ and all f_j are known, it is easy to test each plot in turn. The difficulty is rather to know what critical value to give F. Clearly it has one and $(f-1)$ degrees of freedom, but what significance level should be used? If the answer is 0.05, then on average one datum in twenty will be called into question. Perhaps that is what the experimenter wants. Among 100 plots he is ready to re-examine five data. Perhaps it is not—he must decide—but once F_c has been chosen such that a plot value will be questioned if the right-hand-side of (7.6.2) exceeds F_c, it may be easier to say that dubious plots are those for which

$$\rho_j^2 > \frac{F_c\psi_{jj}f\hat{\sigma}^2}{F_c\psi_{jj} + f - 1}. \tag{7.6.3}$$

If $\psi_{jj} = f/n$, that becomes

$$\rho_j^2 > \frac{F_c f^2 \hat{\sigma}^2}{F_c f + n(f-1)}. \tag{7.6.4}$$

However, there are occasions when values are in doubt for some objective reason. For example, a storm after sowing may have left water standing on some plots and have given rise to fears that germination would be affected. It is then reasonable to ask if the inundated plots differ from the rest in later characteristics, such as crop. A possible way of finding out is to calculate an analysis of variance with the suspected values included and so obtaining an error sum of squared deviations of S with f degrees of freedom. If the analysis is repeated with the doubtful values regarded as missing, the new error sum of squares may be S' with $(f-m)$ degrees of freedom, which gives a test of the extent to which the flooded plots differ from the rest, the criterion being

$$F = \frac{S - S'}{S'} \frac{f - m}{m}$$

with m and $(f-m)$ degrees of freedom. An alternative approach is to form a pseudo-variate that has the value one for each plot suspected of damage and zero for the rest. A covariance adjustment on the pseudo-variate will show if the mishaps have had any effect. The difference between the two approaches is clear. Treating data as missing is equivalent to using m pseudo-variates and so allowing each plot to have had its own reaction to the disturbing factor. In the second method, the enquiry concerns the mean effect, the assumption being that all plots have reacted to much the same extent and only one pseudo-variate is used. Each is valid in its own way. If the damage was evenly spread over an area, it may seem more reasonable to test for a general effect, that being the more sensitive method. If, on the other hand, the damage was patchy in its incidence, the first approach may be preferred. Another advantage of the second method lies in its making use of the differences between affected plots. Once it is assumed that they have all been affected equally by the mishap, their data are available for estimating the error variance. If, however, they are believed to have been affected to differing and unknown degrees, they have nothing to impart and need to be excluded altogether.

Such a case, i.e. one in which there is a clear objective reason for questioning some data, is fairly simple to deal with. At the other extreme is the case where there is nothing to be said against the data except that the experimenter dislikes their rejection of his favourite hypothesis. In between there is difficulty and the only fair advice is to act honestly and to be frank in any subsequent report. Broadly speaking there are three possibilities:

(1) *There are good objective reasons for thinking that the data may be wrong and a test suggests that they are.* Here there is no serious problem except perhaps in deciding the significance level at which to reject. If there are really sound reasons for questioning the data, little confirmation may be called for.

(2) *There are good objective reasons for thinking that the data may be wrong but inspection does not suggest that they are different from the rest.* Here it is better to leave them in, possibly noting the circumstances in any report, because no experiment is improved by the loss of data.

(3) *There are no objective grounds for thinking the data to be different, but they certainly appear to be.* As has been said, any body of data has its outliers and nothing is gained by identifying them. Nevertheless, if some plots are obviously aberrant, the correct approach is that of a biologist, who goes and looks at them, not that of the mathematician, who is concerned only to get rid of them. A lot may be found that way. The great Sir Ronald Hatton's watchword was 'Treasure your exceptions'. If some plots behave one way and some another, that may ruin the current experiment but the long-term value can be immense if it leads to a useful line of thought. Incidentally, a skilled manipulator can find grounds for rejecting almost anything. The inspection of aberrant data may well reveal that they came from plots that were indeed different, e.g. they were on a strip of gravel or nearer the river, in which case it would be reasonable to reject them and to avoid similar misjudgements in future, but everyone should beware of facile reasons for disposing of unwelcome data. An experimenter should face his own prejudices and inspect the site knowing that he has them. He should then make all doubtful decisions concerning rejection against himself. Only in that way can he present his eventual conclusions in good conscience.

Covariance also provides a way of dealing with deaths in plots that contain only a few large plants. In extreme cases there could be only one. Clearly the loss of a plant is serious because an appreciable part of the plot will have gone. In the extreme case it will all have gone, which does at least give a relatively simple situation. The complication comes rather from the need to replant and fill the spaces, as would be done in a commercial plantation.

A lot depends upon the supposed cause of the losses. If the experimenter believes that they are an effect of the particular treatment applied, it would not be right to make excuses for it and cover up its bad consequences by statistical adjustments to the data. The only real remedy, though not always a practicable one, is to use larger plots—eight or twelve plants at the least to a plot—and to replant as in commerce. The variate for analysis would then be the total crop etc. for the plot, the replants being included with the rest.

More usually, however, it is believed that the loss occurred on account of something in the plant itself e.g. it was diseased from the start or was damaged at planting. In that case a small plot is all right. The best procedure is to let such losses occur and when they have finished to replant all at the same time, so that the new plants—often called 'recruits'—are comparable among themselves just as the surviving original plants are comparable. It will, of course, take some time for the recruits to settle down and adapt themselves to the conditions of the plot. When they have done so, it is possible to proceed but covariance adjustments are called for. One is almost essential, namely, upon the number of recruits in each plot. Another may well be desirable, namely, upon the number of original plants to have a recruit as neighbour. Obviously such a plant has had an advantage in the reduced competition, first from the gap and then from the smaller neighbour. (If a plant has two recruits as neighbours, it can be given a value of two in the concomitant variable and so on.) Single plant plots raise no special problems. The first concomitant variable then has only two possible values, zero for an original plant and one for a recruit.

7.7 MIXED-UP PLOTS

In the hurry of crop recording it is quite easy for a container to be lost and then, when it is found, for no one to be certain which of two plots it came from. That is to say, the total for the plots is known, but no one can say how to apportion it. A similar problem arises when samples are collected and two labels are lost. By a process of elimination it is possible to say which two plots the unlabelled samples must have come from, but there may be no way of allocating them beyond that. A critic may make noises about 'reprehensible carelessness' or the like, but practical men know that such mistakes do occur, especially in time of crisis, and methods are needed for dealing with the defective data created.

Before looking at the mathematics of the problem it will be as well to consider it in a simple way. The basis must be the missing plot values for the plots affected. Imagine that the two plots are missing and on that basis would be assigned values of 21.4 kgr and 30.7 kgr. Suppose now that the known crops of the two plots are respectively 22.6 kgr and 24.0 kgr with 5.9 kgr undecided. There can be little doubt which plot has lost the unassigned container. (Such an approach will not appeal to the sophisticated but in this instance common sense can be better than algebra.) In cases that are unclear it is feasible to work out three error sum of squared deviations, i.e. with the two plots missing and with the doubtful value assigned to them in both the possible ways. First the data are analysed using missing-plot values to give an error sum of squared deviations, S, with f degrees of freedom. Then, the missing container is assigned to the first plot, while the second is given only its certain crop, which leads to S being inflated to S_1 and f to $(f+2)$. An F-test will show if the two degrees of freedom for the deviations of the assigned values from the missing-plot values are significant. If they are, the doubtful value has been assigned in a way that appears unlikely. Finally the container is assigned instead to the second plot to give an error sum of squared deviations of S_2, leading to a further test to see if that is any better. Such an approach may well resolve the problem.

On the other hand, it may not. Perhaps both allocations look reasonable or perhaps neither does. (In the latter case, is everyone sure that the loose container really does belong to one of the two plots? It would be a mistake to ask the question aggressively, because that might elicit a wrong answer, but it is just possible that it comes from a discard area or the like.) The best procedure in case of genuine doubt is to assign the whole crop for the two plots, which is not in question, to one of the plots and zero to the other, to make a covariance adjustment on a pseudo-variate that has $+1$ for the plot that has been assigned everything and -1 for the other, with zero for the remainder. As a result a quantity θ, being the regression coefficient, will be transferred from the plot with everything to the one with nothing. The advantage over working with two missing plots lies in the use of all the available information, a fact that is reflected in the additional degree of freedom for error. If more plots have been involved in the muddle, more pseudo-variates are needed. Thus, if there are three, all the crop should be assigned to one plot and two pseudo-variates will be needed. In each most elements are zero, but the first has $+2$ for the plot with everything and -1 for the two that have been credited with nothing; it therefore serves to reduce the favoured plot to its true level. The second also contains mostly zeroes, but it has $+1$ and -1 for the two plots with zero crops and therefore serves to apportion their share between them. The method can readily be extended if more plots are involved in the muddle. It is illustrated in Example 7F.

Iterative methods are available for mixed-up plots similar to those used when plots are missing. As Preece and Gower (1974) have pointed out, if values are guessed subject to the limitation that their sum shall be correct, it is possible to find residuals for all the mixed-up plots. If now those residuals are referred to their own mean as origin, their sum will be zero. Consequently, subtracting each one from the guessed value to which it refers will give an improved approximation. The iteration can proceed until convergence is achieved. Further, the process can be accelerated as in (7.4.7), the limits for ζ being as before.

Also, John and Lewis (1976) have noted another parallel to the methods used with missing plots. Given m mixed-up plots they propose using $(m-1)$ pseudo-variates and proceeding as Rubin did for the missing plot situation.

7.8 THE LOSS OF A TREATMENT

Sometimes a treatment fails so badly that it has to be excluded from the experiment altogether. It is absurd to retain its data if they consist only of a string of zeroes. Again, even if failure is not as bad as that, an unsuccessful treatment can still cause trouble by giving variable results, attaining near success in some parts of the field and failing almost totally elsewhere. Although it is usually better to keep an experiment in its intended form there are times when an investigator has to accept changes.

Often little harm is done. If the design is in randomized blocks, or indeed uses any orthogonal block design, the loss of a treatment leads to a reduction in the number of error degrees of freedom. That could be awkward but not necessarily disastrous. The covariance matrix of the parameters for the remaining treatments would be unaltered apart from a reassessment of the error variance. If the omitted treatment was giving erratic results, that could even be an advantage.

In other block designs there can be a loss of information concerning the other treatments and never a gain. That conclusion is fairly obvious but it can be shown formally using the following lemma.

Lemma 7.8.A *If two designs have coefficient matrices, $\mathbf{C}_1$ and $\mathbf{C}_2 = \mathbf{C}_1 - \mathbf{LL}'$, where $\mathbf{L}$ is some matrix with v rows, and if treatment replications are the same in both designs, then any contrast estimated by the second design is estimated as well or better by the first.*

Proof: Because all treatments have the same replications, then, defining $\boldsymbol{\Omega}$ as in (3.5.11) to allow for contrasts confounded in the first design,

$$\boldsymbol{\Omega}_2^{-1} = \boldsymbol{\Omega}_1^{-1} - \mathbf{LL}'.$$

So

$$\boldsymbol{\Omega}_2 = (\mathbf{I}_v + \boldsymbol{\Omega}_1\mathbf{LL}' + \boldsymbol{\Omega}_1\mathbf{LL}'\boldsymbol{\Omega}\mathbf{LL}' + \cdots)\boldsymbol{\Omega}_1.$$

Hence for any contrast, $\mathbf{c}$, $\mathbf{c}'\boldsymbol{\Omega}_1\mathbf{c} \leqslant \mathbf{c}'\boldsymbol{\Omega}_2\mathbf{c}$. One special case may be noted. If the series on the right-hand side diverges, the contrast is confounded in the second design but not in the first, but that is within the terms of the lemma. ■

Applied to a design from which the first treatment is to be removed, let

$$\mathbf{N} = \begin{bmatrix} \boldsymbol{\alpha}' \\ \mathbf{N}_0 \end{bmatrix} \quad \text{and} \quad \mathbf{r} = \mathbf{N}\mathbf{1}_b = \begin{bmatrix} \rho \\ \mathbf{r}_0 \end{bmatrix}, \quad \text{say,}$$

then

$$\mathbf{C} = \begin{bmatrix} \rho - \boldsymbol{\alpha}'\mathbf{k}^{-\delta}\boldsymbol{\alpha} & -\boldsymbol{\alpha}'\mathbf{k}^{-\delta}\mathbf{N}_0' \\ -\mathbf{N}_0\mathbf{k}^{-\delta}\boldsymbol{\alpha} & \mathbf{Z} = \mathbf{r}_0^{\delta} - \mathbf{N}_0\mathbf{k}^{-\delta}\mathbf{N}_0' \end{bmatrix}.$$

From Lemma (4.3.C) it appears that the analogue of $\mathbf{Z}$ in $\mathbf{C}^-$ is a generalized inverse of a matrix, $\mathbf{Z}_1$, which may be written $\mathbf{Z}_1 = \mathbf{Z} - \mathbf{L}_1\mathbf{L}_1'$. If now the first treatment is omitted, $\mathbf{Z}$ becomes

$$\mathbf{Z}_2 = \mathbf{r}_0^{\delta} - \mathbf{N}_0(\mathbf{k} - \boldsymbol{\alpha})^{-\delta}\mathbf{N}_0' = \mathbf{Z} + \mathbf{L}_2\mathbf{L}_2', \quad \text{say.}$$

Hence

$$\mathbf{Z}_2 - (\mathbf{L}_1 \mid \mathbf{L}_2)(\mathbf{L}_1 \mid \mathbf{L}_2)'$$

and the lemma applies, albeit with a small extension that does not affect the argument.

The difficulties are greater with a row-and-column design, because the omission of a treatment renders the rows and columns non-orthogonal. Where the design as a whole is orthogonal, as with a Latin square, that is not too much of a difficulty. Following the notation of Section 6.6, the design becomes of Type T: OO. Dualizing the treatments and the columns, it becomes one of Type O:OX, where X is some non-orthogonal configuration. However, the dualization will not affect the error sum of squared deviations, which can readily be found. Also, since the treatments are orthogonal to both rows and columns, their sum of squared deviations can be found using the summation terms as described in Section 4.1. However, the loss of a treatment from a non-orthogonal row-and-column design can be very awkward. Where some treatments may prove unsuccessful, it is better to be cautious and avoid such designs.

Some have asked what happens in a factorial context if a treatment combination fails. In practice the problem is not of much importance. If a certain level of Factor A, which is ordinarily successful, fails on being combined with a certain level of Factor B, that is an interaction so marked that it scarcely needs further establishment. Also, the existence of an interaction directs attention to the particular effects, which are here unimpaired, so no great difficulties of interpretation arise.

Examples 7

A. Rayner (1969) has given the following data from an experiment to compare six turnip varieties in a Latin square

E, 9.0;	F, 14.5;	D, 20.5;	A, 22.5;	B, 16.0;	C, 6.5;
B, 17.5;	A, 29.5;	E, 12.0;	C, 9.0;	D, 33.0;	F, 12.5;
F, 17.0;	B, 30.0;	C, 13.0;	D, 29.0;	A, 27.0;	E, 12.0;
A, 31.5;	D, 31.5;	F, 24.0;	E, 19.5;	C, 10.5;	B, 21.0;
D, 25.0;	C, 13.0;	B, 31.0;	F, 26.0;	E, 19.5;	A, m_1;
C, 12.2;	E, 13.0;	A, 34.0;	B, 20.0;	F, m_2;	D, m_3.

Data represent fresh weight (root *plus* tops) in pounds per plot (15 feet × 15 feet). The three plots in the corner had been attacked by vandals and on that account were regarded as missing.

Analyse the data using a computer that has a program to analyse complete data but no procedures for estimating missing plot values or residuals and which cannot adjust for concomitant variables.

Answer: The computations will be lengthy but need not involve a lot of programming. They depend upon the passage below (7.4.6). All that is needed is the writing of a few instructions which, in effect, use the main program as a sub-routine.

The method will make use of the fact that if a missing value is replaced by different values, the error sums of squared deviations when graphed against those values will give a parabola. Further, if the values ascribed to a missing plot are $0.9a$, a and $1.1a$ and if they give error sums of squared deviations of S_1, S_2, and S_3, then the parabola that results will have a minimum when the missing plot is given the value

$$a\left[1+\frac{S_1-S_3}{20(S_1-2S_2+S_3)}\right].$$

That can easily be confirmed.

Accordingly, m_1, m_2, and m_3 will each take an initial value equal to the means of other plots with the same treatment, i.e. $m_1 = 28.90$, $m_2 = 18.80$, $m_3 = 27.80$; then the error sum of squared deviations is $243.56 = S_2$. Now, marking down m_1 to $0.9\,(28.90) = 26.01$ but keeping m_2 and m_3 as before, the error sum of squared deviations becomes $254.08 = S_1$, while marking m_1 up to $1.1\,(28.90) = 31.79$ makes it $242.23 = S_3$. Hence the minimum would be found by setting m_1 equal to

$$28.90\left[1+\frac{11.85}{20(9.19)}\right] = 30.76.$$

The change of $(30.76-28.90) = 1.86$ in m_1 is obviously important but, as minimizing values are approached, the question of ending the iteration will arise. If it is agreed that treatment means should be correct to within about one per cent of a standard deviation, that implies limits of about ± 0.03. (The standard error is of the order of $\sqrt{(242/22)} = 3.3$ or rather less.) Since a missing plot value included in a treatment will be divided by six when means are formed, that requires a precision of about ± 0.2 for such values. The iteration can be expected to converge fairly quickly, but it would be wise to take additional cycles until no value had moved by more than ± 0.05 since it was last evaluated. However it is chosen, there should be a quantity, p, that represents acceptable stability in a value and the computer must count how many changes lie within that limit. Initially the count should be zero and it should be increased by one whenever a missing plot value is adjusted by an amount smaller than p, but returned to zero whenever an adjustment exceeds that limit. The iteration can end when the count equals the number of missing plots, i.e., all their last adjustments have been acceptably small.

To continue the calculation, it is now necessary to take $m_1 = 30.76$, $m_3 = 27.80$ and to try values of m_2 of $0.9\,(18.80) = 16.92$, 18.80 and $1.1\,(18.80) = 20.68$. In that way an adjusted value of m_2 will be found of 20.024. A further cycle will amend m_3 to 24.997. The full results as given by the computer are:

m_1	m_2	m_3	*S-S*	*Count*
28.90	18.80	27.80	243.56	0
30.76	18.80	27.80	241.64	0
30.76	20.02	27.80	240.81	0
30.76	20.02	25.00	236.44	0

30.08	20.02	25.00	236.18	0
30.08	19.53	25.00	236.05	0
30.08	19.53	24.76	236.02	0
30.08	19.53	24.76	236.02	1
30.08	19.48	24.76	236.02	2
30.08	19.48	24.75	236.02	3

It cannot be claimed that the method is economical of computer time, but it does save programming labour, which at some institutes is very short, especially if an expert knowledge of statistics is needed as well. Essentially all that is required is an analysis of variance program appropriate to the design. Even if no one understands it very well, there should be no difficulty about turning it into a sub-routine, which is going to be needed in two forms. In the shorter all that is asked of it is the error sum of squared deviations, which is returned to the main program. Hence everything after the derivation of that quantity can be suppressed, as can all print instructions. In the larger version it serves the same purposes as it did as a main program and needs no amendment apart from reducing the number of degrees of freedom for error. The best way of implementing all this will depend upon the computer language and the operating system but there can be few installations where it would be difficult. The main program, which needs to be written after the method of implementation is agreed, can be quite simple. Further, any number of analysis programs can be adapted for use with it.

B. Assuming that estimates of the residuals are readily available, analyse the data of Example 7A.

Comment: Often there is no difficulty about finding estimates of the residuals. There are two obvious cases. (1) The main analysis program uses a method, such as the Kuiper–Corsten iteration or sweeps, that gives residuals explicitly. (2) The design is so simple that a formula for residuals can readily be derived.

Thus, for a Latin square a residual equals (datum) – (the appropriate row mean) – (the appropriate column mean) – (the appropriate treatment mean) + 2(the general mean). That can be derived from (6.1.6), which gives the value of $\boldsymbol{\phi}$, and from (6.6.4) which shows that $\mathbf{r}^{-\delta} = \mathbf{C}^{-}$ for an orthogonal design. Hence $\boldsymbol{\psi}$ follows and so $\boldsymbol{\psi}\mathbf{y}$.

Answer: The obvious approach is that of Rubin. It is described in Section 7.4. First set $m_1 = m_2 = m_3 = 0$, and find residuals for the three missing plots, calling them ρ_1, ρ_2, ρ_3, then

$$\rho_1 = 0 - 19.08 - 8.67 - 24.08 + 36.79 = -15.04$$

$$\rho_2 = 0 - 13.00 - 17.67 - 15.67 + 36.79 = -9.75$$

$$\rho_3 = 0 - 13.20 - 8.67 - 23.17 + 36.79 = -8.25.$$

The next step is to take a pseudo-variate in which the value of the first missing plot is put equal to one and that of all other plots to zero. The residuals now give the first column of $\mathbf{\Lambda\psi\Lambda'}$, i.e.,

$$1-\tfrac{1}{6}-\tfrac{1}{6}-\tfrac{1}{6}+\tfrac{2}{36}=\tfrac{10}{18}$$
$$0-0-0-0+\tfrac{2}{36}=\tfrac{1}{18}$$
$$0-0-\tfrac{1}{6}-0+\tfrac{2}{36}=-\tfrac{2}{18}.$$

Repeating with two similar pseudo-variates for the other missing plots, it appears that

$$\mathbf{V}=\mathbf{\Lambda\psi\Lambda'}=\tfrac{1}{18}\begin{bmatrix}10 & 1 & -2\\ 1 & 10 & -2\\ -2 & -2 & 10\end{bmatrix},\qquad \mathbf{V}^{-1}=\tfrac{1}{17}\begin{bmatrix}32 & -2 & 6\\ -2 & 32 & 6\\ 6 & 6 & 33\end{bmatrix},$$

and

$$\begin{bmatrix}\hat{m}_1\\ \hat{m}_2\\ \hat{m}_3\end{bmatrix}=-\tfrac{1}{17}\begin{bmatrix}32 & -2 & 6\\ -2 & 32 & 6\\ 6 & 6 & 33\end{bmatrix}\begin{bmatrix}-15.04\\ -9.75\\ -8.25\end{bmatrix}=\begin{bmatrix}30.08\\ 19.50\\ 24.76\end{bmatrix}.$$

The results are not exactly as in Example 7A, the rounding errors having made a small difference.

C. Repeat the calculations of Exercise 7B using an approximation to $\mathbf{\Lambda\psi\Lambda'}$ and iterating.
Comment: In most situations, supposing that estimates of residuals are readily available, this is usually quite the easiest method. The relevant equations are (7.4.7) to (7.4.11).
Answer: Various approximations have been suggested in the text, i.e. (7.4.10) and its modification with Ω_{jj} replaced by $1/r_j$ and the use of $(f/n)\mathbf{I}_m$. In the present instance, which is a very simple one, they all amount to setting $\mathbf{\Lambda\psi\Lambda'}$ equal to $\tfrac{5}{9}\mathbf{I}_3$.
The iteration can be started as was the one in Example 7A by taking $m_1 = 28.90$, $m_2 = 18.80$ and $m_3 = 27.80$. That gives corresponding residuals as follows:

$$\rho_1 = 28.90-23.90-18.12-28.90+40.98 = -1.04$$
$$\rho_2 = 18.80-20.97-20.80-18.80+40.98 = -0.79$$
$$\rho_3 = 27.80-20.97-18.12-27.80+40.98 = +1.89.$$

Multiplying each residual by $\tfrac{9}{5}(=1.8)$ and subtracting gives revised missing plot values of

$$m_1 = 28.90-1.8(-1.04) = 30.77$$
$$m_2 = 18.80-1.8(-0.79) = 20.22$$
$$m_3 = 27.80-1.8(+1.89) = 24.40.$$

Using these values the residuals become

$$\rho_1 = 30.77 - 24.20 - 17.86 - 29.21 + 40.98 = +0.48$$
$$\rho_2 = 20.22 - 20.64 - 21.04 - 19.04 + 40.98 = +0.48$$
$$\rho_3 = 24.40 - 20.64 - 17.86 - 27.23 + 40.98 = -0.35.$$

Continuing in that way, the computer gives successive approximations of

28.90	18.80	27.80
30.76	20.21	24.38
29.94	19.34	25.03
30.15	19.55	24.70
30.06	19.46	24.78
30.09	19.49	24.75

The stopping rule was as before in Example 7A. It will be noticed that the computer, which carries more decimal places, has given slightly different values from quite an early stage. However, it is probably going to give a faster convergence. In an iterative method like this, mistakes and rounding error can slow convergence but should not lead to a false result.

D. Calculate the missing plot values for the data in Example 7A by equating residuals to zero.

Answer: The method relies on (7.4.12). It requires the estimation of residuals for the missing plots, equating each to zero and solving the simultaneous equations.

$$m_1 - (114.5 + m_1)/6 - (52.0 + m_1 + m_3)/6 - (144.5 + m_1)/6$$
$$+ (662.2 + m_1 + m_2 + m_3)/18 = 0$$
$$m_2 - (79.2 + m_2 + m_3)/6 - (106.0 + m_2)/6 - (94.0 + m_2)/6$$
$$+ (662.2 + m_1 + m_2 + m_3)/18 = 0$$
$$m_3 - (79.2 + m_2 + m_3)/6 - (52.0 + m_1 + m_3)/6 - (139.0 + m_3)/6$$
$$+ (662.2 + m_1 + m_2 + m_3)/18 = 0.$$

Hence

$$10m_1 + m_2 - 2m_3 = 270.8$$
$$m_1 + 10m_2 - 2m_3 = 175.4$$
$$-2m_1 - 2m_2 + 10m_3 = 148.4,$$

and so $m_1 = 30.1$, $m_2 = 19.5$, $m_3 = 24.8$. In fact, the method is a variant of that in Example 7B. Anyhow, all methods agree to the first place of decimals, which is all the precision required.

By this time there is little doubt what the missing plot values are. It remains only to introduce them into the gaps and to analyse as usual, changing only the degrees of freedom for error. (If further lines are included, say, for rows and columns, which do not really belong to the stratum, the total degrees of freedom will need adjustment as well.) The result is as follows:

Source	*d.f.*	*s-s*	*m-s*	*F*
Rows	5	327.12		
Columns	5	99.21		
Treatments	5	1579.14	315.83	22.75
Error	17	236.02	13.88	
Total	32	2241.49		

Although the above analysis looks satisfactory it is open to some criticism. For one thing the missing plot values have been chosen to minimize the error sum of squared deviations, so that line is all right. However, when the treatments are merged with error with a view to assessing the combination different missing plot values may well be needed, $\boldsymbol{\phi}$ now being used instead of $\boldsymbol{\psi}$, but the matter has not even been glanced at. Further, the covariance matrix of treatment means, i.e. $\mathbf{C}^{-}\sigma^2$, must have been affected by all this fitting of missing values, but that awkward subject also has been avoided.

E. Analyse the data of Example 7A making use of covariance adjustments on pseudo-variates.

Answer: Three pseudo-variates are required. All have 35 zero elements, the other being equal to one and corresponding to a missing plot, i.e., in the first pseudo-variate to Treatment A in Column VI, in the second to Treatment F in Column V and in the third to Treatment D in Column VI. It is immaterial what values are assigned to the missing turnip yields but here they will all be set equal to zero. For one thing, that is the simplest computationally; for another, it will point the similarities to the method set out in Example 7B.

Adjustments by covariance have been presented in Section 2.3. Here, where there are several concomitant variables, expressions (2.3.10) to (2.3.14) apply. First it is necessary to decide what standard values are to be used and the answer is zero for each concomitant variables (i.e. pseudo-variate), i.e., all plots are to be regarded as present. That being so, $\mathbf{Z}$, as defined above (2.3.10), holds the pseudo-variates in its three columns. It is in fact the same as $\boldsymbol{\Lambda}$ in (7.4.1). Hence, in (2.3.10)

$$\mathbf{V} = \mathbf{Z}\boldsymbol{\psi}\mathbf{Z}' = \boldsymbol{\Lambda}\boldsymbol{\psi}\boldsymbol{\Lambda}'$$

$$\mathbf{U} = \mathbf{Z}\boldsymbol{\psi}\mathbf{y} = \boldsymbol{\Lambda}\boldsymbol{\psi}\mathbf{x}.$$

There is now an interesting relationship between (2.3.10) and (7.4.4) from which it appears that $-\hat{\theta}$ is a vector of missing plot values. Reflection shows that there is nothing strange about that. If the pseudo-variate for the jth missing plot has the element 1, which is changed to a standard value of 0, the dependent variate will be increased by θ_j. Since it was initially zero, it will in consequence come to equal θ_j. The two methods are, in fact, identical.

The advantage of using the analysis of covariance lies in the greater facilities available. For example, a computer program will make the second minimization as part of its ordinary procedure. That is to say, it will merge the treatment and error lines and minimize again using $\boldsymbol{\phi}$ instead of $\boldsymbol{\psi}$. Thus, here

$$\mathbf{V}_0 = \boldsymbol{\Lambda\phi\Lambda}' = \tfrac{1}{36}\begin{bmatrix} 25 & 1 & -5 \\ 1 & 25 & -5 \\ -5 & -5 & 25 \end{bmatrix}, \qquad \mathbf{U}_0 = \boldsymbol{\Lambda\phi}\mathbf{x} = \begin{bmatrix} -9.36 \\ -12.48 \\ -3.48 \end{bmatrix},$$

$$\mathbf{V}_0^{-1} = \tfrac{3}{50}\begin{bmatrix} 25 & 0 & 5 \\ 0 & 25 & 5 \\ 5 & 5 & 26 \end{bmatrix}, \qquad \text{so} \qquad -\mathbf{V}_0^{-1}\mathbf{U}_0 = \begin{bmatrix} 15.08 \\ 19.76 \\ 11.98 \end{bmatrix}.$$

Just as residuals could be found for the full design by a simple formula, so they can be found here by one that is even simpler. Ignoring treatments a residual equals

(datum) − (the corresponding row mean)

− (the corresponding column mean) + (the general mean).

It is now possible to adjust the treatment sum of squared deviations found in the last example. The two estimates of missing plot values differ by

$$\begin{bmatrix} 30.08 - 15.08 \\ 19.50 - 19.76 \\ 24.76 - 11.98 \end{bmatrix} = \begin{bmatrix} 15.00 \\ -0.26 \\ 12.78 \end{bmatrix},$$

so the overestimation equals

$$(15.00 \quad -0.26 \quad 12.78)\, \mathbf{V}_0 \begin{bmatrix} 15.00 \\ -0.26 \\ 12.78 \end{bmatrix} = 210.35.$$

Hence the true value for the treatment sum of squared deviations is not 1579.14, as found in Example 7D, but 1361.96, and F should equal 19.62.

However, a much better way of finding the sum of squared deviations for treatments is to work out the analysis of variance ignoring the treatments and using the missing plot values found above, namely, 15.08, 19.76 and 11.98. That gives an error sum of squared deviations of 1596.82, which exceeds the former value by 1360.80. The discrepancy arises from the need to carry many more decimal places if the missing plot values of (7.4.17) are to be used effectively.

Also, the covariance matrix of the treatment means, as given by (2.3.13) is $(\mathbf{r}^{-\delta} + \mathbf{HV}^{-1}\mathbf{H}')\sigma^2$, where $\mathbf{H}$ gives the deviations of the treatments means of the concomitant variables from their standard values. That is quite easy. The standard values are all zero and that is the treatment mean in most cases. The exceptions occur when the variable relates to a missing plot that should have received Treatment j, in which case the treatment mean for the variable is $\frac{1}{6}$, i.e.

$$\mathbf{H} = \frac{1}{6}\begin{bmatrix} 1 & 0 & 0 \\ 0 & 0 & 0 \\ 0 & 0 & 0 \\ 0 & 0 & 1 \\ 0 & 0 & 0 \\ 0 & 1 & 0 \end{bmatrix}, \qquad \mathbf{H}\mathbf{V}^{-1}\mathbf{H}' = \frac{1}{612}\begin{bmatrix} 32 & 0 & 0 & 6 & 0 & -2 \\ 0 & 0 & 0 & 0 & 0 & 0 \\ 0 & 0 & 0 & 0 & 0 & 0 \\ 6 & 0 & 0 & 33 & 0 & 6 \\ 0 & 0 & 0 & 0 & 0 & 0 \\ -2 & 0 & 0 & 6 & 0 & 32 \end{bmatrix},$$

and the covariance matrix of treatment means is

$$\frac{13.88}{612}\begin{bmatrix} 134 & 0 & 0 & 6 & 0 & -2 \\ 0 & 102 & 0 & 0 & 0 & 0 \\ 0 & 0 & 102 & 0 & 0 & 0 \\ 6 & 0 & 0 & 135 & 0 & 6 \\ 0 & 0 & 0 & 0 & 102 & 0 \\ -2 & 0 & 0 & 6 & 0 & 134 \end{bmatrix}.$$

Where losses have occurred variances are much increased. Take the elementary contrast between Treatments A and B, for example. For the full design, which is orthogonal with six-fold replication, the variance would have been $\frac{1}{3}\sigma^2$. Anyone who noted that Treatment A had only five replicates might have guessed $(\frac{1}{5}+\frac{1}{6})\sigma^2 = 0.367\sigma^2$ and he would have done much better. The true figure is $\frac{236}{612}\sigma^2 = 0.386\sigma^2$. For a contrast like that between Treatments A and D the position is worse. A guess of $(\frac{1}{5}+\frac{1}{5})\sigma^2$ would have been fairly good, but the true value is $\frac{257}{612}\sigma^2 = 0.420\sigma^2$.

F. Suppose that in obtaining the data of Example 2A the samples from the first two plots had become mixed together and no one had been sure how to separate them. After such a mishap, instead of being able to assign the datum 9 to the plot of F3 and 12 to the plot of O it would have been possible to have said only that together they added to 21. Analyse the data that would have resulted, using the approach in Section 7.7.
Answer: Attribute the whole, 21, to the first plot and zero to the second, other plots receiving their true values, which are known. The resulting vector will be called **y**. Treatment means will be the same as in Example 2A except that the mean for F3 will now be 12.50 instead of 9.50, while that for O will be 21.13 instead of 22.63. The analysis of variance will read:

Source	*d.f.*	s.s.
Treatments	6	711.35
Error	25	1599.87
Total	31	2311.22

A pseudo-variate, $\mathbf{x}$, is needed to allocate the combined figure of 21 correctly between the two plots from which it came. Its first element should be $+1$, its second -1, and the rest should be zero. Since there is only one concomitant variable, expressions (2.3.3) to (2.3.9) apply. Then

$$U = \mathbf{x}'\boldsymbol{\psi}\mathbf{y} = \mathbf{x}'(\mathbf{I}_n - \boldsymbol{\Delta}'\mathbf{r}^{-\delta}\boldsymbol{\Delta})\mathbf{y}.$$

Since the treatment means, $\mathbf{r}^{-\delta}\boldsymbol{\Delta}\mathbf{y}$, are known, that presents no difficulty, especially as only the first two elements of $\boldsymbol{\psi}\mathbf{y}$ are needed, namely,

$$\begin{aligned} 21 - 12.50 &= \quad 8.50 \\ 0 - 21.13 &= -21.13 \end{aligned}$$

and $U = 29.63$.

It is even easier to find $\mathbf{V} = \mathbf{x}'\boldsymbol{\psi}\mathbf{x}$. The means for $\mathbf{x}$ for Treatments F3 and O are respectively $\frac{1}{4}$ and $-\frac{1}{8}$, so the first two elements of $\boldsymbol{\psi}\mathbf{x}$ are $\frac{3}{8}$ and $-\frac{7}{8}$ and $V = \frac{13}{8}$. Accordingly $\theta = U/V = 18.2$. That quantity should be subtracted from the value of 21 ascribed to the first plot to give 2.8 and added to the value of 0 ascribed to the second plot to give itself, namely 18.2.

After adjusting the analysis of variance by $\mathbf{x}$, the error sum of squared deviations will be reduced by U^2/V ($= 540.27$) with one degree of freedom to give 1059.60. Although it would be possible to use the same allocation between the two plots for the treatment line, overestimation will again result. It is better to repeat with error and treatments merged. Residuals then equal the difference between a plot value and the general mean, because $\boldsymbol{\phi} = \mathbf{I}_n - \mathbf{1}_n\mathbf{1}_n'/n$ and the quantities sought are $U_0 = \mathbf{x}'\boldsymbol{\phi}\mathbf{y}$ and $V_0 = \mathbf{x}'\boldsymbol{\phi}\mathbf{x}$, namely

$$\begin{aligned} U_0 &= (21 - \text{general mean}) - (0 - \text{general mean}) = 21 \\ V_0 &= (1 - 0) - (-1 - 0) = 2 \end{aligned}$$

and the sum of squared deviations for error and treatments together should be reduced by $U_0^2/V_0 = 220.5$ to give 2090.72 with 30 degrees of freedom. The adjusted analysis now reads

Source	*d.f.*	*s.s.*	*m.s.*	*F*
Treatments	6	1031.12	171.85	3.89
Error	24	1059.60	44.15	
Total	30	2090.72		

Also treatment means are O, 23.40; S3, 16.75; F3, 7.95; S6, 18.25; F6, 15.50; S12, 14.25 and F12, 5.75. The vector, $\mathbf{h}$, which shows the shift in $\mathbf{x}$ required for the various treatments is very simple. For O the change is $+\frac{1}{8}$, for F3 it is $-\frac{1}{4}$ and for other treatments it is zero. Accordingly, the covariance matrix of the treatment means, $(\mathbf{r}^{-\delta}+\mathbf{hh}'/V)\sigma^2$, is

$$44.15\begin{bmatrix} 0.141 & 0.000 & -0.031 & 0.000 & 0.000 & 0.000 & 0.000 \\ 0.000 & 0.250 & 0.000 & 0.000 & 0.000 & 0.000 & 0.000 \\ -0.031 & 0.000 & 0.288 & 0.000 & 0.000 & 0.000 & 0.000 \\ 0.000 & 0.000 & 0.000 & 0.250 & 0.000 & 0.000 & 0.000 \\ 0.000 & 0.000 & 0.000 & 0.000 & 0.250 & 0.000 & 0.000 \\ 0.000 & 0.000 & 0.000 & 0.000 & 0.000 & 0.250 & 0.000 \\ 0.000 & 0.000 & 0.000 & 0.000 & 0.000 & 0.000 & 0.250 \end{bmatrix}.$$

The chief sufferer among the contrasts, at might be expected, is that between Treatments O and F3, where the variance is $0.491\sigma^2$ instead of $0.375\sigma^2$. Treatments not involved in the muddle are unaffected.

G. An experiment is designed in v blocks, each of $(v-1)$ plots, there being v treatments. In each block a different treatment is omitted, the others all occurring once. Derive the coefficient matrix, $\mathbf{C}$, from the known properties of a design in randomized complete blocks.

Answer: The chosen design can be regarded as having v missing plots. As a result of the loss of the plot with the first treatment, the coefficient matrix for the complete design, $\mathbf{C}_c$, is such that

$$\mathbf{C}_c = \begin{bmatrix} v-1 & -1 & \dots & -1 \\ -1 & v-1 & \dots & -1 \\ \vdots & \vdots & & \vdots \\ -1 & -1 & \dots & v-1 \end{bmatrix}$$

is reduced by $\mathbf{e}_1\mathbf{e}_1'$ in (7.3.4), where

$$\mathbf{e}_1' = (v-1,\ -1,\ -1, \dots -1)/\sqrt{[v(v-1)]}.$$

Since the other losses occur in different blocks they have independent effects, which accumulate. In total

$$\sum_i (\mathbf{e}_i\mathbf{e}_i') = \mathbf{C}_c/(v-1).$$

Hence for the design under consideration,

$$\mathbf{C} = \mathbf{C}_c - \sum_i (\mathbf{e}_i\mathbf{e}_i') = (v-2)\mathbf{C}_c/(v-1).$$

Hence,

$$\mathbf{C}^+ = (v-1)\mathbf{C}_c^+/(v-2).$$

Suggestion: The same approach can be tried for other designs in balanced incomplete blocks and for many in supplemented balance.

Chapter 8

Interactions and the confounding of interactions

8.1 THE ANATOMY OF INTERACTIONS

Throughout this text interactions have tacitly been found by taking the elements of two orthogonal contrasts and multiplying them each to each. Thus, in (1.6.3) the interaction of

$$(-1, \quad 1, \; -1, \quad 1)'/2$$

and

$$(-1, \; -1, \quad 1, \quad 1)'/2$$

was found to be

$$(1, \; -1, \; -1, \quad 1)'/2. \tag{8.1.1}$$

That exemplifies a general rule. If elements of the first two vectors are multiplied each to each, the result is

$$(\tfrac{1}{4}, \; -\tfrac{1}{4}, \; -\tfrac{1}{4}, \; \tfrac{1}{4})$$

which, after standardization, gives (8.1.1). Leaving aside questions of scaling the interaction of two orthogonal contrasts such that $\mathbf{c}_1'\mathbf{c}_2 = 0$, is thus given by

$$\mathbf{c}_{12} = \mathbf{c}_1^\delta \mathbf{c}_2 = \mathbf{c}_2^\delta \mathbf{c}_1. \tag{8.1.2}$$

The interaction, $\mathbf{c}_{12}$, is itself necessarily a contrast because

$$\mathbf{c}_{12}'\mathbf{1}_v = \mathbf{c}_2'\mathbf{c}_1^\delta \mathbf{1}_v = \mathbf{c}_2'\mathbf{c}_1 = 0 \tag{8.1.3}$$

or, after scaling

$$\mathbf{c}_{12} = \mathbf{c}_2^\delta \mathbf{c}_1/(\mathbf{c}_1'\mathbf{c}_2^{2\delta}\mathbf{c}_1)^{1/2} = \mathbf{c}_1^\delta \mathbf{c}_2/(\mathbf{c}_2'\mathbf{c}_1^{2\delta}\mathbf{c}_2)^{1/2}. \tag{8.1.4}$$

Further, $\mathbf{c}_{12} = \mathbf{c}_{21}$. Also, if $\mathbf{c}_3$ is a third contrast, orthogonal to the first two, $\mathbf{c}_{123}$ may be written

$$\mathbf{c}_{123} = \mathbf{c}_3^\delta \mathbf{c}_2^\delta \mathbf{c}_1 = \mathbf{c}_2^\delta \mathbf{c}_1^\delta \mathbf{c}_3, \quad \text{etc.} \tag{8.1.5}$$

and so on for higher-order interactions.

It soon appears that interactions are not of much use outside a factorial context. For example, if

$$\mathbf{c}_1 = \frac{1}{\sqrt{12}}\begin{bmatrix} 3 \\ -1 \\ -1 \\ -1 \end{bmatrix}, \quad \mathbf{c}_2 = \frac{1}{\sqrt{6}}\begin{bmatrix} 0 \\ 2 \\ -1 \\ -1 \end{bmatrix}, \quad \mathbf{c}_3 = \frac{1}{\sqrt{2}}\begin{bmatrix} 0 \\ 0 \\ 1 \\ -1 \end{bmatrix}, \tag{8.1.6}$$

then $\mathbf{c}_{ij} = \mathbf{c}_i$, where $j < i$. The result is not particularly helpful. If, however, three orthogonal contrasts are taken to show the linear, quadratic and cubic effects of four equally spaced treatments, i.e.

$$\mathbf{c}_1 = \frac{1}{\sqrt{20}}\begin{bmatrix} -3 \\ -1 \\ 1 \\ 3 \end{bmatrix}, \quad \mathbf{c}_2 = \tfrac{1}{2}\begin{bmatrix} 1 \\ -1 \\ -1 \\ 1 \end{bmatrix}, \quad \mathbf{c}_3 = \frac{1}{\sqrt{20}}\begin{bmatrix} 1 \\ -3 \\ 3 \\ -1 \end{bmatrix},$$

the interactions give rise to a new set

$$\mathbf{c}_{23} = \frac{1}{\sqrt{20}}\begin{bmatrix} 1 \\ 3 \\ -3 \\ -1 \end{bmatrix}, \quad \mathbf{c}_{13} = \tfrac{1}{6}\begin{bmatrix} -3 \\ 3 \\ 3 \\ -3 \end{bmatrix}, \quad \mathbf{c}_{12} = \frac{1}{\sqrt{20}}\begin{bmatrix} -3 \\ 1 \\ -1 \\ 3 \end{bmatrix}.$$

Finding the interactions has led back to something like the original contrasts but $\mathbf{c}_{23}$, $\mathbf{c}_{13}$ and $\mathbf{c}_{12}$ provide an alternative partition that has few uses.

Interactions are chiefly useful in the context of a factorial set of treatments. To take an example, let one factor have levels A, B, C, and D and the other X, Y, and Z. Taking treatment combinations in the order AX, AY, AZ, BX, . . . , DZ, the main effect of the first factor is summed up by three contrasts,

$$\begin{aligned}
\mathbf{c}_1 &= (3,\ 3,\ 3,\ -1,\ -1,\ -1,\ -1,\ -1,\ -1,\ -1,\ -1,\ -1)'/6 \\
\mathbf{c}_2 &= (0,\ 0,\ 0,\ 2,\ 2,\ 2,\ -1,\ -1,\ -1,\ -1,\ -1,\ -1)'/\sqrt{18} \\
\mathbf{c}_3 &= (0,\ 0,\ 0,\ 0,\ 0,\ 0,\ 1,\ 1,\ 1,\ -1,\ -1,\ -1)'/\sqrt{6}.
\end{aligned}$$

The main effects of the other is similarly summed up by two,

$$\begin{aligned}
\mathbf{c}_4 &= (2,\ -1,\ -1,\ 2,\ -1,\ -1,\ 2,\ -1,\ -1,\ 2,\ -1,\ -1)'/\sqrt{24} \\
\mathbf{c}_5 &= (0,\ 1,\ -1,\ 0,\ 1,\ -1,\ 0,\ 1,\ -1,\ 0,\ 1,\ -1)'/\sqrt{8}.
\end{aligned}$$

Between them they give rise to six contrasts for the interaction, obtained by taking each of $\mathbf{c}_1, \mathbf{c}_2$, and $\mathbf{c}_3$ in association with each of $\mathbf{c}_4$ and $\mathbf{c}_5$, multiplying out and standardizing, i.e.,

$$\begin{aligned}
\mathbf{c}_{14} &= (6,\ -3,\ -3,\ -2,\ 1,\ 1,\ -2,\ 1,\ 1,\ -2,\ 1,\ 1)'/\sqrt{72} \\
\mathbf{c}_{15} &= (0,\ 3,\ -3,\ 0,\ -1,\ 1,\ 0,\ -1,\ 1,\ 0,\ -1,\ 1)'/\sqrt{24} \\
\mathbf{c}_{24} &= (0,\ 0,\ 0,\ 4,\ -2,\ -2,\ -2,\ 1,\ 1,\ -2,\ 1,\ 1)'/6 \\
\mathbf{c}_{25} &= (0,\ 0,\ 0,\ 0,\ 2,\ -2,\ 0,\ -1,\ 1,\ 0,\ -1,\ 1)'/\sqrt{6} \\
\mathbf{c}_{34} &= (0,\ 0,\ 0,\ 0,\ 0,\ 0,\ 2,\ -1,\ -1,\ -2,\ 1,\ 1)'/\sqrt{12} \\
\mathbf{c}_{35} &= (0,\ 0,\ 0,\ 0,\ 0,\ 0,\ 0,\ 1,\ -1,\ 0,\ -1,\ 1)'/2.
\end{aligned}$$

It will be found upon examination that the interaction contrasts are orthogonal to one another and to those of the main effects, so a complete partition has been found of the eleven degrees of freedom. The result is no coincidence, as may be seen by considering the general case. Let the first factor have a levels and the second b, so that there are ab combinations, all represented in the design. The main effect of a factor can always be summed up by a series of orthogonal vectors, as in the example with four levels ($a = 4$) given in (8.1.6). Let there be ab treatment combinations in all and let $\mathbf{c}_p$ be the pth contrast of the main effect of the first factor. Then it has $b(p-1)$ elements equal to zero, b equal to $(a-p)$ and $b(a-p)$ equal to -1. (For the moment scaling is being ignored. Although other partitions are possible, they are not needed here.)

Now, for the interaction of the pth contrast from the first factor and the qth from the second, the contrast vector will have one element equal to $(a-p)(b-q)$, $(a-p)$ equal to $-(b-q)$, $(b-q)$ equal to $-(a-p)$ and $(a-p)(b-q)$ equal to one. Multiplying it out by one of the vectors that gave rise to it, say, the pth of the first factor, there is one element equal to $(a-p)^2\ (b-q)$, $(a-p)$ equal to $-(b-\bar{q})$, $(b-q)$ equal to $-(a-p)^2$ and $(a-p)\ (b-q)$ equal to one. In short, any contrast of an interaction is orthogonal to any contrast of a main effect contained in it. The situation is not made more complicated in principle by the introduction of further factors, though the algebra is much increased. In short, given a complete set of interactions produced by the multiplication rule, they are all orthogonal and provide a complete representation of the treatment effects. In any particular design, however, they do not necessarily give a satisfactory partition of the treatment sum of squared deviations because they may not be estimated independently. The skill of a designer often lies in contriving that all main effects and all their dependent interactions shall correspond to natural contrasts or, alternatively, to basic contrasts, as described in Section 3.7. A further refinement arises when general balance (Section 4.9) is achieved.

The value of interactions in a factorial context lies in their providing a description of the way in which different levels of the various factors combine in their effect. If one factor has the same effect at all levels of another, there is no interaction between them. Conversely, if there is, then some levels of one factor create conditions that alter the action of the other. For example, it sometimes happens that soil applications of salts of potassium have no effect unless enough magnesium is present for the plants to make use of what they are given. It can then be said that the two interact, meaning that the effect of potassium has to be assessed according to the amount of magnesium available. Equally, the effect of applications of magnesium will depend upon the amount of potassium applied.

When using interactions it should be remembered that they qualify the main effects and may well show their estimation to be unnecessary. Similarly, a higher-order interaction will qualify judgements about one of lower order contained within itself. Thus, if A × B × C is large, it means that no general statement can be made about A × B; everything depends upon the level of C. Equally A × C depends upon the level of B and so on. In the same way if A × B appears to be significant, nothing can be said about the effect of either factor without specifying the level of the other. All that exemplifies the powerful paper by Nelder (1977) in which such matters are discussed. It stands in contrast to the view, commonly held but mistaken, that the essence of the conclusions lies in the main effects, which are very important, the interactions being some sort of optional ornamentation.

To take a simple example, let there be two factors each at two levels, giving means of *1, *A, *B and *AB. If the factors interact, the important contrasts are those like (*A − *1) and (*AB − *B), which show the particular effects of A at the two levels of B. If, on the other hand, there is no interaction, a mean can be taken of the two particular effects, namely $\frac{1}{2}$(*A + *AB − *1 − *B), to give the main effect. (Similar comments apply, of course, to the effects of B.)

The analysis will give two main effects

$$M_A = -{*1} + {*A} - {*B} + {*AB}$$
$$M_B = -{*1} - {*A} + {*B} + {*AB}$$

and an interaction,

$$Int = {*1} - {*A} - {*B} + {*AB}.$$

The particular effects of A are $\frac{1}{2}(M_A - Int)$ and $\frac{1}{2}(M_A + Int)$; its main effect is $\frac{1}{2}M_A$. Since either particular or main effects may be required according to how the interaction turns out, the example serves as a warning against schemes in which the interaction is poorly determined as compared with the main effects. A well-estimated interaction is needed for two reasons. First, a lot is going to depend upon the decision whether to accept it or not. Second, if particular effects are going to be required, they will involve the interaction and will themselves be poorly estimated if it has a large variance. Also, if an interaction is believed to exist, it must not be confounded, because its estimate will be needed to find the particular effects. Further, its very existence makes the study of those effects imperative.

Nevertheless, when there are many factors there will be interactions that everyone is ready to discard. Whether those interactions are confounded or merged with error is a problem of design rather than of analysis. If groups of them can be confounded, smaller blocks will result and that might well be advantageous. On the other hand, their degrees of freedom will be lost to error and that could be disadvantageous. (However, an experiment with many factors probably has a sufficiency of degrees of freedom and could manage with fewer.) If an interaction is thought possible, it is unwise to discard it.

Some thought should be given to 'hidden replication'. If a set of treatments is being studied on two varieties, it is sometimes claimed for a factorial design that their replication is thereby doubled. If they do not interact with varieties, that is true. However, the most likely justification for introducing the varieties is an expectation that there will be such an interaction, that being another reason for using a factorial structure. There is the danger of the whole argument becoming two-faced.

Various attempts have been made to study results from a two-way table of means without resorting to the additive model implied by a partition into main effects and interactions. Thus, Mandel (1971) considered such a matrix, $\mathbf{M}$, of means in this way. Let θ_i^2 be an eigenvalue of $\mathbf{MM}'$ i.e. $\mathbf{MM}'\mathbf{u}_i = \theta_i^2\mathbf{u}_i$, and let $\mathbf{v}_i = \mathbf{M}'\mathbf{u}_i/\theta_i$. Then a possible solution for $\mathbf{M}$ is

$$\mathbf{M} = \sum_i (\theta_i \mathbf{u}_i \mathbf{v}_i'),$$

where $\mathbf{v}_i$ is scaled so that $\mathbf{v}_i'\mathbf{v}_i = 1$.

Such an examination of $\mathbf{M}$ could clarify what is happening, especially if there were only one appreciable θ_i, because that would indicate that the main effects combined in a multiplicative manner rather than an additive. However, treatment effects are but a specification of differences that exist anyway over the experimental area. If they combine in a non-additive way it is not reasonable to suppose that the effects of blocks and treatments will be additive, or that the residual errors can just be added to a sum of parameters. Where the model appears to be multiplicative, it might be better to transform the data to their logarithms and achieve additivity in that way.

Another suggestion is that of E. J. Williams (1952), who supposed that there was an effect of one factor expressed by the vector, $\boldsymbol{\gamma}$. The effect of the other, expressed by $\boldsymbol{\delta}$, depended in its intensity upon the level of the first, given by another vector, $\boldsymbol{\theta}$, i.e.

$$\tau_{ij} = \gamma_i + \theta_i \delta_j \tag{8.1.7}$$

where i and j are respectively levels of the two factors. In some contexts the model can be very appealing and could usefully be considered. It will be noted that the two factors are not symmetric in (8.1.7).

Although both the above approaches can be associated with significance tests, their main use must lie in describing an interaction once its existence has been established. The general rule remains that an interaction, if it exists at all, rules out any facile description of the data in terms of main effects only. Useful as it is to have a description, it must be remembered that sometimes there can be several alternative models and the data may not differ significantly from any of them. A description may therefore be of value chiefly because it implies some plausible underlying biological cause, which, once suggested, can be investigated further.

8.2 COMPUTATIONS WITH A FACTORIAL SET OF TREATMENTS

Where the treatments form a factorial set there is not usually much difficulty about the separate computation of sums of squares for main effects, etc. Once contrast vectors have been found to sum up the main effects, those for interactions can be derived from them and require no special computation. Expressions (3.6.1), (3.6.2), (3.6.9) and (3.6.12) apply as for any other contrast vector.

Indeed, the danger is rather of the unthinking adoption of some standard procedure even though the problem does not call for it in that particular instance. The fact that a set of treatments is factorial does not mean that interactions, etc.

are required; sometimes they have been taken for granted. For example, if a virus is eradicated from a number of varieties and a factorial set is formed of varieties each with and without virus, it is unlikely that anyone wants an interaction to see if different varieties are equally affected by virus. It is more likely that such an interaction can be assumed. The enquiry will concern rather which varieties are worth the trouble of producing a virus-free strain. The point may be illustrated by some data studied by Federer (1955). They came from an experiment intended to investigate the nutritional needs of paddy. One factor was of three varieties, namely, Red Aus, Kashiphul and Dudkalma: the other was the absence of nitrogenous fertilizer (O) as compared with the application of Ammophos (A_p) and Ammonium Sulphate (A_s). The treatment totals each based on six plots, were

	O	A_p	A_s
Red Aus	698	730	708
Kashiphul	520	498	466
Dudkalma	456	705	609

The error variance was 567.8 with 40 degrees of freedom. Federer was confronted with the question whether different varieties of paddy have the same fertilizer requirements, i.e., whether there is an interaction. In that case the correct partition, as he pointed out, is

Varieties	2	11 867.7
Fertilizer	2	1 878.9
Interaction	4	3 713.2

The interaction mean-square is 928.3 and the value of F is 1.63, so the conclusion was rightly reached that there was insufficient evidence to establish that varieties and fertilizers interacted. Suppose, however, that someone had collected the data convinced that such an interaction existed, his intention being to investigate if any of the three varieties gave a response to nitrogen. In that case he would have partitioned differently, like this

Varieties	2	11 867.7
Fertilizer with Red Aus	2	89.3
Fertilizer with Kashiphul	2	245.7
Fertilizer with Dudkalma	2	5 257.0

He would conclude that Dudkalma responded to added nitrogen but the others did not. The difference between the two partitions is not that one is right and the other is wrong; the point is that either could be right according to the experimenter's preconceptions. If he is enquiring about an interaction, he must be given an appropriate test. If he decides as a result that his factors interact or if he never doubted that they would do so, it is the particular effects that should be examined.

If a factorial partition is required, in the case of an orthogonal design or one that is completely randomized, there is usually no difficulty in effecting it, especially if the treatment combinations are equi-replicate, for then the covariance maxtrix of the treatment parameters equals $\mathbf{I}/r$, where r is a constant. The best procedure usually is to write down data totals for all treatment combinations and to find a 'summation term', by squaring the totals, adding the squares and dividing by the number of data in each. The approach has already been described in Section 4.1. Then factors are omitted in turn and further summation terms are formed. Now, if there are three factors, A, B, and C, their sums of squared deviations are respectively

$$S_A - S_0$$
$$S_B - S_0$$
$$S_C - S_0,$$

where S_A, S_B and S_C are summation terms when the data are classified only by A, only by B and only by C, while $S_0\,(=G^2/n)$ is the summation term when the data are unclassified. Similar terms can be found for two or more factors. Thus, if the data are classified by A and B ignoring C, that would give a summation term, S_{AB}. To find the interaction of A and B, it should be noted that the total sum of squared deviations due to A and B together is $(S_{AB} - S_0)$, of which $(S_A - S_0) + (S_B - S_0)$ has already been accounted for by main effects, which leaves

$$S_{AB} - (S_A + S_B) + S_0$$

for the two-factor interaction. The others similarly equal

$$S_{AC} - (S_A + S_C) + S_0$$
$$S_{BC} - (S_B + S_C) + S_0.$$

Subtracting the main effects and two-factor interactions from $(S_{ABC} - S_0)$ gives

$$S_{ABC} - (S_{BC} + S_{AC} + S_{AB}) + (S_A + S_B + S_C) - S_0$$

for the three-factor interaction and so on. The method is illustrated numerically in Example 8A.

A special method, known as 'Yates's algorithm' exists for the 2^p case. It is more easily illustrated than explained, so Example 8B will set out the method.

8.3 THE 2^p-FACTORIAL PARTITION

One case of special interest and importance is that of the 2^p-factorial, i.e. there are p factors each at two levels. It will first be noted that a main effect is given by a single contrast vector in which each element is either $+1/\sqrt{v}$ or $-1/\sqrt{v}$, the positive and negative signs indicating the two levels of the factor. The contrast vector for any interaction is then readily found. Where a treatment combination occurs that has an odd number of factors in the specification of the interaction it takes a negative sign, but a positive one if the number is even. Thus, given treatments:

$$1 \quad A \quad B \quad AB \quad C \quad AC \quad BC \quad ABC$$

the contrast vector for the three-factor interaction, $A \times B \times C$, is

$$(1, -1, -1, 1, -1, 1, 1, -1)'/\sqrt{8}.$$

That is to say, treatment combinations like 1, AC or BC, which are specified by an even number of letters, take a positive sign, whereas those like A or ABC take a negative one. Similarly the vector for $A \times B$ is

$$(1, -1, -1, 1, 1, -1, -1, 1)'/\sqrt{8},$$

the factor C having no effect. It is very easy in that way to write down any interaction. Also, since all treatment combinations enter into any effect, it is possible to confound any main effect or any interaction without difficulty. All that is needed is an allocation of treatment combinations with a positive sign to one set of blocks and all with a negative sign to the other. Such confounding can also take place between the rows and columns of a row-and-column design and in many other contexts.

Further, there need be no difficulty about confounding two chosen effects in four blocks. Treatment combinations can then be classified as $++$, $+-$, $-+$ and $--$, the two signs relating respectively to the first and second effects. Thus, for the example given, the four groups are:

$++$	$+-$	$-+$	$--$
1	AC	C	A
AB	BC	ABC	B

Obviously, $A \times B \times C$ is confounded between the groups $++$ and $+-$ as compared with $-+$ and $--$, while $A \times B$ is similarly confounded between $++$ and $-+$ on the one hand against $+-$ and $--$ on the other. There is, however, a third comparison possible, namely that of $++$ and $--$ against $+-$ and $-+$, and, rather disconcertingly, it confounds the main effect of C. Reflection shows that some such effect is inevitable. First, the confounding of two effects between four blocks necessarily raises the question of what happens to the third degree of freedom of the multiple disconnection. Secondly, if a treatment combination gives the same sign in both the original contrasts, i.e. it has a positive sign in both or is negative in both, it must have a positive sign in the interaction. On the other hand, if it is positive in one but negative in the other, it will be negative in the interaction. In the interaction in the example, $(A \times B \times C) \times (A \times B) = C$. If, however, $A \times B$ and $B \times C$ are confounded, the third effect confounded is $A \times C$, since both A and C occur in one of $A \times B$ and $B \times C$ but not in the other. The point is that if a factor occurs twice in an extended interaction, it cancels itself because $(-1)^2 = +1$.

The rule can readily be extended, but some care is needed. For example, the designer of an experiment may wish to confound seven effects between eight blocks or eight groups of blocks. He thinks of the seven effects he is most ready to dispense with and starts off with A × B × C and B × C × D. That commits him to confounding A × D also, though it is not on his list. He then adds A × C × D × E and is committed to B × D × E, A × B × E, and C × E, which completes his allowance of seven effects. He may protest that he is now confounding effects that he wished to keep and that his list is not exhausted, but he must accept facts. It is not possible to make an arbitrary list and be sure that it will fit a scheme of confounding. Further, it is not possible to confound only high-order interactions unless there are many factors. For example, if there are only four in all, any three-factor interaction implies as its complement the main effect of the fourth factor. If another three-factor interaction is to be confounded, that implies another main effect. If both interactions are confounded, that leads to the confounding of the two-factor interaction of their complements. Thus the confounding of both A × B × C and A × B × D implies the confounding of C × D, unpalatable as that may be.

The problem of evolving designs in which all main effects and all two-factor interactions are retained is of some importance. They are especially useful when a range of factors is to be given preliminary study to see which can usefully be examined further. Too often the first experiments of a project are poorly designed and poorly executed, the excuse being that they are 'only preliminary'. If their purpose is to discover what lines of research can most usefully be pursued, a lot depends upon them and they need to be designed and carried out with the greatest of care, otherwise the most fruitful approach may be abandoned at an early stage for no good reason. They can be the most critical in the whole project. On the other hand detailed information about higher-order interactions is not really called for. The need is rather to define the field in which interactions are to be sought and to obtain warnings about combinations of factors that might well need to be studied in association. Thus, if A, B, and C all gave evidence of main effects and if A × B and A × C both appeared to be appreciable, that would be an argument for a further experiment that studied A, B, and C, perhaps at more than two levels each.

Although 2^p designs are usually equi-replicate they do not have to be. In fact, the whole approach can be quite flexible if need be. Thus, with a 2^3 treatment structure, say, factors X, Y and Z, each at two levels, it may well be decided to confound X × Y × Z but to keep everything else. If blocks of four plots were readily available, there would be no problem, but what if they are not? All such blocks may have been used already. The suggestion has been made (Pearce, 1970b) that if blocks of three or five plots could be found, the following design would be available:

Block	I	YZ,	XZ,	XY;	Block	V	X,	X,	Y,	Z,	XYZ;
	II	1	XZ,	XY;		VI	X,	Y,	Y,	Z,	XYZ;
	III	1	YZ,	XY;		VII	X,	Y,	Z,	Z,	XYZ;
	IV	1	YZ,	XZ;		VIII	X,	Y,	Z,	XYZ,	XYZ.

Taking treatments in the order 1, YZ, XZ, XY, X, Y, Z, XYZ, **C** equals

$$\frac{2}{15}\begin{bmatrix} 15 & -5 & -5 & -5 & 0 & 0 & 0 & 0 \\ -5 & 15 & -5 & -5 & 0 & 0 & 0 & 0 \\ -5 & -5 & 15 & -5 & 0 & 0 & 0 & 0 \\ -5 & -5 & -5 & 15 & 0 & 0 & 0 & 0 \\ 0 & 0 & 0 & 0 & 27 & -9 & -9 & -9 \\ 0 & 0 & 0 & 0 & -9 & 27 & -9 & -9 \\ 0 & 0 & 0 & 0 & -9 & -9 & 27 & -9 \\ 0 & 0 & 0 & 0 & -9 & -9 & -9 & 27 \end{bmatrix}.$$

Accordingly, a possible solution for $\mathbf{C}^-$ is

$$\frac{1}{7200}\begin{bmatrix} 2825 & 125 & 125 & 125 & 0 & 0 & 0 & 0 \\ 125 & 2825 & 125 & 125 & 0 & 0 & 0 & 0 \\ 125 & 125 & 2825 & 125 & 0 & 0 & 0 & 0 \\ 125 & 125 & 125 & 2825 & 0 & 0 & 0 & 0 \\ 0 & 0 & 0 & 0 & 1413 & -87 & -87 & -87 \\ 0 & 0 & 0 & 0 & -87 & 1413 & -87 & -87 \\ 0 & 0 & 0 & 0 & -87 & -87 & 1413 & -87 \\ 0 & 0 & 0 & 0 & -87 & -87 & -87 & 1413 \end{bmatrix}.$$

For a contrast like the main effect of X, i.e. $\mathbf{c} = (-1, -1, 1, 1, 1, -1, -1, 1)'$, or indeed any of the main effects or two-factor interactions, $\mathbf{c}'\mathbf{C}^-\mathbf{c}\sigma^2 = \frac{7}{3}\sigma^2$. If it had been practicable to use blocks each of four plots, the figure would have been $2\sigma^2$, so there has been a loss of information, though not a disastrous one.

8.4 THE 3^p-FACTORIAL PARTITION

Other factorial systems do not have the neatness of the one just considered, which becomes apparent when passing from the 2^p partition to the 3^p. To take the simplest example, let there be only two factors, giving treatment combinations, PX, PY, PZ, QX, QY, QZ, RX, RY, RZ. Then, main effects are given by the vectors

$$\mathbf{c}_1 = (1, 1, 1, 0, 0, 0, -1, -1, -1)'/\sqrt{6}$$
$$\mathbf{c}_2 = (1, 1, 1, -2, -2, -2, 1, 1, 1)'/\sqrt{18}$$

for PvQvR and

$$\mathbf{c}_3 = (1, 0, -1, 1, 0, -1, 1, 0, -1)'/\sqrt{6}$$
$$\mathbf{c}_4 = (1, -2, 1, 1, -2, 1, 1, -2, 1)'/\sqrt{18}$$

for X v Y v Z. Anyone who concluded that the interaction with four degrees of freedom could be found from $\mathbf{c}_1 \times \mathbf{c}_3$, $\mathbf{c}_1 \times \mathbf{c}_4$, $\mathbf{c}_2 \times \mathbf{c}_3$ and $\mathbf{c}_2 \times \mathbf{c}_4$ would be quite right but, unfortunately, those vectors do not readily lend themselves to confounding.

The only one that uses all treatment combinations is $\mathbf{c}_2^\delta \mathbf{c}_4$, namely

$$\mathbf{c}_2^\delta \mathbf{c}_4 = (1, -2, 1, -2, 4, -2, 1, -2, 1)'/6$$

but it is difficult to imagine anyone wanting a block made up of PX, PZ, QY (four times), RX and RZ for comparison with another made up of PY, QX, QZ, RY all twice. Again, a vector like

$$\mathbf{c}_1^{\delta}\mathbf{c}_3 = (1, 0, -1, 0, 0, 0, 1, 0, -1)'/2$$

also implies multiple replications within a block, though it is known that precision of comparison requires the dispersal of treatments among blocks. A much more convenient expression of the interaction is found by writing

$$\begin{aligned}
\mathbf{c}_5 &= (\ \ 1,\ -1,\ \ \ 0,\ \ \ 0,\ \ 1,\ -1,\ -1,\ \ \ 0,\ \ \ 1)'/\sqrt{6}\\
\mathbf{c}_6 &= (\ \ 1,\ \ \ 1,\ -2,\ -2,\ \ 1,\ \ \ 1,\ \ \ 1,\ -2,\ \ \ 1)'/\sqrt{18}\\
\mathbf{c}_7 &= (-1,\ \ \ 0,\ \ \ 1,\ \ \ 0,\ \ 1,\ -1,\ \ \ 1,\ -1,\ \ \ 0)'/\sqrt{6}\\
\mathbf{c}_8 &= (\ \ 1,\ -2,\ \ \ 1,\ -2,\ \ 1,\ \ \ 1,\ \ \ 1,\ \ \ 1,\ -2)'/\sqrt{18}.
\end{aligned}$$

The first pair, $\mathbf{c}_5$ and $\mathbf{c}_6$, represent the contrasts between the groups (PX, QY, RZ), (PY, QZ, RX), and (PZ, QX, RY). Similarly $\mathbf{c}_7$ and $\mathbf{c}_8$ express the differences between the groups (PX, QZ, CY), (PY, QX, RZ), and (PZ, QY, RX). In fact, the same result is being applied as was used in Section 5.8, namely, that a square array of treatments m by m can be given $(m+1)$ classifications, such that any two levels of different classifications have only one cell in common. Here, $m = 3$, so there are four groupings, one by each main effect and two by the interaction. The classifications are in fact those of the Graeco-Latin square

$A\alpha$	$B\beta$	$C\gamma$		PX	PY	PZ
$C\beta$	$A\gamma$	$B\alpha$	corresponding to	QX	QY	QZ
$B\gamma$	$C\alpha$	$A\beta$		RX	RY	RZ

The rows and columns correspond to the two main effects and the Latin and Greek letters together make up the interaction.

It is clear that the pairs of interaction contrasts lend themselves readily to confounding. Thus, if three blocks are taken

Block I	PX	QY	RZ
II	PY	QZ	RX
III	PZ	QX	RY

then $\mathbf{c}_5$ and $\mathbf{c}_6$ are confounded between them. That is called the I-component. If another three are taken

Block IV	PX	QZ	RY
V	PY	QX	RZ
VI	PZ	QY	RX

they confound $\mathbf{c}_7$ and $\mathbf{c}_8$, which constitute the J-component. The trouble is that the whole design is giving half efficiency to all interaction contrasts instead of confounding them, i.e.

$$\mathbf{C} = \tfrac{1}{3}\begin{bmatrix} 4 & 0 & 0 & 0 & -1 & -1 & 0 & -1 & -1 \\ 0 & 4 & 0 & -1 & 0 & -1 & -1 & 0 & -1 \\ 0 & 0 & 4 & -1 & -1 & 0 & -1 & -1 & 0 \\ 0 & -1 & -1 & 4 & 0 & 0 & 0 & -1 & -1 \\ -1 & 0 & -1 & 0 & 4 & 0 & -1 & 0 & -1 \\ -1 & -1 & 0 & 0 & 0 & 4 & -1 & -1 & 0 \\ 0 & -1 & -1 & 0 & -1 & -1 & 4 & 0 & 0 \\ -1 & 0 & -1 & -1 & 0 & -1 & 0 & 4 & 0 \\ -1 & -1 & 0 & -1 & -1 & 0 & 0 & 0 & 4 \end{bmatrix}.$$

All the proposed contrasts are eigenvectors. The corresponding eigenvalues, i.e. the effective replications, are 2 for $\mathbf{c}_1$, $\mathbf{c}_2$, $\mathbf{c}_3$ and $\mathbf{c}_4$ and 1 for all the rest. The design, incidentally, is of Type F, i.e., it has the factorial balance described in Section 5.5. There is no possibility of confounding $\mathbf{c}_5$, $\mathbf{c}_6$, $\mathbf{c}_7$, and $\mathbf{c}_8$ simultaneously.

Although the above approach is usual, it may be questioned whether there is much value in estimating an interaction with half efficiency. For reasons already given, an interaction should be well estimated if it is needed and discarded if it is not. Indeed, there is a case for confounding $\mathbf{c}_1$ and $\mathbf{c}_2$ in three blocks, i.e.

Block I	PX	PY	PZ
II	QX	QY	QZ
III	RX	RY	RZ

and confounding $\mathbf{c}_3$ and $\mathbf{c}_4$ in three more, i.e.,

Block IV	PX	QX	RX
V	PY	QY	RY
VI	PZ	QZ	RZ

The interaction is then preserved at full efficiency. Alternatively, if the interaction is not needed, there is little point in retaining vestiges of it. What is unconfounded would be better merged with error.

If there are more factors, the process can be continued. Thus with a 3^3 factorial set of treatments there are two degrees of freedom for each of three main effects and four degrees of freedom for each of three two-factor interactions. The eight degrees of freedom left over correspond to a three-factor interaction. They fall into four groups of two, each referring to a separate component. For example, there were three groups associated with the I-component of the 3^2 set, namely I_1 = PX, QY, RZ, I_2 = PY, QZ, RX and I_3 = PZ, QX, RY. Combining them with A, B and C, the three levels of a third factor, gives

AI_1	BI_1	CI_1
AI_2	BI_2	CI_2
AI_3	BI_3	CI_3

Taking an I-component of that table gives groups, (AI_1, BI_2, CI_3), (BI_1, CI_2, AI_3),

(CI_1, BI_3, AI_2). The first of these may be expanded to read

APX, AQY, ARZ, BPY, BQZ, BRX, CPZ, CQX, CRY

and the others likewise. Together the three make up a complete set of treatments and the two degrees of freedom between them are part of the three-factor interaction. A further two are found by taking a J-component of the above table. The other degrees of freedom for the interaction are found by taking I- and J-components of

$$\begin{array}{ccc} AJ_1 & BJ_1 & CJ_1 \\ AJ_2 & BJ_2 & CJ_2 \\ AJ_3 & BJ_3 & CJ_3 \end{array}$$

The process can be continued indefinitely for more factors. Now, if each component of the 3^3 design is confounded in one set of three blocks, it will be of full efficiency in three other sets, giving an efficiency of $\frac{3}{4}$, which is at least better than $\frac{1}{2}$. Of course, an unwanted interaction or some component of it can always be confounded to reduce block size, the rest of it being merged with error, and that is perhaps the chief use of the techniques just described.

Where the three levels of a factor represent three equally spaced applications of a fertilizer or concentrations of a spray, they can be broken down into a linear and quadratic effect in the usual way. However, in the 3^p case that does not usually help much in confounding.

8.5 THE 4^p-FACTORIAL DESIGN

Given a 4^p-factorial set of treatments, it can be approached in much the same way as for a 3^p set. Thus, it is possible to regard a 4^2 array as having five orthogonal classifications like the rows, columns, Roman letters, Greek letters and numbers in Section 1.1. If the rows and columns represent main effects, each with three degrees of freedom, the other classifications give three components, each with three degrees of freedom, which make up the nine degrees of freedom for interaction. If there are more factors, the process can be continued as with the 3^p case. Nevertheless, schemes with many factors, each at four levels, can be very cumbersome.

However, 4^p schemes can be useful when the levels represent increasing applications of some substance, precisely the circumstances in which a 3^p scheme often fails. Let there be levels a_0, a_1, a_2 and a_3 and let them be regarded notionally as a 2^2 factorial set, i.e.,

$$a_0 = 1, \quad a_1 = W, \quad a_2 = X, \quad a_3 = WX.$$

Then the main effect of X is given by the vector $(-1, -1, 1, 1)'$, which is not an ideal representation of the linear effect of increasing applications but it will serve. Also $W \times X$ gives the vector $(1, -1, -1, 1)'$, which provides a good measure of the quadratic effect. That leaves the main effect of W, which can well be confounded, i.e. $(-1, 1, -1, 1)'$. If now there is a second factor at four levels, say,

$$b_0 = 1, \quad b_1 = Y, \quad b_2 = Z, \quad b_3 = YZ,$$

similarly no harm will be done by confounding the main effect of Y. In the following 2^4 scheme W, Y and W × Y are all confounded:

Block I	1	X	Z	XZ
II	W	WX	WZ	WXZ
III	Y	XY	YZ	XYZ
IV	WY	WXY	WYZ	WXYZ

It can equally well be written in the form of a 4^2 scheme thus:

Block I	00	20	02	22
II	10	30	12	32
III	01	21	03	23
IV	11	31	13	33

The two digits show the respective levels of a and b. Of course, that leaves no degrees of freedom for error so the whole scheme would need to be replicated.

8.6 FRACTIONAL REPLICATION

Once it is accepted that unwanted interactions can be merged with the error, the next step is to form an error entirely of such interactions, there being no replication. Finney (1945b, 1946c, 1950) took the matter even further. Suppose that there is a 2^7 factorial scheme of treatments; that would need 128 plots for even a single replicate, but would it not be possible to use only 64? He showed that it can be done, but a smaller and more manageable example will be afforded by taking a 2^5 scheme and asking if it can be accommodated on 16 plots. Finney's method involves selecting a 'defining contrast' and using only those treatment combinations that have the same sign in it. Thus, here it will be convenient to take A × B × C × D × E, which gives a choice of two sets of 16 treatment combinations, i.e. either those with a positive sign in the defining contrast, namely

1, AB, AC, AD, AE, BC, BD, BE, CD, CE, DE, BCDE, ACDE, ABDE, ABCE, ABCD,

or those with a negative sign, namely

A, B, C, D, E, ABC, ABD, ABE, ACD, ACE, ADE, BCD, BCE, BDE, CDE, ABCDE.

The choice may be made either way. With a set of treatment combinations restricted in that way any effect has an 'alias'. Thus, supposing the first set to have been chosen, someone might want to estimate the main effect of A. He therefore proposes to use the contrast vector

$$(-1, +1, +1, +1, +1, -1, -1, -1, -1, -1, -1, -1, +1, +1, +1, +1)',$$

thus picking up those treatment combinations that do contain A and contrasting them with those that do not. However, his proposed vector would serve as well for the estimation of B × C × D × E. The reason is that the only treatment combinations available are those represented by an even number of letters. If a selection is made of those that contain A, then B, C, D, E together must come to an odd number. Similarly, those that do not contain A must have an even number of B, C, D, and E, which leads to B × C × D × E as an alternative expression of A. That is why the one contrast is said to be an alias of the other. They are the same thing under two different names. Also, a two-factor interaction, like A × B, has a three-factor alias, i.e. C × D × E.

As far as the main effect is concerned, if the contrast should be marked, most people would be in no doubt that the cause lay in the effect of A and would discount the four-factor interaction, but they would be less certain about preferring an interaction with two factors to one with three. That is why a 2^7 scheme is often regarded as the smallest feasible, though 2^6 is a possibility.

Mention has already been made of the need of good designs for an exploratory situation, where a number of factors need to be introduced with a view to identifying those that can usefully be studied further. A design with partial replication that will examine main effects and two-factor interactions can be ideal. It will admittedly not be reliable about higher-order interactions, but as has been said in Section 8.3, at a preliminary stage that may not matter much.

By using two defining contrasts, the size of the experiment can be reduced even further to a quarter-replicate. Of course, the two defining contrasts will give rise to a third, as is the way with confounding. As a consequence each effect will have three aliases, so some skill is needed in choosing the defining contrasts aright. The problem does not arise, however, unless there are many factors.

The introduction of blocks is permissible, but it necessarily calls for further confounding both of a contrast and its alias. For example, in the 2^5 example, if it were decided to have two blocks, discarding A × B and C × D × E in the process, the design would work out like this:

Block I	AC,	AD,	AE,	BC,	BD,	BE,	BCDE,	ACDE
II	1,	AB,	CD,	CE,	DE,	ABDE,	ABCE,	ABCD

With such designs the method of sweeping, which was explained in Chapter 6, has notable advantages because the only contrasts that need to be considered are those that will appear in the final analysis. Also, all unconfounded contrasts are of full efficiency, which can lead to useful simplifications.

8.7 QUALITY–QUANTITY INTERACTIONS

Sometimes two factors are related, one concerning the form in which some substance should be supplied, e.g. potassic fertilizer as chloride or as sulphate, and the other concerning the amount. A problem arises if zero applications are included, because not applying potassium chloride is the same as not applying sulphate. The problem is related to the merging of treatments considered in Section 4.2, but it is of some importance and merits study in its own right. Such a treatment set was presented in Example 2A.

The classical exposition is that of Yates (1937), who considered only orthogonal, equi-replicate designs. Suppose that the six treatments are thought of as OA, OB, 1A, 1B, 2A, and 2B: a possible start is to transfer the elementary contrast of OA and OB to error. Any further contrasts considered will have to be orthogonal to that one and in all Yates suggested

$$\frac{1}{\sqrt{2}}\begin{bmatrix}1\\-1\\0\\0\\0\\0\end{bmatrix},\quad \frac{1}{\sqrt{10}}\begin{bmatrix}0\\0\\-1\\1\\-2\\2\end{bmatrix},\quad \frac{1}{\sqrt{2}}\begin{bmatrix}-1\\-1\\0\\0\\1\\1\end{bmatrix},$$

$$\frac{1}{\sqrt{12}}\begin{bmatrix}1\\1\\-2\\-2\\1\\1\end{bmatrix},\quad\text{and}\quad \frac{1}{\sqrt{10}}\begin{bmatrix}0\\0\\2\\-2\\-1\\1\end{bmatrix}.$$

The first is to be assigned to error; the second represents the difference between the two forms, giving double weight to the double application. The third and fourth set out the linear and quadratic effect for the effect of increasing level of application. That leaves a final vector which represents the interaction of single and double applications with the two substances. The single applications have perforce to have double weighting if everything is to be orthogonal, but it is quite logical for them to be treated in that way since they are assumed to be having only half an effect.

A difficulty with that partition is the attributing of double effectiveness to a double application, which is unlikely except at very low levels of application. Also, in the form given it is not easy to generalize the approach. The following vectors are perhaps more useful:

$$\frac{1}{\sqrt{2}}\begin{bmatrix}1\\-1\\0\\0\\0\\0\end{bmatrix},\quad \tfrac{1}{2}\begin{bmatrix}0\\0\\-1\\1\\-1\\1\end{bmatrix},\quad \tfrac{1}{2}\begin{bmatrix}-1\\-1\\0\\0\\1\\1\end{bmatrix},\quad \frac{1}{\sqrt{12}}\begin{bmatrix}1\\1\\-2\\-2\\1\\1\end{bmatrix},\quad\text{and}\quad \tfrac{1}{2}\begin{bmatrix}0\\0\\1\\-1\\-1\\1\end{bmatrix}.$$

The partition of the treatment sum of squared deviations is then

Source	*d.f.*
Quantity	2
*Quality	1
*Interaction	1
Treatments	4

Where the design is orthogonal, or if there are no blocks, the method of summation terms is available as presented in Section 4.1, though the lines marked with an asterisk should be calculated ignoring the treatment combinations with a zero application, while the others make use of them. That is a general method, however many forms of the substance are used and however many levels of application. Let those numbers be respectively f and l, then in general the treatment partition is

Source	*d.f.*
Quantity	$l-1$
*Quality	$f-1$
*Interaction	$(l-2)(f-1)$
Treatments	$f(l-1)$

The asterisks have the same meaning as before. The partition is applied to the data of Example 2A in Example 8G at the end of the chapter.

8.8. SYMMETRIC INTERACTIONS

Special problems arise when the levels of two factors are the same. For example, with diallel crosses the same varieties are used as pollen parents and as seed parents in all combinations. However, as Martin (1980) has remarked, the class is really wider. Such interactions arise whenever the means on the leading diagonal of the table can be regarded as essentially different from the rest and whenever there is a relation between a treatment and its reflection in that diagonal. Thus, with diallel crosses the treatment combinations featured down the leading diagonal represent self-fertilization and with some species they could well be less satisfactory than those in which the pollen comes from a different variety. Also, the genetical composition when X is pollinated by Y should be much the same as when Y is pollinated by X. Nevertheless, in some instances the point could well be investigated.

The problem was first mentioned by Vyvyan (1955). He had three apple varieties M.IV, M.IX, and M.XII, and grafted them in all nine combinations, using each variety both as rootstock and as scion. He therefore had a table of nine means, thus:

		Scion		
		M.IV	M.IX	M.XII
	M.IV	A	B	C
Rootstock	M.IX	D	E	F
	M.XII	G	H	J

Here there are nine means; in general there will be p^2.

Vyvyan first took out the two main effects and then proceeded to partition the interaction, which remained. One obvious contrast is that between like unions and unlike, i.e.

$$2(A+E+J)-(B+C+D+F+G+H)$$

That is Martin's η-component; it represents the effect of a parameter that equals η on the leading diagonal and $-\eta/(p-1)$ elsewhere.

Vyvyan then made a comparison between a set of graft unions in which all varieties appeared both as rootstock and scion and compared it with a set in which roles were reversed, i.e.,

$$(B+F+G)-(C+D+H).$$

He called that the 'equivalence effect'. In Martin's generalization it becomes the κ-component, the corresponding parameters being such that $\kappa_{ii}=0$, $\kappa_{ij}=-\kappa_{ji}$ $(i\neq j)$ with the constants, $\Sigma_i\kappa_{ij}=0=\Sigma_j\kappa_{ij}$. In general, it has $\frac{1}{2}(p-1)(p-2)$ degrees of freedom, which equals one when $p=3$.

The remaining degrees of freedom were assigned by Vyvyan to a 'union effect', in which he compared groups of unions distinguished by the varieties in the unlike crosses, i.e. he took two degrees of freedom between the groups

$$(A+F+H),\quad (C+E+G),\quad \text{and}\quad (B+D+J).$$

That gives Martin's ω-component, which has $\frac{1}{2}(p+1)(p-2)$ degrees of freedom. Its parameters, ω_{ij}, are such that $\omega_{ij}=\omega_{ji}$, subject to the constraints that $\Sigma_i\omega_{ij}=0=\Sigma_j\omega_{ji}$, and that $(p-1)\Sigma\omega_{ii}$ equals the sum of off-diagonal elements, the last being required to secure orthogonality with η. In the simple case where $p=3$, there is an obvious similarity to the partition in Section 8.4, derived from a Graeco-Latin square.

8.9 INTERACTIONS INVOLVING BLOCKS

The assumption that the parameters for blocks and treatments can be added is implicit in most of what has already been said, starting at (3.1.1). Nevertheless, such a law is not self-evident and is indeed open to question, so some care is needed to avoid an assumption that may be quite unjustified.

Since all error is in some sense the result of non-additivity, no test can do more than examine one particular pattern and pronounce on the evidence for its existence. One long-established approach is that of Tukey (1949). Let (3.1.1) be written as

$$\mathbf{y}=\alpha\mathbf{1}_n+\mathbf{D}'\boldsymbol{\beta}+\boldsymbol{\Delta}'\boldsymbol{\tau}+\boldsymbol{\eta}$$

or

$$y_{ij}=\alpha+\beta_i+\tau_j+\eta_{ij}, \tag{8.9.1}$$

where $\mathbf{k}'\boldsymbol{\beta}=0=\mathbf{r}'\boldsymbol{\tau}$; then it is quite possible that additivity will fail because the effect of treatments will depend upon the inherent fertility of the blocks. Thus, in a fertilizer trial it could be that treatment differences will show up more clearly on poor soil than on good. Equally, in a trial of measures to control a pest the treatments may have a larger effect where infestations are high than where it is low. Other similar examples can readily be suggested. To test whether the treatments and blocks do interact in some such way, Tukey proposed the model

$$y_{ij}=\alpha+(\beta_i+\tau_j)+\lambda(\beta_i+\tau_j)^2+\eta_{ij}. \tag{8.9.2}$$

He was thinking in terms of randomized replicate blocks in which $\mathbf{N} = \mathbf{1}_v\mathbf{1}_b'$. For negative λ, if β_i and τ_j both have the same sign the expectation of y_{ij} is decreased. That corresponds to the case where the effects of treatments are more marked at lower levels of the data, e.g., the fertilizer effects that show up more on poor soil. For positive λ, however, the opposite occurs as with the sprays that are more effective when infestation levels are high. Either way the model is realistic. As Tukey remarks, the function is 'slightly non-linear'. To that extent the ordinary least squares solution is not exact, but within reason that need not cause much alarm. It is, at least, better to introduce a necessary non-linear term frankly and to face the consequences rather than to pretend that there is no need of it. Federer (1955) has some interesting comments on non-additivity in general and alternatives to (8.9.2) in particular.

To explain Tukey's test it will be convenient to write the data, $\mathbf{Y}$, as a matrix of b rows and v columns and not as a vector, $\mathbf{y}$, as formerly. That raises the question of duplicate values of y_{ij}, but like voids they will not be considered here. It will be assumed that each block contains each treatment once as with randomized replicate blocks. Now, writing $G = \mathbf{1}_b'\mathbf{B} = \mathbf{1}_v'\mathbf{T}$ and $n = bv$ as before, let

$$K = \mathbf{B}'\mathbf{Y}\mathbf{T} - G^3/n$$
$$H = \mathbf{B}'\mathbf{B}/v - G^2/n$$
$$J = \mathbf{T}'\mathbf{T}/b - G^2/n.$$

It will be noted that H and J are respectively the sums of squared deviations for blocks and treatments. Then Tukey's method requires the separation from the error of a quantity

$$\frac{[K - G(H + J)]^2}{bvHJ} \tag{8.9.3}$$

with one degree of freedom. It represents, though only approximately, the effect of λ. With its removal a new error variance is found with one degree of freedom less than before, which provides an F-test for λ to see if it can reasonably be regarded as non-zero. As Tukey points out, if non-additivity is established, the analysis cannot just proceed as if the new error made everything right. Perhaps the need is for a transformation that will give a combination of parameters that fits the data better; perhaps one or two data are seriously at fault and need to be reconsidered. Care is needed and there is no automated solution. The calculations are shown in Example 8D.

The expression in (8.9.3) can usefully be examined further. Let

$$\mathbf{Y} = \alpha\mathbf{1}_b\mathbf{1}_v' + \boldsymbol{\beta}\mathbf{1}_v' + \mathbf{1}_b\boldsymbol{\tau}' \tag{8.9.4}$$

with no interaction or error and let $\mathbf{1}_b'\boldsymbol{\beta} = \mathbf{1}_v'\boldsymbol{\tau} = 0$. Then

$$B = v\alpha\mathbf{1}_b + v\boldsymbol{\beta}$$
$$T = b\alpha\mathbf{1}_v + b\boldsymbol{\tau}$$
$$G = bv\alpha$$

$$H = v\boldsymbol{\beta}'\boldsymbol{\beta}$$
$$J = b\boldsymbol{\tau}'\boldsymbol{\tau}$$
$$K = bv\alpha(v\boldsymbol{\beta}'\boldsymbol{\beta} + b\boldsymbol{\tau}'\boldsymbol{\tau})$$
$$K - G(H + J) = 0$$

as might be expected.

Given orthogonality a similar approach may be used with a row-and-column design. Thus, in a Latin square, treatments are applied so as not to affect row or column totals and it would therefore be possible to use Tukey's test to explore whether rows and columns did or did not combine in an additive manner or according to (8.9.2) and so for other situations (Tukey, 1955; Abraham, 1960).

Where there are groups of plots that have both block and treatment in common, it is always possible to measure the differences between them; they will be free of any interaction of blocks and treatments. If the sum of squared deviations is subtracted from that for error, there will remain a residue in which an interaction, if one exists, is concentrated. An F-test between the two components of error will gauge the weight of evidence that an interaction exists. The method is illustrated in Example 8E. It has been used by Trail and Weeks (1973).

The test differs from Tukey's in accepting any pattern, not just the one in (8.9.2). Also, instead of one degree of freedom for additivity there will usually be many, ordinarily $(b-1)(v-1)$, though sometimes it will not be possible to have all combinations of blocks and treatments represented.

Examples 8

A. The following table has been published (Pearce, 1953). It sets out crops in pounds over a ten-year period from 160 pear trees. There were eight randomized blocks, each of which contained four trees of five varieties, Beurré d'Amanlis, Beurré Hardy, Conference, Fertility and Pitmaston Duchess. To the four trees of one variety in one block, a factorial set of pruning treatments was applied, i.e. either hard pruned or lightly pruned with either few or many leaders. The full set of 20 treatment combinations was therefore of structure $2^2 \times 5$. Totals were:

		Amanlis	*Hardy*	*Conference*	*Fertility*	*Pitmaston*
Few	Hard	5129	3416	3326	4382	2625
	Light	5181	4045	5022	4875	3480
Many	Hard	6063	3764	3967	4430	2809
	Light	6626	4516	4909	5906	3664

Work out the full factorial partition of the sum of squared deviations.

Answer: Calling the factors V for varieties, N for number of leaders and D for degree of pruning, the figures given enable S_{VND} to be calculated directly, namely

$$S_{\text{VND}} = (5129^2 + 3416^2 + \cdots + 3664^2)/8 = 51\,289\,262.$$

So many significant figures are not really needed, but they arise directly from the squared data and may as well be retained. The method of using summation terms is available on account of the orthogonality.

Factors should now be omitted one by one. First, summing over varieties gives

	Hard	Light
Few	18 878	22 603
Many	21 033	25 621

Hence,

$$\begin{aligned} S_{\mathrm{ND}} &= (22\,603^2 + 18\,878^2 + 25\,621^2 + 21\,033^2)/40 \\ &= 49\,152\,431 \end{aligned}$$

Now summing over number of leaders and over degree of pruning, two more summation terms can be found.

	Amanlis	*Hardy*	*Conference*	*Fertility*	*Pitmaston*
Hard	11 192	7 180	7 293	8 812	5 434
Light	11 807	8 561	9 931	10 781	7 144
Few	10 310	7 461	8 348	9 257	6 105
Many	12 689	8 280	8 876	10 336	6 473

They give

$$S_{\mathrm{VD}} = (11\,192^2 + 7\,180^2 + \ldots + 7\,144^2)/16 = 50\,985\,515$$

$$S_{\mathrm{VN}} = (10\,310^2 + 7\,461^2 + \cdots + 6\,473^2)/16 = 50\,731\,245$$

The next summation terms needed are those for single classifications, i.e. two fewer than the original. It will be noticed that each can be obtained in two ways. For example, either of the two tables immediately above this will give varietal totals, which are respectively

22 999 15 741 17 224 19 593 12 578

Other totals are:

Number of leaders	41 481	46 654
Degree of pruning	39 911	48 224

Finally, the grand total can be found in three ways as 88 135. Hence

$$S_{\mathrm{V}} = (22\,999^2 + 15\,741^2 + \cdots + 12\,578^2)/32 = 50\,484\,093$$

$$S_{\mathrm{N}} = (41\,481^2 + 46\,654^2)/80 = 48\,715\,863$$

$$S_{\mathrm{D}} = (39\,911^2 + 48\,224^2)/80 = 48\,980\,526$$

$$S_0 = 88\,135^2/160 = 48\,548\,614$$

Hence the partition is

Source	*d.f.*	s.s.
Varieties (V)	4	1 935 479
Number (N)	1	167 249
Degree (D)	1	431 912
N × D	1	4 656
V × D	4	69 510
V × N	4	79 903
V × N × D	4	51 939
Treatment	19	2 740 648

The error line had 133 degrees of freedom and gave a variance of 9096, so it appears that no interactions are significant. That allows attention to be concentrated on the main effects and here all are significant at a level, $P < 0.01$. (Of course, there is no point really in testing the effect of V, because the five varieties were chosen to be as different as possible, though all at the time were of commercial importance.) It may be concluded that for a wide range of varieties light pruning with many leaders gave the best crops. That lesson has now been learnt and the spur pruning used in the experiment is no longer a commercial method.

B. The following data (Andrews and Herzberg, 1980) show yields per plot in kilograms from an experiment, encoded LFAN in their list. In each plot 16 square yards were recorded. There was a 2^3 factorial set of treatments, made up of nil and double levels of three nutritional elements, nitrogen, phosphorus and potassium. In each block there were also two plots that received the usual application of each element. Analyse the data (i) ignoring the additional plots and considering only the factorial set and (ii) using all the data (1 square yard = 0.00051 hectare)

	Block				
Treatment	I	II	III	IV	Total
1 1 1	2.92	5.40	4.94	6.02	33.66
1 1 1	2.94	4.14	3.72	3.58	
0 0 0	1.60	—	2.90	—	4.50
0 0 2	—	1.72	—	1.48	3.20
0 2 0	—	3.94	—	3.18	7.12
0 2 2	1.68	—	1.22	—	2.90
2 0 0	—	5.62	—	3.95	9.57
2 0 2	4.44	—	5.26	—	9.70
2 2 0	4.74	—	5.00	—	9.74
2 2 2	—	5.78	—	5.94	11.72

Answer: It will be seen that the three-factor interaction of N × P × K has been confounded between Blocks I and III on the one hand and Blocks II and IV on the other.

Table 8B
Calculations using Yates's algorithm for the 2^3 factorial set in Example 8B.

Treatments	*Totals*	(A)	(B)	(C)	*S.S.*
000	4.50	7.70	17.72	58.45	213.5252
002	3.20	10.02	40.73	3.41	0.7268
020	7.12	19.27	5.52	−4.51	1.2713
022	2.90	21.46	−2.11	−1.07	0.0716
200	9.57	1.30	−2.32	−23.01	33.0913
202	9.70	4.22	−2.19	7.63	3.6386
220	9.74	−0.13	−2.92	−0.13	0.0011
222	11.72	−1.98	1.85	−4.77	1.4221

Omitting Treatment 111 the design has a 2^3 factorial set and is readily approached using Yates's algorithm, but first the summation terms will be helpful:

Total	256.7933
Blocks	216.1955
Treatments	253.7477
Correction	213.5252

To use the algorithm it is first necessary to set out treatments in the order given above. That is to say, the last factor is allowed to change, the others being kept constant, i.e., the list starts with 000 and 002. Then the last factor but one is changed, giving 020 and 022. Then the last but two is changed to give 200, 202, 220, 222. If there were more factors the process could be continued. The remaining calculations are set out in Table 8B, which starts by listing the treatments and their data totals. The upper half of Column (A) is formed by successive sums of values taken in pairs, i.e. $4.50 + 3.20$, $7.12 + 2.90$, $9.57 + 9.70$, and $9.74 + 11.72$. The lower half is similarly formed from differences, i.e. $4.50 - 3.20$, $7.12 - 2.90$, $9.57 - 9.70$, and $9.74 - 11.72$. Column (B) is then formed from Column (A) in the same way and Column (C) from Column (B). The process ends when the summing and differencing has taken place as many times as there are factors.

It will be seen on examination that the algorithm has generated in Column (C) the following vectors in order,

$$(+1,\ +1,\ +1,\ +1,\ +1,\ +1,\ +1,\ +1)'$$
$$(+1,\ -1,\ +1,\ -1,\ +1,\ -1,\ +1,\ -1)'$$
$$(+1,\ +1,\ -1,\ -1,\ +1,\ +1,\ -1,\ -1)'$$
$$(+1,\ -1,\ -1,\ +1,\ +1,\ -1,\ -1,\ +1)'$$
$$(+1,\ +1,\ +1,\ +1,\ -1,\ -1,\ -1,\ -1)'$$
$$(+1,\ -1,\ +1,\ -1,\ -1,\ +1,\ -1,\ +1)'$$
$$(+1,\ +1,\ -1,\ -1,\ -1,\ -1,\ +1,\ +1)'$$
$$(+1,\ -1,\ -1,\ +1,\ -1,\ +1,\ +1,\ -1)'.$$

Bearing in mind that a contrast is not affected by reversal of all signs they represent respectively the grand total, the main effect of K, the main effect of P, the interaction of P and K, the main effect of N, the interaction of N and K, the interaction of N and P and the three-factor interaction. (The treatments in the first column give an indication of the contrasts eventually reached, as may readily be noted.) Squaring each quantity in Column (C) and dividing it by the number of plots assigned to the factorial set, i.e. by 16, will give sums of squared deviations, which are set out in the last column. The first is the correction term; the others are for the treatment effects. In fact, the analysis is:

Source	*d.f.*	*s.s.*	*m.s.*	*F*
Blocks	3	2.6703		
N	1	33.0913		110.46
P	1	1.2713		4.24
K	1	0.7268		2.43
P × K	1	0.0716		0.24
N × K	1	3.6386		12.15
N × P	1	0.0011		0.00
Error	6	1.7974	0.29957	
Total	15	43.2681		

There is a rounding error of 0.0003.

The blocks have been obtained from summation terms in the usual way, i.e. as 216.1955 − 213.5252, as has the total, i.e. 256.7933 − 213.5252. The partitioned treatment effects come from the algorithm, while the error is found as usual but with the addition of the confounded interaction, i.e.

$$(256.7933 - 216.1955 - 253.7477 + 213.5252) + 1.4221.$$

There is a rounding error of 0.0003.

A feature of the analysis is the clear warning that the effects of nitrogen and potassium are interacting, though with little evidence of a main effect of potassium. Examination of the means suggests a conclusion

		K	
		0	2
N	0	2.91	1.53
	2	4.83	5.36

It looks as if potassium in the absence of nitrogen may even have done harm. The difference of 1.38 (= 2.91 − 1.53) has a variance of $(\frac{1}{4}+\frac{1}{4})$ 0.29957 $= (0.387)^2$, so it does appear to be significant ($P = 0.025$).

Now if the median treatment, 111, is included, nothing much is altered. The contrast added is that of 111 versus the rest, which shows whether evenness of application of fertilizer application is important. The contrast is orthogonal, so the obvious way of finding the sum of squared deviations is to use summation terms, i.e.

$$\left(\frac{33.66^2}{8}+\frac{58.45^2}{16}\right)-\frac{92.11^2}{24}=1.6391.$$

Other summation terms are now

$$S(\text{total}) = 407.5617$$
$$S_b(\text{blocks}) = 359.5411$$
$$S_t(\text{treatments}) = 395.3722$$
$$S_o(\text{correction}) = 353.5105$$

There is, however, one complication. The design is no longer disconnected, so there is some intra-block information about the three-factor interaction. Since it will be poorly determined and since it was discarded before, there is perhaps no reason to consider it now. However, it will here be taken into account to illustrate the method. It is not difficult to find the coefficient matrix, **C**, whence it appears that the contrast vector for the three-factor interaction is an eigenvector with an effective replication equal to $\frac{2}{3}$. Since the actual replication is 2, that gives an efficiency factor of $\frac{1}{3}$. Accordingly only one-third of the sum of squared deviations given by the algorithm will be ascribed to it, namely 0.4740 = 1.4221/3. The analysis now reads

Source	*d.f.*	*s.s.*	*m.s.*
Blocks	3	6.0306	
(111)v the rest	1	1.6391	
N	1	33.0913	
P	1	1.2713	
K	1	0.7268	
P × K	1	0.0716	
N × K	1	3.6386	
N × P	1	0.0011	
N × P × K	1	0.4740	
Error	12	7.1070	0.59225
Total	23	54.0512	

Again there is a small rounding error.

There are several things worth noting about the analysis. One is the sharp rise in error variance from 0.29957 to 0.59225. Examination of the data suggests that the trouble comes from the duplicate values in Block IV, namely 6.02 and 3.58, which contribute $\frac{1}{2}(6.02-3.58)^2 = 2.9768$ with one degree of freedom to error. That leaves a variance of 0.37547, which is much closer to what was found before, so there is perhaps something to investigate here. As to the conclusions themselves they are much as before. There does not appear to be a three-factor interaction. Also, Treatment 111 has given a mean reasonably close to that of the rest combined, which suggests that all three fertilizer elements are being applied at a low level where the response curve is still straight, so dressings could perhaps be advantageously increased. (The experiment did in fact include three supplementary treatments, 311, 131, and 113, in case it should appear that the applications were appreciably below the optimal values.)

C. Take the 3 × 3 factorial set used in Section 8.2 and work out main effects and interactions using the method of components, as described in Section 8.4. *Answer:* The data totals were

	O	A_p	A_s
Red Aus	698	730	708
Kashiphul	520	498	466
Dudkalma	456	705	609

The method associates the table with the Graeco-Latin square.

$A\alpha$	$B\beta$	$C\gamma$
$C\beta$	$A\gamma$	$B\alpha$
$B\gamma$	$C\alpha$	$A\beta$

For the rows, i.e., the main effect of varieties, totals are 2136, 1484 and 1770, the grand total being 5390. The sum of squared deviations is therefore

$$(2136^2 + 1484^2 + 1770^2)/18 - 5390^2/54 = 11867.7.$$

For the columns, i.e., the main effect of fertilizers, the corresponding figure is

$$(1674^2 + 1933^2 + 1783^2)/18 - 5390^2/54 = 1878.9.$$

For the Roman letters and the Greek letters, which together make up the interaction, the figures are

$$(1805^2 + 1652^2 + 1933^2)/18 - 5390^2/54 = 2199.1$$

$$(1869^2 + 1859^2 + 1662^2)/18 - 5390^2/54 = 1514.0.$$

The total for the interaction is therefore 3713.1.

D. Rayner (1969) presents the following data from an experiment in five randomized blocks. The data represent yields of maize in pounds per plot of 0.01 morgen (1 lb = 453.6 gr, 1 morgen = 0.86 hectare). Unfortunately for present purposes, the five treatments, A, B, C, D, and E, are unspecified.

	Treatment				
	A	B	C	D	E
Block I	33.7	37.1	33.7	34.3	32.6
II	36.5	40.7	42.7	42.7	51.3
III	37.1	44.4	51.0	45.8	56.5
IV	37.1	40.4	50.2	44.0	50.4
V	37.3	46.2	49.0	25.8	47.0

Apply Tukey's test for non-additivity, described in Section 8.9.

Answer: It may be assumed that the ordinary analysis of variance has already been carried out. It reads

Source	*d.f.*	*s.s.*
Blocks	4	$457.49 = H$
Treatments	4	$430.45 = J$
Error	16	420.66
Total	24	1308.60

Estimation of the data suggests that there could well be an interaction of the kind considered by Tukey. Thus, the least fertile block is I with a mean of 34.3 compared with a general mean of 41.9, while the worst treatment is A with a mean of 36.3. Hence Treatment A in Block I could be expected to give a yield of

$$34.3 + 36.3 - 41.9 = 28.7.$$

In fact it does much better. Similarly, the best treatment in the best block, i.e. E in III, could be expected to give

$$47.6 + 47.0 - 41.9 = 52.7.$$

Again, it does better. Certainly, there could be an interaction of the sort given by a positive λ, i.e. low values are avoided and high ones are exaggerated.

The data are presented so as to give the matrix, $\mathbf{Y}$, where $b = v = 5$ and $G = 1047.5$. Also,

$$\mathbf{B}' = (171.4 \quad 213.9 \quad 234.8 \quad 222.1 \quad 205.3)$$
$$\mathbf{T}' = (181.7 \quad 208.8 \quad 226.6 \quad 192.6 \quad 237.8),$$

so

$$K = \mathbf{B}'\mathbf{Y}\mathbf{T} - G^3/n = 952\,953$$
$$K - G(H + J) = 22\,836,$$

and the contribution of λ to error is

$$\frac{22\,836}{25 \times 430.45 \times 457.49} = 105.92,$$

so the partition of error reads

Source	*d.f.*	*s.s.*	*m.s.*	*F*
Non-additivity	1	105.92	105.92	5.04
Residue	15	314.74	20.98	
Error	16	420.66		

There is, in fact, some evidence for such an interaction.

E. Examine the data of Example 3A to see if there is evidence of a block × treatment interaction. (See Section 8.9)

Answer: Altogether there are eight pairs of comparable plots. Thus in Block I there are two of Treatment A, which gives values of 166 and 163 and two of Treatment O with 177 and 190. The contribution is therefore

$$\tfrac{1}{2}(166-163)^2+\tfrac{1}{2}(177-190)^2=\tfrac{1}{2}(3^2+13^2).$$

Other blocks give similar pairs, so that there are eight degrees of freedom in all, corresponding to error variation that is free of interaction, their total sum of squared deviations being

$$\tfrac{1}{2}(3^2+13^2+13^2+18^2+21^2+10^2+14^2+12^2)=776.$$

Hence, there is a variance of $776/8 = 97.00$ with eight degrees of freedom that represents random error and another of $(3460-776)/(20-8) = 223.67$ with twelve degrees of freedom which represents random error possibly inflated by an interaction of blocks and treatments. That makes F equal to $223.6/97.00 = 2.31$ with twelve and eight degrees of freedom, which certainly raises suspicion of an interaction but without establishing that one exists. It will be helpful to look at block-treatment means, i.e.

	A	B	C	D	O
I	165	133	166	107	184
II	136	126	152	104	155
III	117	132	129	103	181
IV	130	103	147	102	179

There is no obvious pattern in the table so the question is better left.

F. The following data represent the yield of dry matter in pounds for three green manure crops (A = Sunnhemp, B = Woolly Pyrol, C = Sword bean) in six randomized blocks (Federer, 1955). Actually there were two factors, three species and two levels of nitrogen, but the other factor has been ignored and the figures show the yield of two plots, each of $\frac{1}{36}$ acre. (1 pound = 453.6 gr, 1 acre = 0.405 hectare.)

	I	II	III	IV	V	VI
A	535	565	552	613	735	593
B	183	247	202	370	327	182
C	354	388	399	459	379	577

Examine the data for additivity.
Answer: An attempt to use Tukey's method revealed that the non-additivity, which is considerable, certainly did not follow the form implied in his approach. Indeed, the sum of squared deviations removed, namely, 10 out of a total of 50 355, was remarkably low. (This is perhaps the place to comment that some forms of non-additivity, so far from inflating Tukey's component, appear to suppress it.) Examination of the data reveals a phenomenon that is sometimes encountered. The first three blocks taken separately give an error variance of 230; the other three one of 12 321. Only examination of the site

can explain what happened, but the example shows how experiments sometimes give information about the land on which they are conducted as well as about the effects of treatments.

G. Take the data of Example 2A and partition the treatment sum of squared deviations to give main effects and the quality–quantity interaction.
Answer: Using the method of Example 8A and Section 8.7 there are six summation terms needed:

(1) Using all data
  A. Quantity (i.e. weight of applications)
  B. Correction
(2) Ignoring data from the control
  C. Quantity and Quality (i.e. Season of application)
  D. Quantity
  E. Quality
  F. Correction.

They can readily be found

$$A = (181^2 + 105^2 + 135^2 + 80^2)/8 = 8551.38$$
$$B = 501^2/32 = 7843.78$$
$$C = (38^2 + 62^2 + \ldots + 57^2)/4 = 4721.00$$
$$D = (105^2 + 135^2 + 80^2)/8 = 4456.25$$
$$E = (123^2 + 197^2)/12 = 4494.83$$
$$F = 320^2/24 = 4266.67.$$

The partition is therefore

Source	*d.f.*	*s.s.*
Weight	3	707.60
*Season	1	228.16
*Interaction	2	36.59
Treatments	6	972.35

Lines marked with an asterisk are derived ignoring data from the untreated control.

Chapter 9

Some special topics

9.1 SYSTEMATIC DESIGNS

So far all the emphasis has been on randomized designs, which do indeed represent the norm, but there are occasions when randomization is impracticable or unwise. An obvious example is afforded by spacing trials. They are difficult anyway, but chiefly because extreme treatments must not adjoin. If they do, on account of wind shadows and the like, a plot will not behave as it would do if surrounded by an area of its own treatment. Very large discard areas would be needed to overcome the difficulty and they would be added to plots that were large anyway. (Thus, row spacings of 80 cm, 100 cm and 120 cm really need plots 12 metres wide to contain 15, 12 and 10 rows respectively. Further, if outside rows are discarded to avoid edge effects, that leaves plots of unequal area. It all gets very complicated.) The difficulties are, however, much reduced once it is recognized that the need is simply to compare each treatment with its neighbours in the series, contrasts between extremes being of little interest. It then becomes appropriate to use one of the 'fan designs' described by Nelder (1962) and illustrated in Fig. 3. Basically all involve planting on rays emanating from a centre, the angle between successive rays usually being about five degrees. Apart from that they do not necessarily have much in common. Thus, if the object is to compare inter-row spacings, planting density along a row can well be constant as in the top example in Fig. 3. Perhaps, however, intra-row spacings are to change proportionately with the inter-row to maintain rectangularity of plot shape as in the middle example. Another possible aim is to maintain a constant plot area but to use a range of rectangularities, in which case a design like that in the bottom example may be used, the intra-row spacing decreasing as the inter-row distance increases. Then at one end of the fan the planting is in rows close together but with wide spacing in the row; at the other the plants are packed close in rows far apart. An example is afforded by Freeman (1964b).

As a mathematical pastime designing fan trials is quite entertaining and more useful than some better-known alternatives. Anyone who tries his hand at it should remember that the product of his thoughts will have to be laid down on land that is not an exact plane. Over a small area that may not matter a lot, but

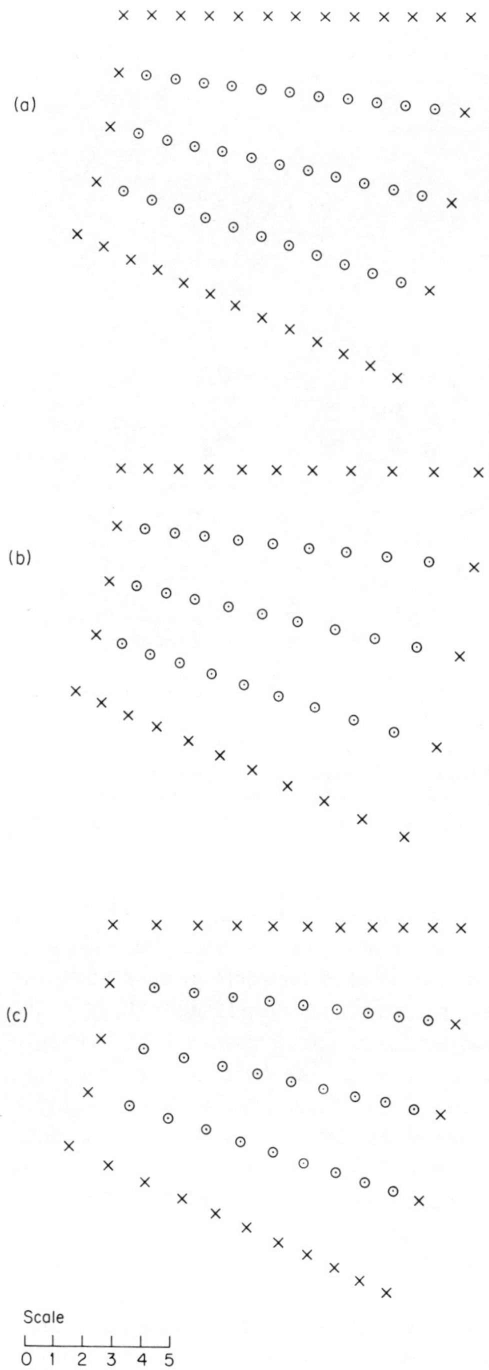

Figure 3. The three fan designs illustrated are similar but adapted to different purposes. In each × indicates an external ground plant and ⊙ an experimental plant, i.e. one that is internal. Each such internal plant is assumed to occupy the land half-way to its neighbour in any direction. It therefore grows in a space that is approximately a rectangle and can be described by two variables, namely, A, the area, and R, the rectangularity (the ratio of the longer side of the rectangle to the smaller).

In Fan(a) the spacing within rows is constant at one unit. Consequently both A and R are variable, their values ranging from $A = R = 2.0$ at the inner end to $A = R = 3.0$ at the outer.

In Fan(b) as the rows become farther apart, the spacing within rows is increased to keep the rectangularity constant. In this example $R = 2.0$. Here A increases from 2.00 at the inner end to 4.45 at the outer.

In Fan(c) the spacing within rows is decreased as that between rows is increased. That is done so as to keep the area constant. Here $A = 3.0$ throughout. In consequence R varies from 1.40 at the inner end to 3.00 at the outer.

Calculations for all fans were based on the formulae given by Pearce (1976a, Section 66)

Figure 4. The fan trials described in Section 9.1 are often extended to form a complete circle. The experiment on Brussels sprouts shown in this picture (Finch, Skinner and Freeman, 1976) was intended to study the effect of plant density on insect population and was designed to give a range of spacings with constant rectangularity. It is unusual in having a sector 120° wide on which insecticide was applied, the remaining 240° being untreated. Reproduced by permission of the National Vegetable Research Station

larger experiments will call for the services of a surveyor and he will be more pleased if he can fix the perimeter of the fan and work inwards, thus providing himself with checks, than if he has to start from somewhere in the middle and work outwards. Also, farm supervisors sometimes insist on straight alleys for the movement of farm implements. The writer has presented his own solution to such problems elsewhere (Pearce, 1976a, Section 66), but he is not the only one (Bleasdale, 1967b; Freyman and Dolman, 1971; Rogers, 1972). There is plenty of guidance in the literature, but many experimenters produce their own schemes and there is no reason why they should not. Plots will be chords crossing the rays. All plants of a plot can be recorded collectively, because rays of the same fan do not provide independent testimony as to spacing effects.

Whatever kind of fan is used, the data should not be regarded as coming from a randomized experiment, because they do not. What they express is a relationship between growth (or crop or disease incidence) and the character that is constrained to alter along the rays, whether that is area, inter-row spacing or rectangularity. The aim is to find the maximum, or perhaps the minimum, of the measured quantity. Hence the appropriate statistical tool is the fitting of a curve.

Here a number of papers will be found useful for indicating the sort of relationship that could be considered (Holliday, 1960; Bleasdale and Nelder, 1960; Bleasdale, 1967ab; Farazdagh and Harris, 1968). The general problem of spacing has been usefully examined by Mead (1966, 1967) and Berry (1967).

Like any other kind of spacing trial fans give rise to changes in microclimate but, being systematic, their effects are more obvious. In particular wind is a nuisance. If it blows into a fan it is funnelled and its effects are intensified, especially at the narrow end; if it blows the other way it is divided and the whole fan is unduly calm, again especially at the narrow end. For that reason it is often convenient to have fans in pairs side by side, the narrow end of one adjoining the wide end of the other. Not only does that give an approximately rectangular shape, but it is to be expected that biasses in the two fans, whether due to microclimate or to fertility trends, will be in opposite directions. (It is not reasonably to be expected that they will cancel, only that they will to some extent compensate for one another.) Also, it is not unusual for fans to have enough rays to complete a circle. Since it is difficult to regard a pair of compensating fans or a circle as being independent of a similar figure alongside, there is much to be said for dispersing them. As remarked in Section 1.2, any experiment confined to one site gives results that are difficult to generalize, whereas dispersal widens the inferential base and it does so whatever the design.

Similar problems arise with fertilizer trials. If the aim is to compare different sources of the same element, e.g. potassium applied as chloride or as sulphate, everything can be quite straightforward, but if the task is to find optimal levels of an element the same difficulty arises if extreme treatments come together. To arrange for a heavily fertilized plot to adjoin one that is virtually unfertilized is to ask for trouble. Roots of plants in the starved area will search for food. As soon as one finds its way into the neighbouring plot, it will develop at the expense of those that have been less successful and robbing can occur on a large scale. It is true that plots can be made large and the outsides discarded. Alternatively, it is possible to place discard strips between the plots and to assign them an intermediate level of fertilizer. Either way the difficulties of an excessive experimental area begin to appear in the same way as with the spacing trials. Deep ploughing of alleys between plots will cut roots that have strayed, but that may only delay matters. As soon as a deep root gets across, the trouble begins. One solution that is effective but expensive is to cut deep trenches and to place sheets of plastic in them before refilling (Goode and Marchant, 1962; Marchant and Boa, 1962). It is no wonder that a common solution is the adoption of a systematic design that keeps extreme treatments apart. Again the object is to find an optimal input, the contrasts of chief interest being those between a treatment and those most like it, so curve fitting should give all that is needed. The object, however, is not necessarily the finding of a maximum, so it is useful to know about the relationship between fertilization and cropping advanced by Mitscherlich (1934) and ways of fitting it (Pimental-Gomes, 1953), and of bringing in economic considerations (Finney, 1953). A suitable design is obtained by dividing a rectangular area into long, narrow plots laid side by side. The lowest fertilizer application is given at one end and increased by one step for each plot until the maximum is given at the other end. A second adjacent rectangle is needed with the fertilizer pattern reversed to compensate for trends that could lead to bias. If there are two elements under study, say, nitrogen and phosphorus, a factorial design is possible. The unit plots are now very small, being formed by the intersection of strips at right angles. One element is applied in graded amounts to successive rows and the other in the same way to columns. The result is the growing of a response surface for all to see. Here four rectangles are needed to form a self-compensating set; any further quartets are better placed at a distance to achieve independence. The whole approach has been discussed by Cleaver *et al.*, (1970).

So far the designs considered have implied either a rectangular or a radial format of plants. In the first, plots have four sides at right angles; in the latter they form trapezia, as in a fan trial. However, where individual plants form the plots there is often much to be said for a triangular format, in which each plant lies at the centre of a regular hexagon, three adjacent plants forming an equilateral triangle, as shown in Fig. 5. Rows are a distance, $\sqrt{3}a/2$, apart and plants are a distance, a, apart in rows. The feature that distinguishes a triangular format from a rectangular one is the displacement of each row by $\frac{1}{2}a$ relative to those on either side of it. That gives a distance between neighbours of a. For example, if such a format is adopted and one diseased plant is introduced somewhere in the middle useful information should become available about the spread of the disease, whether it is downwind, along the direction of cultivation, strictly to adjacent plants or by the sporadic appearance of fresh foci, the last being characteristic of infection from mobile insects. Since each plant has six neigh-

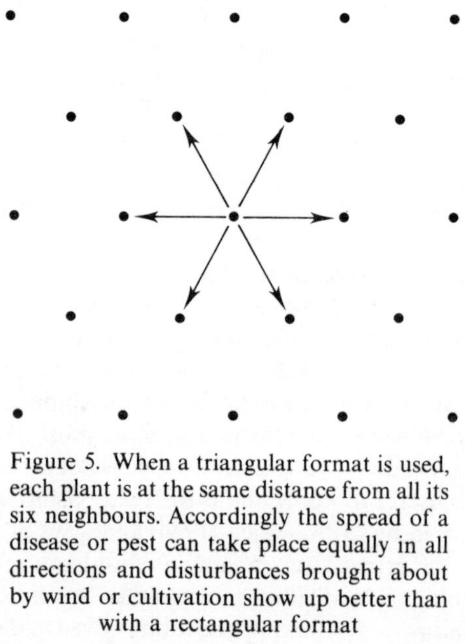

Figure 5. When a triangular format is used, each plant is at the same distance from all its six neighbours. Accordingly the spread of a disease or pest can take place equally in all directions and disturbances brought about by wind or cultivation show up better than with a rectangular format

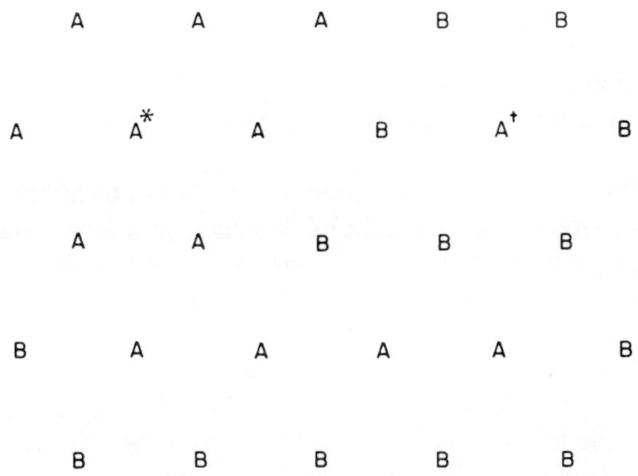

Figure 6. In a beehive design two kinds of plant, A and B, are intermingled on a triangular format in such a way that there is a wide range of competition. Thus, there are here nine internal plants of A. The one marked * has all six neighbours of its own kind, whereas that marked † has all neighbours of B. All intermediate cases are represented

bours, set on six equally spaced radii, spread can take place in all directions with nearly equal ease.

Triangular formats have an especial usefulness in competition experiments. It was Martin (1973) who suggested designs like those in Fig. 6. They are called 'beehive' designs. Here A and B represent two kinds of plant. Excluding those on the perimeter there are eleven in all, eight being of type A. Of those eight, one is surrounded by its own kind, one has one neighbour of B, another has two such neighbours and so on up to one that has all neighbours of B. The only break in the series comes when half the neighbours are A and half are B because there are two plants of A so placed. It is therefore possible to plot performance of those eight plants against the number of similar neighbours and so determine competition effects.

The design suggested is unsatisfactory in several respects. For one thing there are not really enough plants to determine performance under any degree of competition. For another, the plants with few A-neighbours are at one end and those with many at the other. Finally, a better design might enable the B-plants to be studied also, but with more plants those objections can be overcome.

Those matters were discussed by Martin in his original paper and by Veevers and Boffey (1975, 1979). Mead (1979) has considered competition experiments in general.

9.2 COMPLETE LATIN SQUARES AND SIMILAR DESIGNS

For many years agronomists have sought to obtain genuine 'polycross designs' in which there are v treatments disposed over a field in such a way that blossoms of any variety will be pollinated equally by the other $(v - 1)$. Only in that way can the varieties be fairly compared as seed parents, because if Variety A were always pollinated by B and C and Variety B always by C and D it would not be clear whether the comparison was being made on the basis of potentiality as a seed or as a pollen parent.

Freeman (1967, 1969) has drawn attention to the problem of design and has suggested various solutions, but first he made a distinction between the directional case, like that of wind-borne pollen where the direction of the wind can be foreseen, and the non-directional, when pollen might spread in all directions. (Of course, the distinction cannot be complete. During the critical period the wind may blow predominantly from one direction, but there could be occasions when it would blow from another and those occasions could be anything from rare to frequent according to location.) Also, as Dyson and Freeman (1968) point out, there are a number of other difficulties, e.g. the varieties may not produce pollen in equal amounts or differences in time of blossoming may have a disturbing effect. All that must be recognized, but those difficulties exist however the varieties are distributed; the optimal design still needs to be sought.

Freeman has considered design problems further in other papers (Freeman, 1979a, 1979b, 1981). If a systematic Latin square is taken, e.g.

A	B	C	D
B	C	D	A
C	D	A	B
D	A	B	C

then A always has B to the east and south of itself, and D to the west and north, while other varieties are surrounded in the same way. Nevertheless, if the wind came mostly from the west in a certain season, A would chiefly be pollinated by D, B by A, C by B, and D by C, so the confounding of seed parentage and pollen parentage would arise, as it would if the wind blew from any other quarter. However, designs known as 'complete Latin squares' exist in which the directional pattern is eliminated, e.g. as in

A	B	C	D
B	D	A	C
C	A	D	B
D	C	B	A

Here, A has three neighbours to the north, and they prove to be one each of B, C and D. So it is for the other directions and for all other varieties and directions. The second design is therefore admirable for the purpose of pollen dispersal. With some numbers of varieties, rows and columns, perfect designs are not possible, but often it is possible to arrange for a variety to have a wide range of neighbours if it does not always adjoin each of the other varieties. Further, by selecting seed only from certain plants, the others serving solely as pollen parents, other possibilities appear.

It may be objected that such work is not concerned with experimentation in the ordinary sense but with raising the material for experiments, and that is true. It is nonetheless relevant. The complete Latin square, however, is interesting on another account. One difficulty with the method of Papadakis, as set out in Section 1.5, is the way in which the concomitant value for a plot depends upon the treatments applied to its neighbours. Do complete Latin squares help the situation? Also, is there a realistic model to be built that is concerned with correlations over plot boundaries? If so, a design in which all pairs of treatments adjoin could be invaluable. The future may, therefore, hold some excitement. Of course, there is a danger of building some superb system of experimentation upon a postulate about fertility patterns which, however convenient, may be unreliable. To some extent the established theory of block designs has done that, but the use of randomization to break the effect of spatial correlations has provided a safeguard. Systematic designs might entail more risks.

A similar problem can arise in two dimensions. Dyke and Shelley (1976), for example, considered the case of four barley varieties, A, B, C and D, with different susceptibilities to mildew. On account of practical difficulties like the spraying of discard areas they wanted to arrange their varieties so that each occurred with all 36 possible combinations of neighbours, i.e. A B A, A B C, etc., but not A B B. By using a computer they obtained a number of solutions, e.g.

ABACDABCDBDACBCADADBCBDCACABDBADCDCBAB

As they remark, there are a number of modifications that could well be made to accommodate other problems. Such designs cannot strictly speaking be regarded as validly randomized. Nevertheless, a reasonable approximation is possible; further, a covariance adjustment on position in the row is quite easy to arrange.

9.3 BIVARIATE ANALYSIS OF VARIANCE

Originally the analysis of variance was thought of in purely univariate terms. Several quantities might be measured, say, crop both as fresh weight and as dry weight, plant height, straw produced etc., but each would be analysed separately. However, multivariate tests were soon devised (e.g. Wilks, 1932) and have received a lot of attention since but without finding much favour in an agricultural context.

In any context a lot of silliness results if people just throw unrelated variates into a computer to be processed by some magnificient program that no one really understands, but in agriculture there are occasions when the problem is in fact multivariate in character. Perhaps that goes too far but 'bivariate' does not. The amendment is important because bivariate tests are sometimes exact, whereas more general ones rarely are. Also, conclusions from them can be represented on a two-dimensional sheet of paper; more variates are more difficult to set out.

To take an example, the size of strawberry plants is commonly measured either by height or spread or both. It is true that the two measures are ordinarily so closely related that one will serve for the other, but their relationship can be revealing. When a plant is infested with virus it may to some extent collapse, in which case spread is increased and height decreased. Neither the gain in spread nor the loss in height may appear significant by itself, but the fact that the variates have moved in opposite directions when usually there is a strong positive correlation between them may show up markedly in a bivariate analysis. Much the same situation can arise in an irrigation trial where the availability of water may cause plants to revive with a decrease in spread. To take another example, there are some species that give two crops. Thus, millet and sorghum provide fodder as well as grain, a feature of some importance where draft oxen have to be fed as well as people. A treatment might appear to have no effect on either crop, but a bivariate analysis might reveal that the movement of the two variates went against their usual relationship. At this point someone might object that two negligible changes are no more important than one, but non-significant treatment effects can be quite considerable and two such effects can amount to something quite important. (The objection does, however, draw attention to the need for full assessment of a bivariate conclusion, a point that will be considered later.) A third example, and perhaps the most important, concerns intercropping experiments in which two species, such as sorghum and cowpea or maize and cassava, are grown in alternate rows or in some other geometric configuration. Ordinarily a gain for one species in a plot is a loss for the other on account of competition, but if some treatment, like heavier fertilization, leads to an increase in both crops it is not to be lightly dismissed. It may be that neither increase is significant when examined by itself in a univariate analysis, but together, acting as they do against the usual correlation, they may well be significant in a bivariate one. Further, two small gains taken together may represent a very acceptable increase in productivity.

A bivariate analysis is not difficult to carry out. As explained in Chapter 3 a univariate analysis is based upon two minimizations, one representing the residual sum of squared deviations, $\mathbf{y}'\boldsymbol{\phi}\mathbf{y}$, from a model (3.3.10) that does not contain any treatment parameters and the other, $\mathbf{y}'\boldsymbol{\psi}\mathbf{y}$, that arises when treatment parameters are included (3.1.1). The criterion, F, is

$$F = \frac{e(\mathbf{y}'\boldsymbol{\phi}\mathbf{y} - \mathbf{y}'\boldsymbol{\psi}\mathbf{y})}{h\mathbf{y}'\boldsymbol{\psi}\mathbf{y}}$$

with h and e degrees of freedom, where h is the number of independent unconfounded treatment contrasts and e is the number of degrees of freedom for $\mathbf{y}'\boldsymbol{\psi}\mathbf{y}$. In a bivariate analysis there is a second variate x. Writing

$$D = \det\begin{bmatrix} \mathbf{y}'\boldsymbol{\phi}\mathbf{y} & \mathbf{x}'\boldsymbol{\phi}\mathbf{y} \\ \mathbf{y}'\boldsymbol{\phi}\mathbf{x} & \mathbf{x}'\boldsymbol{\phi}\mathbf{x} \end{bmatrix} \tag{9.3.1}$$

and

$$E = \det\begin{bmatrix} \mathbf{y}'\boldsymbol{\psi}\mathbf{y} & \mathbf{x}'\boldsymbol{\psi}\mathbf{y} \\ \mathbf{y}'\boldsymbol{\psi}\mathbf{x} & \mathbf{x}'\boldsymbol{\psi}\mathbf{x} \end{bmatrix}, \tag{9.3.2}$$

the bivariate F is found as

$$F = \frac{1-\sqrt{\Lambda}}{\sqrt{\Lambda}}\,\frac{e-1}{h} \tag{9.3.3}$$

with $2h$ and $2(e-1)$ degrees of freedom (Rao, 1952a), where $\Lambda = E/D$. The off-diagonal elements of the matrices need cause no difficulty. If they cannot readily be obtained in any other way, it is sufficient to analyze $\mathbf{z} = \theta\mathbf{x} + \mathbf{y}$, where θ is a scaling factor chosen so that the error variance of θx shall be approximately that of $\mathbf{y}$. Then

$$\begin{aligned} \mathbf{y}'\boldsymbol{\psi}\mathbf{x} &= \mathbf{x}'\boldsymbol{\psi}\mathbf{y} \\ &= (\mathbf{z}'\boldsymbol{\psi}\mathbf{z} - \theta^2\mathbf{x}'\boldsymbol{\psi}\mathbf{x} - \mathbf{y}'\boldsymbol{\psi}\mathbf{y})/(2\theta) \end{aligned} \tag{9.3.4}$$

and similarly for $\mathbf{y}'\boldsymbol{\phi}\mathbf{x}$ and $\mathbf{x}'\boldsymbol{\phi}\mathbf{y}$. Of course, the bivariate F-test given by (9.3.3) can be applied to partitions of the treatment effects, e.g. main effects, interactions, linear and quadratic effects, as well as to the whole.

However, as has been said, a bivariate analysis presents special difficulties of assessment. Where only one variate is involved there are no problems of scaling and the best treatment is the one with the largest, or perhaps the smallest, mean. Where there are two variates, say yield of fodder and yield of grain, there are the problems of scaling that arise from having two variances and a correlation coefficient. Also it is necessary to be clear what constitutes excellence in a treatment. If everything can be expressed in terms of money the position is eased, but the solution can be meretricious because prices change and a peasant farmer is often more concerned with supplying his own needs and those of his farm animals than with the monetary value of crops that he is not proposing to sell. If a common basis of assessment is needed, energy or protein content can be useful, but they do not help, e.g. with a millet and cotton intercrop, if the two yields are required for different purposes.

In general a bivariate situation needs a two-dimensional diagram for its assessment. (Indeed, in the case of heights and spreads of strawberry plants, where it is the relationship of the two variates that is in question and not how they should be combined, it is difficult to see any other way.) If a straightforward diagram is produced with one variate plotted along each axis it will be found difficult of interpretation, because treatment points close together may nonetheless be very significantly different, while two points far apart may nonetheless be within limits of random variation of one another. A useful device is to use $\mathbf{X}$ and $\mathbf{Y}$ instead of $\mathbf{x}$ and $\mathbf{y}$, where

$$\mathbf{X} = \mathbf{x}/\sqrt{V_{xx}} \tag{9.3.5a}$$

$$\mathbf{Y} = (\mathbf{y} - V_{xy}\mathbf{x}/V_{xx})/\sqrt{V_{yy}} \tag{9.3.5b}$$

where V_{xx}, V_{xy} and V_{yy} are respectively $\mathbf{x}'\boldsymbol{\psi}\mathbf{x}$, $\mathbf{x}'\boldsymbol{\psi}\mathbf{y} = \mathbf{y}'\boldsymbol{\psi}\mathbf{x}$ and $\mathbf{y}'\boldsymbol{\psi}\mathbf{y}$, each divided by e, the error degrees of freedom, and $V'_{yy} = V_{yy} - V^2_{xy}/V_{xx}$. The variaties thus produced each have unit variance and are uncorrelated, so the ordinary Euclidean distance between two points on the diagram corresponds to the statistical significance of the difference between the two treatments represented by those points. Also, if a circle is drawn round a point with a radius of $1/\sqrt{r}$, where r is the replication of the treatment, that will correspond to putting limits of plus or minus a standard error round a mean in the univariate case. Similarly, if in the univariate case a difference between two treatment means would have a variance of $\theta\sigma^2$, where σ^2 is the variance of an observation, then for the bivariate case the treatments differ significantly at a given level if the distance between their points on the transformed diagram exceeds the square root of

$$\frac{2\theta e F_c}{(e-1)} \tag{9.3.6}$$

where F_c is the critical value of F for 2 and $2(e-1)$ degrees of freedom and the significance level required. Where θ is the same for all elementary contrasts, as with totally balanced designs and those that are equi-replicate and orthogonal, it is possible to draw circles at the critical distance round points. Then if the point for one treatment lies within the circle of the other, the two do not differ significantly. (Of course, that corresponds to working out a least significant difference and is valid only under the usual conditions.) The computations are illustrated in Example 9A.

The advantage of the approach, which has been set out in some detail elsewhere (Pearce and Gilliver, 1978, 1979), lies in its clarifying the agronomic situation before proceeding to assessment. Once such a diagram has been produced, contour lines can be drawn on it for monetary value, protein and much else to see how the treatments compare in those various respects. Further, if monetary values are used, it is possible to draw different sets of contours, one for each of several foreseeable price patterns, in order to judge how far the conclusions reached for the present time might hold in later seasons when the relative prices of crops could have changed.

9.4 USING ORGANISMS AS BLOCKS

In general the plots of a block are isolated from one another so that treatments applied to one cannot affect another, but there are times when the most natural block is made up of plots that necessarily cannot be kept independent. For example, if fruit trees each have three branches, it could be convenient to regard each as a block and to apply different blossom-thinning procedures to the branches as plots. The trouble here lies in treatments possibly having two effects, a local effect with parameters, $\boldsymbol{\gamma}$, acting on the plots to which each is actually applied, and a remote effect with parameters, $\boldsymbol{\delta}$, that acts on other plots of the same block. Thus, the removal of blossoms must have a local effect on the fruitfulness of the branch from which they are taken, but it is quite possible that it will have a remote effect on other parts of the same tree. Further the remote effect may have the same sign as that obtaining locally, as would happen if it were a dilution of the local effect; on the other hand, it would have the opposite sign if, say, nutrients that would have gone to the treated branch go instead to those that retain more blossoms. Although the problem most often arises with experiments on large, woody plants, it occurs also to half leaves injected with, for example, nutrient or pathogen. The technique is useful, the whole leaf being the block and the half-leaf the plot.

The problem has received some attention (Pearce, 1957). The basic solution, however, is simple. Let the treatments be applied non-orthogonally to blocks (trees or whole leaves) in such a way that all contrasts are estimated both intra-block and inter-block. Also, let each block contain the same number of plots, k. It follows that treatment parameters estimated intra-block will represent the local effects, $\boldsymbol{\gamma}$, diminished by $\boldsymbol{\delta}$, i.e. in (3.3.1)

$$\hat{\boldsymbol{\tau}} = \hat{\boldsymbol{\gamma}} - \hat{\boldsymbol{\delta}}. \tag{9.4.1}$$

Inter-block, however, the effect of $\boldsymbol{\gamma}$ is reinforced by $(k-1)$ plots with $\boldsymbol{\delta}$, so in (3.7.4)

$$\hat{\boldsymbol{\tau}}_0 = \hat{\boldsymbol{\gamma}} + (k-1)\hat{\boldsymbol{\delta}}. \tag{9.4.2}$$

Hence

$$\hat{\boldsymbol{\gamma}} = [(k-1)\hat{\boldsymbol{\tau}} + \hat{\boldsymbol{\tau}}_0]/k, \tag{9.4.3a}$$

$$\hat{\boldsymbol{\delta}} = (\hat{\boldsymbol{\tau}}_0 - \hat{\boldsymbol{\tau}})/k. \tag{9.4.3b}$$

Since $\hat{\boldsymbol{\tau}}$ and $\hat{\boldsymbol{\tau}}_0$ are estimated independently (Section 6.4), there is no difficulty about finding the variances of $\hat{\boldsymbol{\gamma}}$ and $\hat{\boldsymbol{\delta}}$.

There are two main design problems. First, it is important to obtain a good estimate of $\hat{\boldsymbol{\tau}}_0$. Consequently designs that might be rejected in more usual circumstances could be valuable here, bearing in mind that efficiency lost from the intra-block analysis appears inter-block instead. The other difficulty concerns block size. If k equals two the model merely assumes that a local effect is matched by a remote effect on the other plot, but if it equals three the two plots that do not receive a treatment must be disposed symmetrically with respect to the one that does; otherwise they cannot be expected to suffer the same remote effect. (In the case of a tree with three branches spaced at about 120°, a common event with a species like apple, there is no difficulty.) With more plots to the block such symmetry is difficult to arrange. However, the need for precise inter-block information militates against large values of k whatever the circumstances.

Of course, it is possible that the local and remote effects may interact. The appropriate analysis for balanced incomplete block design has been considered elsewhere (Pearce, 1957). Goodchild (1971) has examined the availability of covariance adjustments. A useful contribution comes from the paper by Freeman (1973).

In forming plots it is not essential to use the whole of the organism designated as a block. Indeed, it may be difficult to do so, because branches of a tree can be of quite different size and even the two sides of a leaf are not necessarily very similar. In both cases samples may well be taken as plots and there is no reason to the contrary.

A specimen analysis is presented in Example 9B.

9.5 COVARIANCE WITH MEASURED QUANTITIES AS CONCOMITANT VARIATES

The analysis of covariance was introduced in Section 2.2 as a possible way of controlling local variation and again in Sections 7.5, 7.6 and 7.7 to cope with defective or dubious data. Valuable though those applications are, it should not be supposed that the chief use of the technique lies in making adjustments on pseudo-variates. On the contrary, its original application and still its chief use comes in making adjustments on quantities, known as 'independent variates', that have been measured to throw light on the variate under study, known as the 'dependent variate'. Although the calculations look much the same as those in Section 9.3, so (9.3.4) is again useful, it should be emphasized that their purpose is quite different. The aim is not to study two variates, x and y, in conjunction but to study y after eliminating the disturbing effect upon it of x, or perhaps to find out if the two variates are correlated. Of course, there could be more than one concomitant variate. In Section 2.3 formulae were given for p such variates and p may be large, always provided that the error degrees of freedom, $(n-p-v)$ in number, are sufficient for the purposes of the analysis. The expressions given in Section 2.3 referred to experiments without local control, i.e. those that are completely randomized, but they can be used for any others by using the appropriate values of $\boldsymbol{\phi}$ and $\boldsymbol{\psi}$.

To take a practical example, when transplanting it is often useful to measure plant weights, $\mathbf{x}$. (The figures may be inaccurate on account of soil adhering to the roots, but in a sense that is part of the problem. Botanically there may be good grounds for thinking that planting weight will forecast future development; it does not follow that there is any practicable record that will serve that purpose, bearing in mind the undesirability of removing soil from the roots at such a time.) Later, some other quantity, $\mathbf{y}$, such as yield or height or final size is to be measured and the question arises how far it will be related to $\mathbf{x}$. The question may be asked in two contexts. It may be desired to adjust all values of $\mathbf{y}$ to a standard value of $\mathbf{x}$, thus eliminating the effects of initial variability, or there may be interest in the extent to which initial size has had a lasting effect. If it has none, the conclusion may follow that too much attention has been given to standardizing plant size. In that case future rejection standards can perhaps be less strict.

Similarly, if an experiment has to be laid down on variable land, preliminary measures of, say, soil texture on each plot may explain a lot of the variation. Furthermore the analysis of variance will provide a means of allowing for the extraneous variation, if that is needed. To take an interesting variant, Lockwood (1979) made useful adjustments on the basis of trees that were removed for thinning.

Nevertheless care is essential. For one thing, use of (2.3.10) implies the assumption that the regression coefficients, $\boldsymbol{\theta}$, are unaffected by the treatments. Very often the assumption holds, but not always. For example, to take variable soil texture, it might well have less effect on irrigated plots than on unirrigated. Again, an initial record of trunk circumference on trees does not necessarily give the same regression coefficient with trees that are growing well as with those that are dwarfed. Unfortunately, there is no very satisfactory way of dealing with the data in which the regression coefficient is not constant. Computers can indeed effect some sort of least squares solution but interpretation is difficult.

Transformations also need to be considered. In Section 9.12 attention will be given to functions of data that will give stable variances. That is implied in a context of the analysis of covariance also. Consequently, $\mathbf{y}$ should be transformed, if at all, as for an analysis of variance. The problem comes with **x**. No assumptions have been made about its variance, normality or anything else relating to its distribution, so they do not matter, but it has been assumed that it is linearly related to **y**, which has perhaps itself been transformed. Often it is difficult to say what should be done and it is not unusual to use several concomitant variates, say $\mathbf{x}^{1/2}$, **x** and $\mathbf{x}^2$, with reasonable hope that one of them or some linear combination of them will give a useful relationship.

In general, independent variates should be adjusted to standard values equal to their means in the experiment itself. It is clear from Chapter 3 that essentially an experiment estimates contrasts. Consequently very little depends upon the standard values adopted, though conclusions may be more easily understood if they have an intuitive appeal.

It should be remembered that regression coefficients do not express laws of nature but only set out what happened in the body of data under examination. To take an example, plants that are crowded will often make growth in search of light and will be less fruitful on that account. On the other hand, growth induced by better fertilization will probably lead to improved cropping. Consequently, the correlation coefficient between height and crop can be positive or negative according to which effect dominates and there can be occasions when the two cancel so that no correlation appears at all. It is the same with the regression coefficients, $\boldsymbol{\theta}$, in (2.3.10). They are observed in one experiment and made use of there, but it is unwise to assume that they will be repeated in another. On the other hand, it is a strength of the covariance approach that it estimates regression coefficients each time and so always uses appropriate values.

In general, it is best to use only independent variates that have been measured before the imposition of treatments, like initial weight, or are unlikely to have been affected by the treatments, like soil acidity in a spraying trial. The role of the independent variates is then clear. However, there is no need to ban others altogether. For example, in a fertilizer trial it may appear that the differences in crop between the different treatments are eliminated by a covariance adjustment on mean leaf area. The information given by the analysis of variance that better fertilization has indeed led to more crop is important, though it might have been expected. It is, however, usefully supplemented by the further conclusion, given by the analysis of covariance, that the effect on crop appears to derive entirely from the differences induced in leaf size.

A specimen computation is given in Example 9C. It underlines a point already noted in Section 2.3. Contrasts that would be estimated independently in the analysis of variance become related when a covariance adjustment is introduced. When adjustment is made on a pseudo-variate that often does no great harm, but the effects of an arbitrary independent variate can be decidedly awkward. The subject has received careful study by Preece (1980).

9.6 OPTIMALITY OF DESIGN

With response surfaces a great deal of work has been done to establish the best design for a given problem, but it does not transfer well to comparative experiments. When there are several quantities, i.e., contrasts, that can be estimated, some weighting of them is required and that introduces an important complication. If it is assumed that all are of equal importance, a design in total balance is needed (see Section 5.2) and that is obtainable easily enough. Rather a lot of effort has been wasted evolving the optimal design to suit that problem and too much enthusiasm has been expressed when the outcome has been in balanced incomplete blocks, a result that could have been foreseen.

The problem may be expressed formally thus: An experimenter has n available plots and he wants them divided among blocks according to a design matrix, $\mathbf{D}$ (see Section 3.1). (Of course, he may have some alternatives for $\mathbf{D}$ but that does not really affect the problem. Each can be considered on its own.) He now asks what design matrix for treatments, $\boldsymbol{\Delta}$, will best suit his problem and here it is necessary to ask him precisely what that problem is; otherwise nothing can be done. In reply, he may specify some contrasts. Further, if pressed he may be willing to grade them, assigning each a weighting. If he is reluctant to weight them, he may still insist that those are the contrasts he must study and that all matter. In that case each can be given a weight of one, all unspecified contrasts orthogonal to them being given zero weight. It is now reasonable to choose a design such that its coefficient matrix, $\mathbf{C}$, minimizes

$$E_1 = \sum_i (w_i \mathbf{c}_i' \mathbf{C}^- \mathbf{c}_i). \tag{9.6.1}$$

Here the contrasts of interest are given by the $\mathbf{c}_i$, each with its weight, w_i.

Although the minimization of E_1 has its advantages, there can still be reservations. First, it has not been required that the $\mathbf{c}_i$ shall be independent. Secondly, there could be reasons for keeping some treatments in equivalence sets (see Section 4.5), but in (9.6.1) the set of contrasts is unstructured. The reservations do not present insuperable obstacles, but they do suggest that the task of specifying the problem is not quite as simple as may appear.

In fact, E_1 is not the obvious choice of anyone accustomed to the advantages of D-optimality in the response surface problem. He would expect

$$E_2 = \prod_i (w_i \mathbf{c}_i' \mathbf{C}^- \mathbf{c}_i), \tag{9.6.2}$$

but that is to merge all the weights, w_i, so that the contrasts of interest are treated as of equal importance. A third possibility is to maximize

$$E_3 = \sum_i (w_i \mathbf{c}_i' \mathbf{C} \mathbf{c}_i). \tag{9.6.3}$$

The difference from E_1 lies in the extent to which all contrasts of interest are regarded as essential. To take an extreme example, if a design were suggested that confounded one of them (i.e. $\mathbf{c}_j' \mathbf{C}^- \mathbf{c}_j$ was infinite), but effected precise estimation of the rest, E_1 and E_3 would respond very differently. Then E_1 would itself become infinite, thereby disqualifying the design from further consideration, whereas E_3 would merely have had no contribution from $\mathbf{c}_j$. That would be to the disadvantage of the design but it would not prevent due credit being given for excellence at estimating the other contrasts of interest. The choice between E_1 and E_3, in short, depends really upon the kind of optimality that is being sought. It will be noted that E_3 is easier to calculate because it involves **C** instead of $\mathbf{C}^-$.

A practical problem is the fixing of weights. It may be agreed that one contrast is more important than another and there may be agreement which of two designs places the emphasis better, but what does that imply in relative values for weights? Here an empirical study by Freeman (1976a) provides some useful guidance, but more work is needed.

The best approach to optimality of design is to use the computer. It is given **D** together with contrasts of interest with a relative weighting for each. Then, it either starts with a suggested design or works out something for itself and tries to improve what it has. Whenever it finds something better, it starts another cycle and so on until it can effect no further improvement. One particularly useful program is that of Jones (1976) for finding an optimal connected block design. Starting with the definition of Ξ^{-1} given by (3.4.4) he proposed the following approximation to Ξ, following (3.4.8):

$$\Xi = q^{-\delta} - \sum_{i=1}^{H} (\mathbf{1}_v \mathbf{1}_v'/u - \mathbf{q}^{-\delta}\mathbf{W})^i \mathbf{q}^{-\delta}. \tag{9.6.4}$$

Experience shows that $H = 4$ gives a good enough result, though it cannot be relied upon when there are blocks that contain only two plots or when there are only two treatments. The criterion adopted was

$$\begin{aligned} E_1 &= \sum_i (w_i \mathbf{c}_i' \Xi \mathbf{c}_i) \\ &= \operatorname{tr}(\mathbf{X}'\Xi\mathbf{X}\mathbf{w}^\delta) = \operatorname{tr}(\mathbf{X}\mathbf{w}^\delta\mathbf{X}'\Xi) \qquad \text{from Lemma 2.1.A,} \end{aligned} \tag{9.6.5}$$

where **X** holds the various contrasts of interest in its columns. It will be seen that $\mathbf{X}\mathbf{w}^\delta\mathbf{X}'$ exists independently of any design, representing in fact the problem, whereas Ξ is derived entirely from the design that is in process of being optimized.

The computer first examines the design before it and works out

$$M = \frac{N_{ji}}{k_i} - \frac{r_j}{n}$$

for each plot. The quantity M represents the departure from orthogonality as given by (4.2.3) and for present purposes is always given a positive sign. The computer then sums M over each block and chooses the two that depart most from orthogonality. Taking those two blocks it interchanges treatments between them in all possible ways and for each interchange it works out Ξ from (9.6.4) and E_1 from (9.6.5). Having found the interchange that gives the lowest value of E_1, it updates the design and starts again. Although the algorithm does not always find the best design, where that is known, it has a good success rate and the designs it does suggest are rarely far from the ideal. It is also useful for taking a design favoured by the experimenter or biometrician and seeing if any improvement can be suggested. It has been extended to cover row-and-column designs also (Jones, 1979).

A disadvantage of the approach lies in the immutability of the replications. Once they are allocated to the starting design, whether by the methods of Section 2.1 or in some other way, interchanges leave them unaltered. A useful development is therefore afforded by the work of Jones and Eccleston (1980) who supplemented interchanges by an exchange procedure, in which the treatment given to a plot was replaced by another with no compensation elsewhere in the design. First they found the 'weakest observation' of the design by working out

$$\operatorname{tr}(\mathbf{X}\mathbf{w}^{\delta}\mathbf{X}'\mathbf{C}_{n-1}^{-}) - \operatorname{tr}(\mathbf{X}\mathbf{w}^{\delta}\mathbf{X}'\mathbf{C}_{n}^{-}) \tag{9.6.6}$$

for each plot, where $\mathbf{C}_n$ relates to the design under examination and $\mathbf{C}_{n-1}$ to what is left when the plot is removed. Having found the plot whose loss would do least harm to the design, the other treatments were substituted in turn to find which gave the greatest advantage. This exchange procedure was continued until no further improvement could be effected and then an interchange procedure was started instead. The revised algorithm has had some marked successes and has been extended to cover row-and-column designs (Eccleston and Jones, 1980).

9.7 THE CHANGING OF TREATMENTS

Usually the treatments of an experiment are applied over the whole lifetime of the plants but that is not necessarily so with perennial species. Here the plants may last so long that there is opportunity to carry out several experiments in succession.

It is not always easy to assess when an experiment has run its course because sometimes treatments can change their effect with the passage of time. Thus, a fertilizer may initially lead to better growth and cropping but if it is applied over a long period it may have toxic effects. Further, the whole purpose of the experiment may be to find out if that is going to happen. More usually, however, there will come a time when the treatment effects have settled down and their F-value has become stable. It may then be decided that they have told their story and there is no point in going on. Actually, with newly planted material stability may be achieved quite quickly. What happens is this: initially plants tend to react on their nursery performance because the larger ones commonly transplant less well. Consequently variability with respect to size tends to fall at first but to build up later as permanent features, such as soil, begin to have their effect (Pearce, 1960b). Meanwhile the effects of treatments have already started to appear and a stable F may be speedily attained. It is then that a fresh experiment can be considered.

Another occasion arises when differential sprays have been applied to control pests or diseases. In the following year, supposing that the sprays have not been applied a second time, there could still be residual effects arising from spores or eggs left over from the previous season even though the organism may have dispersed, but with a second dispersal such residual effects are unlikely to continue into a third season. Of course, that is a biological judgement and a lot must depend upon the life cycle of the organism. Further, conclusions should be supported by observation. Nevertheless, there will usually come a time when it can safely be assumed that residual effects will have disappeared.

If that assumption can be made with confidence there is no reason to pay any regard to history when designing a fresh experiment. It may however be wise to assume that there could be residual effects, though without any expectation that they will interact with those of treatments subsequently applied. In the case of some treatments, e.g. fertilizers, there could be permanent residual effects on the trees, though again it may be reasonable to discount any interaction with a fresh set. In that case, the experiment has acquired a second blocking system. Where once there were blocks to be allowed for, there are now blocks and former treatments. The situation can, however, be a little complicated. A row-and-column design (Section 6.6) is necessarily of Type O : YZ because rows and columns are always orthogonal. If a block design of Type X is concluded it can often be built up again to one of Type X : YZ with a fresh set of treatments, but X is not necessarily equal to O. Also, dualization (Section 4.6) needs to be considered. In a block experiment it is unusual for anyone to enquire about the effect of blocks, which are ordinarily taken for granted. When, however, an experiment has been built up from blocks and original treatments to a three-way classification of blocks, original treatments and current treatments, it is quite reasonable to enquire if there is still residual effects of the original treatments. For that reason Preece (1966) has suggested that the notation could usefully be extended. Thus, to take the first such design advanced (Pearce and Taylor, 1948):

		Original treatments					
		A	B	C	D	E	F
Block	I	c	d	d	c	b	a
	II	c	a	c	b	b	d
	III	b	a	a	c	c	d
	IV	a	c	d	c	d	b
	V	b	d	b	a	a	c
	VI	d	b	b	d	a	c

Current treatments: a–d

It is of Type O : TT, i.e. the original treatments are disposed orthogonally to blocks, while the treatments of the second set are arranged so that each block and each original treatment receives a complete set of a, b, c and d together with one of the combinations of the four taken two at a time, i.e., ab, ac, ad, bc, bd or cd. The complication comes when study is resumed of the original treatments, which are not disposed in total balance with respect to the current treatments but in a group-divisible design. In Preece's extended notation the design is Type O : TT :: GG : P.

Other useful papers are those of Freeman (1957ab, 1958, 1959).

Those difficulties apart, given an orthogonal block design with b blocks and v treatments, the combinatorial problem of adding a further set of w treatments is no different from that of producing a row-and-column design with w treatments on a $b \times v$ format of plots. If no interaction is expected between the two sets of treatments the analysis of variance also is the same. There are more difficulties starting with a block design of Type X (X $\neq$ O), but those relating to combinatorial properties often admit of quite neat solutions (Hoblyn *et al.* 1954).

Similarly, a row-and-column design can often be given a fourth classification. Thus, a Latin square (Type O:OO) can be built up into a Graeco-Latin square (Type O:OO:OOO) provided it is not of order 6. Other orthogonal solutions have been suggested (Finney, 1945a, 1946a, 1946b), while Freeman (1964a) has studied designs of Type O:OO:SSS. Some attention has also been given to the extension of Youden squares (Type O:OT) to form designs of Types O:OT:TOO and O:OT:OTT (Clarke, 1963; Preece, 1966). Other papers have explored the basis upon which a further set of treatments can be added e.g. those of Freeman (1966), Hedayat *et al.* (1970, 1971), Hedayat and Seiden (1970) and Hedayat (1973, 1975), and of Hall and Williams (1973). There is also the paper of Preece *et al.* (1978) in which some of the issues relating to randomization are considered.

In general, when the final design is of Type O:OX:OYZ the first classification is orthogonal to all the rest. Its sum of squared deviations can therefore be calculated as for an orthogonal design, i.e. as in (4.1.5). The analysis of variance can therefore be carried out ignoring it, the design then being of Type X:YZ. The error sum of squared deviations so obtained should then be partitioned by subtracting the component for the first classification, thus leaving the error for the design as a whole. Designs of Type O:XO:YOZ can be dealt with similarly. Also, with a design of Type O:OX:YOO there are two independent non-orthogonalities, i.e. the third classification relative to the second and the fourth relative to the third, and that simplifies the least squares solution. A similar comment applies to designs of Type O:XO:OYO.

9.8 MULTIPLE CROPPING

In a tropical country where traditional methods prevail the research worker is often confronted with the phenomenon of multiple cropping. He will find that local farmers do not grow one crop at a time but several. In the extreme case they may actually sow a mixture of seed or drop two seeds from different species in each hole. That may be called 'mixed cropping'. Less extreme is the case in which two species are grown in alternate rows, often called 'intercropping'. Either way there is intense interspecific competition. A related practice is 'relay cropping', in which one species is sown between the rows of another that is yet to be harvested. The possibilities are nearly endless and extremely puzzling to anyone trained in so-called 'developed agricultures'.

If anyone is bewildered, he should avoid the temptation to escape the problems by denouncing what he sees as primitive or absurd, because it is neither. On the contrary, it often represents an approach of great subtlety that calls for skills he has never had to acquire and therefore does not understand. Several unconvincing arguments are used against the practices of multiple cropping. It is true that they do not lend themselves to mechanization, but mostly that is not in prospect anyway, so it would be kinder to say that they make full use of the flexibility of manual operations. It is also true that a big change in conditions, like an irrigation scheme or radically improved varieties, leads to the adoption of monocultures, but that only underlines the difficulty of devising a new multiple cropping system *ab initio*. For a poor man with limited resources they are often ideal.

Their first advantage is their reliability. If two crops are sown together, e.g. hill rice and maize as in parts of Central America, the success of one implies partial failure of the other. In that example, if it rains the maize may be killed and if it does not rain the rice will die, but either way the farmer's family has something to eat. Their second advantage comes from better exploitation of scarce resources like nutrients and water. Thus millet and sorghum are similar crops and serve much the same purpose in a family's economy. Nevertheless they differ enough in season for a combination to yield better than either alone, at least in many conditions.

There are, however, two cases to be considered. In the examples given above both crops are cereals and the one is an acceptable substitute for the other. However, it is noticeable that in many traditional combinations the two species serve complementary purposes. For example, in West Africa cassava, which contains virtually no protein, is often intercropped with pulses, which supply the deficiency. In that case the argument of reliability reverses its direction. If one species were to weaken the other a bad situation would arise, because a minimum crop is required from each. The justification must come from the other argument, namely that of better use of land.

It follows that anyone who sets out to design and conduct an experiment on multiple cropping systems has to face several special problems. For example, assessment raises certain difficulties. With a single species the best treatment is the one that gives the largest crop, always supposing that it does not lead to a loss of quality or deplete the soil, but with more species a gain in one may imply a loss in another. The facile solution is to express everything in terms of money but that can be misleading. For one thing the farmer may not intend to sell his produce but to eat it himself. For another, prices are not a constant of nature but vary from place to place and from season to season. If there are two species it is better to plot them for each treatment on a diagram with an axis for each crop. It is then possible to draw sets of contours for the various characteristics of interest, one set for protein, another for energy and so on. For money there could be several sets for different relative prices for the two crops. A treatment can then be assessed in various ways. If there are three or more species in the system, as there could be, the method of Andrews (1972) could have advantages. Given two species bivariate methods are available (Pearce and Gilliver, 1978, 1979) in which the crops are subjected to a transformation to secure desirable statistical properties (see Section 9.3), but that is not essential. Indeed, if the assumptions look doubtful it is better avoided, though it eases interpretation when they are acceptable.

Another problem concerns the need for plots of sole crops. Sometimes they are not needed at all, e.g., if a single system is under test and the only questions concern soil preparation. On the other hand, if the object is to find the gain in yield by growing two crops together instead of apportioning land between them as sole crops, they are essential and in greater numbers than is usual at some institutes.

One difficulty concerns configuration, because a botanical study of competition is not the same as an agronomic study of intercropping. If maize is usually grown in rows 100 cm apart and it is proposed to introduce soya beans between the rows, a botanist would rightly insist on retaining the row spacing for the maize, whereas an agronomist, also rightly from his point of view, would see nothing wrong in putting the maize rows 120 cm apart if that would help the soya beans.

In general, the specification of objectives for a multiple cropping trial causes more trouble than for one with a single crop. One formulation must be resisted. People often want to know if a subsidiary species can be grown under the main species 'without harming it', but the competition must have some effect and the objective should be to minimize the harm, not to prevent it. As a matter of fact that problem can sometimes be better formulated in terms of configurations. For example, suppose that millet is usually grown as a sole crop in rows 80 cm apart but someone wants to introduce ground nuts as well. It could well be that the millet would come to little harm if row spacings were alternately 120 cm and 40 cm. If so, there need be no difficulty about space for the ground nuts. (Incidentally, at the stage of studying configurations for millet, there is no need for any ground nuts. That comes later.)

With any multiple cropping system, however, the first need is an understanding of how it works and that can be a complex matter both botanically and statistically. There is always the danger that a system which gives satisfaction at one place or in one season will prove disastrous in other conditions, usually because more precipitation or less or different soil upsets the balance of species. That relates to another difficulty in studying such systems. The local smallholder appears to have no system for anyone to study. He did one thing last year and is doing something else now, but that illustrates his subtlety of approach. Adaptation to varying conditions is itself part of the system.

The whole subject has a large literature but the following papers deserve special mention on account of their biometrical interest. Federer *et al.* (1976) and Federer (1979) have considered various problems that arise in mixed cropping, just as Mead and Stern (1980) have for intercropping. The paper by Mead and Riley (1981) was discussed at a meeting of the Royal Statistical Society and, taken in conjunction with the extended comments of those present, gives as wide a picture as is presently available. Mead and Willey (1980) have summarized ways of assessing the yield advantages to be gained from intercropping, while an excellent general background is afforded by the survey of Willey (1979ab).

9.9 EXPERIMENTS AT SEVERAL SITES

To continue the theme of Section 1.2, an experiment at a single site raises difficulties of generalization. Strictly speaking the conclusions relate to differences over the area of the experiment. That has been sampled effectively by a process of randomization, but there are no statistical grounds for generalizing the differences even to the rest of the field, to the next field or to the district, let alone to a complete tract. The agronomist may argue that conditions are sufficiently similar to apply the results more widely and that may well be so, but the data as they stand neither confirm nor deny the assertion.

The remedy is to use several sites. Only rarely can they be chosen at random over the tract but that is not usually required. The aim is often to establish that there is no large interaction of site and treatments and that is best accomplished by choosing sites that are different and different in known ways, some wet, some dry, some with deep soil and some with shallow. If now there is replication at each site, so that a valid estimate can be obtained of error, it should be possible to test the interaction. Further, if the treatment effect sometimes disappears, it should be possible to discern the conditions in which that happens.

The official making a bulk purchase, say, of fertilizer or the owners of a large estate may well be interested in the mean effect. If the treatment is unsuccessful here, that does not matter if it does well there, provided that it justifies itself overall. To investigate that question does require a random sample of sites and people with such power may be able to arrange one, but such enquiries are not usual.

There are two difficulties with a multi-site experiment. One is interactions and the other is inequality of error at the various sites. Neither is much trouble without the other. As has been said, if an experimental error can be determined for each site and they all appear to be about the same, there is no great difficulty. Everything can proceed with an additional stratum. Following the thought of Section 6.1 the structure for a series of block experiments is:

$$\text{Plots} \rightarrow \text{Blocks} \rightarrow \text{Sites} \rightarrow \text{Total of experimental land.}$$

Assuming orthogonality at each site, both the main effects of treatments and its interaction with sites is estimated with full efficiency in the first stratum, i.e. plots within blocks.

Equally, there is no difficulty if it is believed that there will be no interaction, but some sites will give a better estimate of the difference than others. In that case the difference is estimated overall by weighting the difference at each site by the reciprocal of its variance, so that sites that give a better determination, whether on account of higher replication or lower error variance, are given greater weight. That is to say, if d_i is the estimate of a difference at the ith site, its variance being $\theta_i \sigma_i^2$, where θ_i is some constant depending on the design at that site and σ_i^2 is the error variance, the estimate of the difference should be

$$\sum_i [d_i/(\theta_i \sigma_i^2)] \sum_i [1/(\theta_i \sigma_i^2)].$$

The whole operation can, however, be rather pointless. If it is believed that there is no interaction, why bother about a range of sites? If the answer is that the absence of an interaction has been confirmed by a test, that brings up the question of how the test could have been carried out when the various sites are giving different error variances.

Before going further it may be noted that there is an alternative approach. The error first used was derived from variation within individual sites but the investigator may be interested in something else. If a recommendation is to be made about the general usefulness of a treatment it would be better to compare the mean squared deviation of its main effect with that for its interaction with sites. If that appeared significant it could be asserted that the treatment difference applied generally.

The best resolution of the problem of a suspected interaction in conjunction with suspected unequal errors at the various sites is to work with single degree of freedom effects, i.e., specified contrasts between treatments. If a contrast is evaluated from data at each of the s sites separately, there is no difficulty about a little analysis in which the mean effect with one degree of freedom is compared with its own interaction with sites, which would have $(s-1)$. Of course, that is to come down firmly on the side of those who want to use the site × treatment interaction as error anyway. Since no use is made of estimates of error internal to the sites, the method is available for the dispersed experiments suggested in Section 1.2. They would have no replication at sites but treatment structures would mostly be fairly simple.

Given a determination of a contrast for each site, there is no difficulty about relating its magnitude to measurable site differences using regression methods. Further, if the sites can be classified, e.g., some on this soil type and some on that, there is no difficulty about carrying out an analysis of variance to see if the contrast is more marked at one sort of site compared with another.

There are several publications that can be recommended to those who would like to know more, among them Yates and Cochran (1938) and Cochran and Cox (1950).

9.10 ROTATION EXPERIMENTS

The design of rotation experiments is a specialized topic of which the author has too little experience. He will do no more than indicate some general principles and refer his readers to others who are better informed.

The cardinal principle is to have all phases of all rotations present in each year. To take a simple example, if a crop A were to be grown in alternate years and the question concerned the crop, whether X or Y, that was to be grown in the other years, it would be necessary to have at least four plots and at least three seasons, the first being required to set up the system; i.e., the absolute minimum would be

		Plot			
		1	2	3	4
Season	0	A	X	A	Y
	1	X	A	Y	A
	2	A	X	A	Y

There would then be a comparison of A in the two rotations both in Season 1 and Season 2. That is obviously not enough. The addition of four more, 5–8, which reproduced 1–4, would obviously help matters, as would the continuance of the rotations over two more seasons, 3 and 4. The effects would, however, be rather different.

The use of more plots would lead to a straightforward increase in replication, no plot being used more than once for A, at least not during the experimental period. The analysis of variance would have seven degrees of freedom in all, one for seasons, one for differences brought about by the other crop, whether X or Y, one for the interaction and four for error. The addition of further plots would improve the experiment without causing any complications. Further, it is possible to form the plots into blocks. Given eight, plots 1–4 would form one block and 5–8 another. It is true that there would then be only three degrees of freedom for error instead of four and that could be a serious loss. Nevertheless, there is nothing wrong about using blocks and with more plots they could well be needed. There is, however, an objection to using an experimental period of only two seasons. It is not necessarily very serious, but it lies in the possibility of those seasons, the character of which is not under the control of the experimenter, being unrepresentative.

Before proceeding it may be noted that no experiment provides an estimate of its own error if it has only one replicate, so it will be assumed henceforth that there must be at least eight plots.

If a wider range of seasons is desired the experiment could continue into Seasons 3 and 4 or perhaps into many more. The data then present analogies with those from a split-plot design (see Section 6.5). Adding data for each plot over all seasons enables a 'main plot' analysis to be calculated like the one given above, the only difference being that the degree of freedom for seasons now represents the difference between odd-numbered seasons and even-numbered. A further analysis on the results of each plot in each pair of seasons is analogous to the 'split-plot' analysis. Given only four experimental seasons there would be one degree of freedom for pairs of seasons and one for the interaction of those pairs with the other species (X or Y). In the absence of blocks that would be enough and it would leave five degrees of freedom for another error, since within the stratum formed by crossing the pairs of seasons with plots there would be seven degrees of freedom in all, (see Section 6.1). It would also be legitimate to remove from error degrees of freedom for the interaction of pairs of seasons and blocks and for the three-factor interaction of pairs of seasons, blocks and treatments.

Of course, there can be many complications. A species may occur twice in a rotation system and cycles are not necessarily all of the same length. Also, treatments such as differential fertilizer applications may be needed and so on. The reader could well consult the publications of Yates (1954) and Patterson (1964). There are others, but further references can be found in those mentioned.

9.11 EXPERIMENTS WITH PERENNIAL PLANTS

When the plants can be expected to live for a long time a number of special problems arise, which have been examined in a previous book (Pearce, 1953, 1976a). Here it will be enough to follow an experiment from inception to completion, pointing out in order the decisions that had to be made as it progressed and drawing conclusions of wider application.

The experiment in question had a 2^3 factorial set of treatments (high and low nitrogen, phosphate and potassium) in eight randomized blocks. It was on apples and was carried out at the East Malling Research Station, where the author worked from 1937, so he knew it for the greater part of its life. It was often pointed out as the first long-term factorial experiment and the claim was probably justified, at least as far as date of planning was concerned, because two years had to elapse before the specially grafted trees could be planted. (Going by date of first application of treatments, however, it lost by a few months to a 2 × 5 fertilizer experiment designed for the Tea Research Institute of Ceylon with the help of R. A. Fisher.)

Planting took place in January 1931 following the plan in Fig. 7. It will be seen that each plot consisted initially of nine experimental trees. In the middle was a tree of Cox's Orange Pippin on the vigorous rootstock, M.XII, which was not

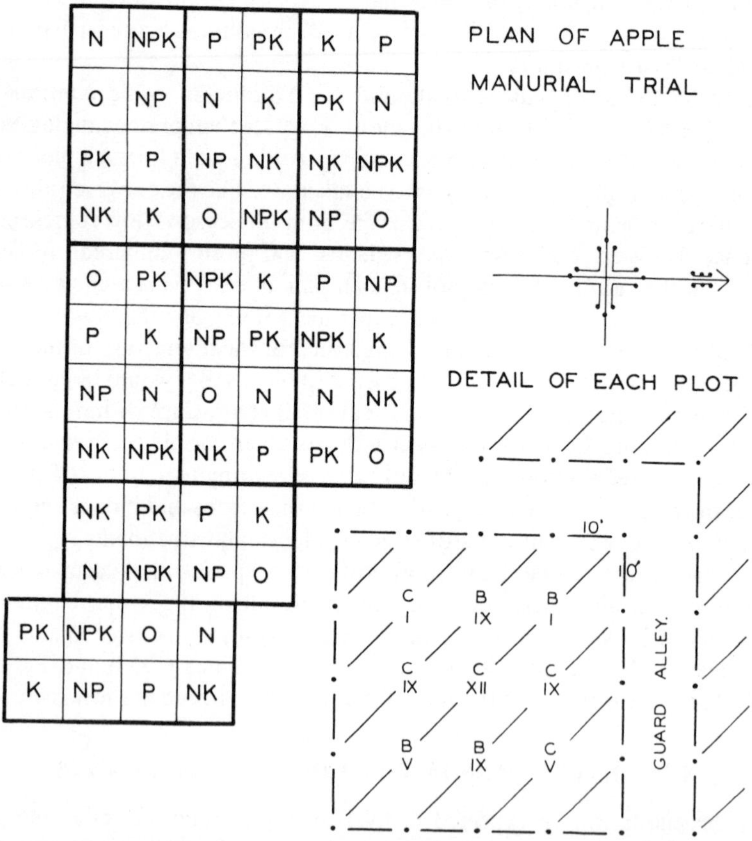

Figure 7. This early field plan of the long-term experiment described in Section 9.11 shows all essential detail, yet it is quite clear, an important matter when staff have to find their way round the trees and plots, perhaps in difficult conditions. Reproduced by permission of East Malling Research Station

expected to crop for some years. The corners were occupied by four trees, which themselves formed a factorial set, there being two each of Cox's Orange Pippin and Beauty of Bath, one of each pair being on the rootstock M.I and the other on M.V. Both rootstocks are semi-dwarfing, M.V being reputed to have difficulty in finding potassium, an element of some importance in East Malling conditions. In the middle of the four sides there were trees on the dwarfing rootstock, M.IX, two being of the variety Cox's Orange Pippin and two of Beauty of Bath. They were expected to come into cropping with little delay.

The intention was to thin progressively as the trees grew into one another, starting with the dwarfs and eventually leaving only the vigorous tree in the middle. In any long-term trial of this sort it is necessary to look ahead and make provision for thinning. However, the act of removing some trees necessarily disturbs those that remain and hopes of calibrating their performance after thinning by what they did before are likely to be unfulfilled (Pearce and Brown, 1960).

In a fertilizer trial it is most important to guard plots against intrusive roots

Figure 8. The experiment described in Section 9.11 went on for about a quarter of a century and went through many phases. It was helped by the essential simplicity of the original design. This aerial photograph, taken in 1935 shows the trees when young before any modifications had been made to the design. Reproduced by permission of East Malling Research Station

Figure 9. With the passing of years a long-term experiment often has to pass through a lot of phases, partly because the increasing size of plants may force the experimenter to remove some of them, partly because treatments sometimes have unexpected effects and need to be modified and partly because problems change while the experiment is in progress. Figures 7 and 8 showed an experiment at its inception. This picture shows it just before it was grubbed in 1965; the intervening changes are summarized in Section 9.11. Reproduced by permission of East Malling Research Station. Photography by Hunting Surveys Ltd

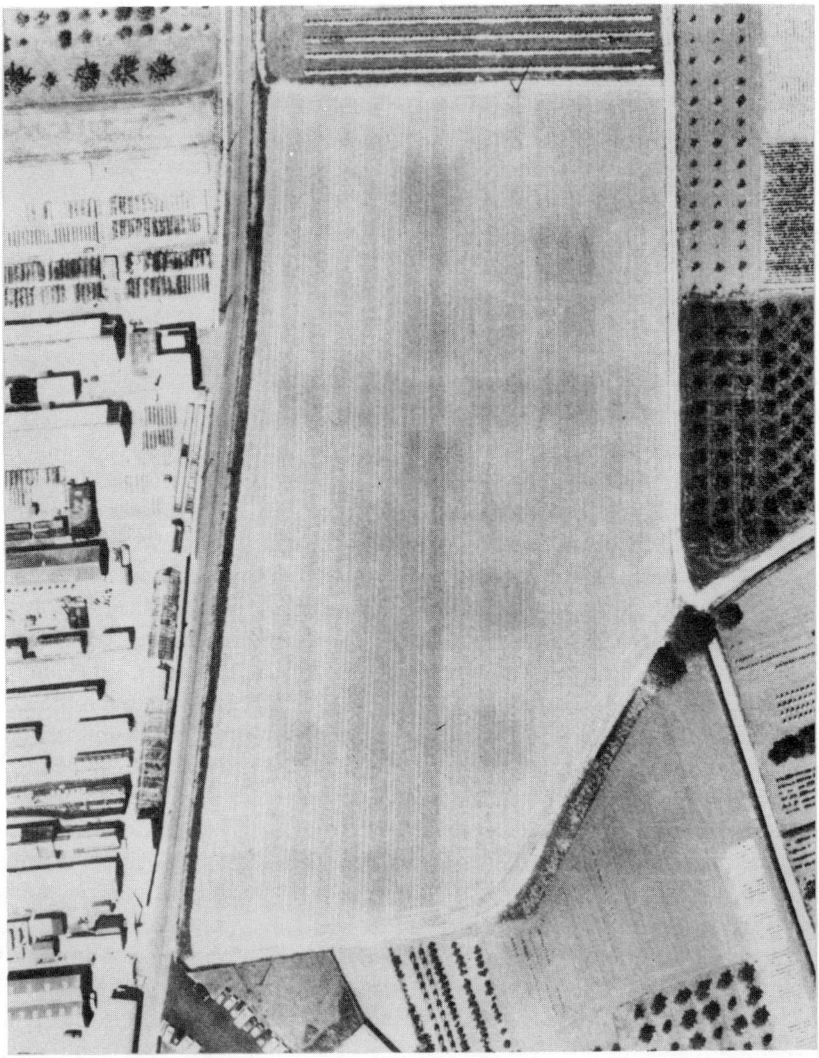

Figure 10. When an experiment has been in the ground for a long time, it leaves differences that are virtually permanent if the treatments have been applied to the soil and long-lasting even if they have been applied to the plants themselves. This aerial photograph of the experiment described in the text, and illustrated in Figures 8 and 9, was taken in 1967 and shows the marked residual effect of former treatments on a subsequent crop. However, that does not mean that the land had become useless for experimental purposes. On the contrary, everything had been surveyed accurately and the site became a most valuable asset, providing as it did areas of known and diverse fertilizer history. Reproduced by permission of East Malling Research Station. Photography by Hunting Surveys Ltd

from neighbours. The device here was to surround each plot with sixteen trees, all on the dwarfing M.IX, alternately of the varieties Worcester Pearmain and Duchess Favourite. (Different varieties were chosen to provide a range of useful pollen, an important matter with apples.) As a result of this arrangement plots were separated by double guards and that proved something of a mistake. Initially the fertilizers were taken only to the guard row for that plot, which meant that most trees were fertilized on only one side or in some instances on only one quadrant. As a result they did not develop root systems large enough to be effective. Later fertilizers were taken out to the middle of the alleys, which led to better growth but partly lost the advantage of double guards. The whole episode illustrates the problem of how to deal with discard areas, to which there is still no satisfactory solution.

At the start, despite the introduction of rootstock M.V, it was not appreciated how disastrous the lower level of potassium application would be. Large differences built up and a decision was taken to apply a basal dressing of that element over the whole area, leaving however the differential due to the factor. That restored the position to some extent but right to the end of its life there was a tendency to think of the experiment in two parts, i.e. those trees with normal development and those that had been initially stunted. That raises an important point. A 2^p design does not enable anyone to study response curves but it does reply to the question of what elements can advantageously be applied. That being so, the time comes when the answer is clear and the design could well be modified. That, however, is a comment from a later age. At the time all this happened no one thought about omitting or adding factors, though in a 2^p design that is quite easy. Given modern knowledge it would have been simple to have omitted the potassium factor, because the element was clearly essential, though some might have wanted to keep it for its interactions. Even if differential levels were discontinued, with so marked an effect there could well be residual effects that might interact, so it would be necessary to keep the factor in the design without continuing treatment applications. A contrary problem arose with the phosphate factor, which never showed much. Here an early acceptance of the fact that phosphate was of little importance could have led to its being discontinued and thereafter disregarded. However, by the time confounding was well understood and the possibility of adding factors was more apparent, the amount of phosphate that had been added to the soil and remained there unused was so high that the factor was continued to the end, though its discontinuance was several times mooted.

As to design the greatest care was taken to make the site uniform by growing cereal crops while the trees were being raised. Nevertheless, the experiment always gave rather high standard errors, partly because of a strip of rich clay that ran down the middle under the loam and partly because the soil was shallow down the north side, i.e., the right-hand of the plan in Figure 7. Again, several points arise. One concerns the advantages possessed by modern experimenters. An older generation did not have the facilities to study the land and decide how best to form blocks. They had to do their best on the basis of what they could see and it is no wonder that their blocking systems, as here, were not always ideal. Nowadays, when facilities are much better, an experimenter has no excuse for making mistakes that formerly were unavoidable. Also, with a long-term experiment it is most important to allow for site characteristics that are going to last for a long time and can therefore build up large differences. It is the same with borrowing money. For a loan of one year interest at 10 per cent is little better than at $10\frac{1}{2}$ per cent, but for compound interest over a long period the difference is important.

Already one change has been noted, namely, the increase in the basal application of potassium. Later, it was realized that more factors were going to be needed and plans were drawn up for adding two more, X and Y, with the consequent confounding of three interactions. In the event only X was used, being equated with S for sward, half the plots being sown with grass. There was some talk of equating Y with applications of magnesium, because some of the trees that received the higher level of potassium were showing signs of deficiency of that element, but that idea was abandoned and Y was never used. Figure 9 shows the experiment towards the end of its life. The plots are now reduced to the single tree in the middle and there are four factors N, P, K and S with $P \times K \times S$ confounded. It came to an end in 1966.

A later photograph, Fig. 10 shows the land afterwards, the crop being mangolds. Two features stand out. One is the residual effect of nitrogen; the other is the current effect of the strip of clay. The need for a preliminary study of soil and site has already been mentioned, but good maps are needed as well as upon these can be noted any permanent features thought to be important and also any that have been added. Even if soil treatments have been applied for only a short period, it is desirable to know exactly where each plot was situated. With a long-term experiment that is not only desirable; it is essential. When research institutes are founded they often have more land than they need, but sooner or later they find themselves using the same area a second time and then they sometimes regret that they did not survey everything properly from the start.

In the case of the experiment just described the existence of residual effects in known locations has proved to be a most useful asset and there are still investigations in progress that depend upon them.

This is not the place to discuss the conclusion drawn from the experiment. Those interested are referred to the two main reports (Hoblyn, 1941; Greenham, 1965).

9.12 TRANSFORMATIONS

When confronted by a body of data the first thing is to cast an eye over it to see if it presents any unusual or disturbing features. Sometimes there will be irregularities that can be cleared up, using perhaps the methods of Section 7.6, though it is unethical as well as unwise to exclude data only because they are unpleasing. Nevertheless, such an examination can contribute to something more important than the immediate question of excluding particular data, namely, a general knowledge of the behaviour of the quantity as a variate in statistical analysis. A high value that stands away from the rest or a zero value may then be seen as part of a general phenomenon whereas it cannot be understood in isolation.

Statisticians assume normality of distribution because it suits them to do so, not because data necessarily accord with their desires. To take an example, fungal lesions can be measured either by an area or a radius; the two resulting variates cannot both be normally distributed but either might be proffered for analysis. The variability of an experimental crop is commonly presented as a coefficient of variation, accompanied perhaps by a comment that the value is higher or lower than usual. The approach argues that standard errors of crops are ordinarily proportionate to the means. If that is so between experiments supposedly it holds within them also. The argument leads to the awkward conclusion that crops should ordinarily be analysed in logarithmic transformation, but how sound is the practice of always thinking in terms of coefficients of variation? The general situation with regard to transformations is fairly clear, as appears from the following argument, which has been presented many times before. Let the standard error of a quantity, x, be σ. If its distribution is normal, σ should be independent of the mean, μ, but suppose that it is not. Let

$$\sigma = f(\mu) \tag{9.12.1}$$

What function, $y = F(x)$, will have constant standard error? Since

$$\sigma_y = \sigma_x \frac{\mathrm{d}y}{\mathrm{d}x} \qquad \text{(approximately)},$$

if $\sigma_y = f(x)\, \mathrm{d}y/\mathrm{d}x$ is to equal some constant, K, then y should be chosen so that

$$\frac{\mathrm{d}y}{\mathrm{d}x} = \frac{K}{f(x)}. \tag{9.12.2}$$

Hence if the standard error is proportionate to the mean, (9.12.1) gives $\sigma = \alpha\mu$, so $y = \log(x)$, i.e. a logarithmic transformation of x should give a more acceptable variate. Alternatively, if the variance is proportionate to the mean, $\sigma^2 = \alpha\mu$, (9.12.2) leads to the use of the square root of x as a suitable variate rather than x itself. To take another case, if x follows a binomial distribution, as a percentage often does,

$$\sigma^2 = \mu(A - \mu),$$

where A is some upper bound. Taking $f(x)$ to be $[x(1-x)]^{1/2}$, where x is expressed as a proportion of A, suggests $y = \theta = \text{arc sin}\sqrt{x}$ as an acceptable variate. If x is not in fact a continuous variate but proceeds by steps of h, whether on account of a large unit of measurement or for some other reason, $y = \ln(x + \frac{3}{8}h)$ and $y = (x + \frac{3}{8}h)^{1/2}$ are preferable (Anscombe, 1948). Similarly, if x represents an integer, k, expressed as a proportion of another integer, n, then

$$\theta = \text{arc sin}\,[(k + \tfrac{3}{8})/(n + \tfrac{3}{4})]^{1/2}$$

is better than

$$\theta = \text{arc sin}\,(k/n)^{1/2}.$$

A full and useful bibliography has been given by Hoyle (1973). There is also a most useful paper by Box and Cox (1964) on the general problem.

The above approach can be very valuable used rightly, but there are dangers in applying transformations in a piecemeal manner, so that one year's results are analysed in one way and the next year's in another. What is needed is well-based knowledge about the way in which important quantities like crop, plant height, etc., do behave, so that a consistent policy can be followed.

Certain general principles can be adduced. For example, $\log(0) = -\infty$. Consequently a logarithmic transformation cannot be correct if zero values can turn up legitimately. (That disposes of any suggestion that crop weights need such a transformation despite the coefficient of variation being commonly supposed constant.) In fact, such a transformation is chiefly of use when studying the result of a logarithmic growth process, where obviously zero is not possible anyway. (A plant may have zero crop or zero number of buds but it cannot have zero weight or it would not exist.) Again, it is not possible to take the square root of a negative number, at least not with real numbers, so quantities that can legitimately be negative are not candidates for a square-root transformation. Thus, the numbers of insects on a leaf may possibly represent a Poisson distribution. Certainly negative values are impossible, so a square-root transformation may be needed, but only study of real data will show. Again, the need is for a body of knowledge upon which to base practice as opposed to a succession of expedients dependent upon the vagaries of the data on the desk at the moment.

The best transformation when found may tell something about the problem under investigation. Thus, if a fungal lesion begins with a single spore, the distance of spread can be the underlying biological phenomenon under study. (It will be so if the experiment is to compare sprays intended to inhibit such spread. It may not be so if the treatments are such as to prevent spores being scattered in the first place.) Again, if heights or weights are measured on two occasions it may be difficult to find a transformation that will consistently benefit the increment, $y = (x_2 - x_1)$. Reflection may suggest that with exponential growth

$$y = \log(x_2) - \log(x_1) = \log(x_2/x_1)$$

might be better and often it is. The advantage of variance stability may well be matched by a closer approach to additivity because the variate measures better the biological feature under investigation. Too often data represent whatever aspect of the phenomenon is easiest to measure, so transformation may be called for, not merely for statistical reasons, but to come nearer the heart of the matter. By the same argument it is better to avoid coefficients of this and indices of the other, so prevalent in biological literature. To take the example already given: the occurrence of lesions is one phenomenon and it needs to be prevented as far as practicable. The growth of those lesions is another phenomenon also to be prevented, but a spray or procedure that helps with one may do little for the other. To combine the two sets of records into some high-sounding 'index of fungal activity' is to confuse the phenomena. One essential of all data analysis is knowing what one is doing and then proceeding without deviation.

Examples 9

A. The following data come from a banana experiment carried out at the Winban Research Station, Roseau, St. Lucia, in the West Indies. They are given here by permission of the Director of the Station. There were five randomized blocks and six treatments, representing the application of five commercial nematicides, A, B, C, D, E, in comparison with one another and with an untreated control, O. The experiment was continued till the first ratoon crop had been harvested as well as the plant crop. The figures represent total bunch weights per plot in kilograms, the plant crop being presented first.

		I	II	III	IV	V
Treatment	A	391 611	398 605	443 623	436 544	368 489
	B	400 699	414 606	366 611	411 628	408 684
	C	401 621	367 637	387 588	401 622	389 586
	D	353 610	380 606	388 647	391 631	367 679
	E	374 634	398 680	405 600	393 619	447 666
	O	292 552	373 522	410 462	365 494	410 609

How should the data be analysed?

Answer: There is no answer that would be accepted by all biometricians. The best response is to set out the possibilities. First, however, as a basis, the following sums of squared deviations and product of deviations are given. Note that the plant crop is here called x and the ratoon crop, y.

Source	*d.f.*	x^2	xy	y^2
Treatments	5	5 847	7 266	53 319
Error	20	15 237	7 386	37 041
Total	25	21 084	14 652	90 360

The block line has been omitted. Conventionally it would be included, but it has no part to play. From Section 6.1 it will be seen that it does not really belong to the intra-block stratum, so its omission is logical whatever convention may require.

It will be noted that the correlation between the two variates judged by the error line is $+0.31$, so they cannot be regarded as independent, nor would anyone expect them to be.

(1) There are those who would argue that variates are often correlated yet no one objects to their being studied in isolation. If that is done here, values of F for the two variates are respectively 1.53 and 5.76, the latter being significant at the level $P = 0.01$. It could therefore be concluded that the application of nematicides had little effect on the plant crop but an appreciable effect on the ratoons.

(2) Others would argue that the two variates are logically related and cannot be separated. They would therefore analyse s, their sum, and d, their difference. The first would show the total crop; the second would reveal any

tendency for the nematicides to build up an effect. Since

$$s^2 = x^2 + 2xy + y^2$$
$$d^2 = x^2 - 2xy + y^2$$

the required analyses are readily derived from what is known. They are

Source	*d.f.*	s^2	d^2
Treatments	5	73 698	44 634
Error	20	67 050	37 506
Total	25	140 748	82 140

The values of F are now respectively 4.40 and 4.76. Both are significant at the level, $P = 0.01$.

It appears then that there has been an effect on total crop and that some treatments are developing the effect more than others.

(3) A third approach is to carry out a bivariate analysis, as described in Section 9.3.

To start with the transformation, it is

$$X = 0.03623x \qquad \text{(from 9.3.4a)}$$
$$Y = -0.01185x + 0.02445y \qquad \text{(from 9.3.4b).}$$

Of these, X is only x rescaled, while Y represents that part of the ratoon crop that cannot be related to the plant crop, i.e., it represents fresh information. Like X it has been rescaled.

At this point a small digression may be helpful. It is not needed in practice but it will illustrate the reason for the transformation. If X and Y are derived and submitted to analysis in the same way as x and y, the result is

Source	*d.f.*	X^2	XY	Y^2
Treatments	5	7.67	3.93	28.49
Error	20	20.00	0.00	20.00
Total	25	27.67	3.93	48.49

In fact, each has unit variance and their covariance is zero. Accordingly, when plotted on a two-dimensional diagram, ordinary Euclidean geometry can be used for distances between points, etc.

To return to the main theme of the calculations, the next task is to find the bivariate F-value. It may be found either from X and Y or from x and y, using the same method in each case. First, to use the transformed variates

$$\Lambda = \frac{\det\begin{bmatrix} 20.00 & 0.00 \\ 0.00 & 20.00 \end{bmatrix}}{\det\begin{bmatrix} 27.67 & 3.93 \\ 3.93 & 48.49 \end{bmatrix}} = 0.3016 = (0.549)^2,$$

so

$$F = \frac{1-0.549}{0.549} \cdot \frac{19}{5} = 3.12.$$

The multipler $\frac{19}{5}$ equals $(e-1)/h$, where $e(=20)$ is the number of degrees of freedom for error and $h\,(=5)$ is the number for treatments. The bivariate F can be referred to tables for $2h$ and $2(e-1)$ degrees of freedom, i.e. 10 and 38. It is again significant at the level, $P = 0.01$.

The same values are found if the original variates are used instead, i.e.

$$\Lambda = \frac{\det\begin{bmatrix} 15\,237 & 7\,386 \\ 7\,386 & 37\,041 \end{bmatrix}}{\det\begin{bmatrix} 21\,084 & 14\,652 \\ 14\,652 & 90\,360 \end{bmatrix}} = 0.3016$$

To help assess the various approaches, the following treatment means are presented.

Treatment	x	y	s	d	X	Y
A	407	574	982	167	14.75	9.22
B	400	646	1045	246	14.48	11.05
C	389	611	1000	222	14.09	10.33
D	376	635	1010	259	13.62	11.06
E	403	640	1043	236	14.62	10.86
O	370	528	898	158	13.41	8.52

The values of X and Y are graphed in Figure 11.

To consider the first method, it disregards any relationship between the two crops and that could be a serious defect. The second variate could merely be a reproduction of the first on a different scale and so could offer no additional information. Alternatively, there could be such a reaction in the ratoon crop against the plant crop as to cancel out any early gain by a subsequent loss. However, to assess the position in the light of the two analyses, a nematode treatment, A–E, is significantly different ($P = 0.05$) from the control if it differs by

$2.086\sqrt{\left(\frac{15237\theta}{20}\right)}$ in respect of x or $2.086\sqrt{\left(\frac{37041\theta}{20}\right)}$ in respect of y

where $\theta = \frac{1}{5} + \frac{1}{5} = 0.4$, i.e. by 36.4 or 56.8. That is to say, Treatment A is just significantly better than the control for the plant crop; the other four treatments are better for the ratoon crop.

Turning to the second approach, corresponding least significant differences ($P = 0.05$) are 76.4 and 57.1 for s and d respectively. It appears then that all treatments showed an improvement as judged by total crop, s, and all with the exception of A gave hope of better crops in the future by showing an increase of yield from plant crop to ratoon crop greater than that of the control. The method is simple and can be used with more crops than two.

(Indeed, with coconuts or coffee there would be many. For that matter, bananas eventually have several.) When that happens the need is for simple combinations of the successive crops to give intelligible measures of plant performance, e.g. total yield, rate of increase in yield, biennial effects, etc. Each can then be analysed, either separately as in the first approach or collectively as in the third.

The third is the most elaborate; some may think too much so. It does, however, tell the most detailed story. From (9.3.6) two treatments differ at the level, $P = 0.05$, if their points are more than 1.65 units apart. A circle of that radius round Point O shows that Treatments B, C, D and E differ markedly from the control, though not all in the same way. Treatment D for example stands apart. With regard to X, which is the plant crop rescaled, it has done poorly; its advantage lies rather in its good value of Y, which represents the excess of the ratoon crop over what would have been expected from the plant crop. One most marked conclusion is the fact that the nematode treatments are sometimes very different from one another. Points A and D, to take an example, are a distance 2.16 apart.

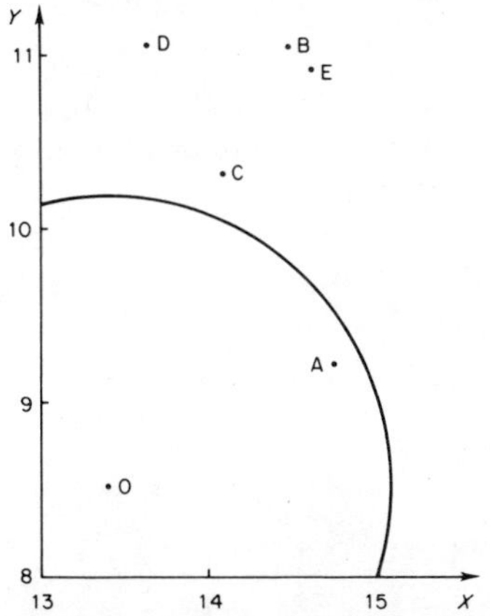

Figure 11. Plotted values of X and Y for the five treatments, A–E, and the untreated control, O, in Example 9A. The circular arc with centre at Point O has a radius of 1.65 to show the least significant difference ($P = 0.05$).

If the author had to choose, he would favour the second approach for most purposes, provided the variates were of the same kind and so suitable for combination. Thus, two crops can be added but it does not make sense to add a weight and a height. The third approach, on the other hand, can cope readily with such a situation. Nevertheless, it is at its best when there is a change of sign between the sums of products of deviations for treatments and error. In the example, both were positive and the bivariate F was 3.12. It would have had the same value if both had been negative, but if they had had different signs it would have been 3.55, which is more.

B. The following data have been presented (Pearce, 1957) from an experiment in which there were eight blocks, each consisting of one tree of Cox's Orange Pippin apple in a pot. On each tree the blossoms were divided into three groups to form plots, which received different pollens. (A = King of the Pippins, B = Ellison's Orange, C = James Grieve, D = Worcester

Pearmain.) The whole formed a totally balanced design. The data show mean fruit diameters in millimetres.

Block = tree	Kind of pollen A	B	C	D
I	—	18.71	17.01	17.23
II	20.42	—	18.02	19.59
III	21.39	20.85	—	18.94
IV	19.49	18.48	17.22	—
V	—	23.16	23.33	23.31
VI	24.67	—	24.54	25.00
VII	24.14	21.78	—	20.86
VIII	26.70	25.75	24.77	—

In order to spread the onerous task of emasculating blossoms and hand-pollinating them, the trees of Blocks I–IV had been brought on by heat treatment. It had been intended to have four more trees later by chilling but their development was so abnormal that the attempt was abandoned.

Analyse the data, allowing for both local and remote effects, as recommended in Section 9.4.

Note: The experiment was initiated in response to reports from fruit growers that Cox's Orange Pippin sets larger fruits when pollinated with King of the Pippins (A). The other varieties represent the usual pollinators.

Answer: The intra-block analysis is

Source	*d.f.*	*s.s.*	
Pollen	3	11.5746	
Error	13	6.7243	$\hat{\sigma}^2 = 0.5173$
Total	16	18.2989	

because

$$\mathbf{Q} = \begin{bmatrix} +5.940 \\ +1.023 \\ -4.243 \\ -2.720 \end{bmatrix}, \quad \mathbf{\Omega} = \tfrac{1}{192}\begin{bmatrix} 35 & -1 & -1 & -1 \\ -1 & 35 & -1 & -1 \\ -1 & -1 & 35 & -1 \\ -1 & -1 & -1 & 35 \end{bmatrix}$$

and $\hat{\boldsymbol{\tau}} = \begin{bmatrix} +1.114 \\ +0.192 \\ -0.796 \\ -0.510 \end{bmatrix}$.

Eliminating the difference between the trees that were advanced by warmth and those that developed normally, the inter-block analysis of variance is

Source	*d.f.*	*s.s.*	
Pollen	3	10.3598	
Error	3	23.7493	$\hat{\sigma}_0^2 = 7.9164$
Total	6	34.1091	

These values were obtained from the following results, which depend upon (3.8.6b), (3.8.7) and (3.8.8)

$$\mathbf{Q}_0 = \begin{bmatrix} +2.030 \\ -1.133 \\ +0.293 \\ -1.190 \end{bmatrix}, \quad \mathbf{\Omega}_0 = \tfrac{1}{6}\begin{bmatrix} 7 & -2 & -2 & -2 \\ -2 & 7 & -2 & -2 \\ -2 & -2 & 7 & -2 \\ -2 & -2 & -2 & 7 \end{bmatrix}$$

and $$\hat{\boldsymbol{\tau}}_0 = \begin{bmatrix} +3.045 \\ -1.700 \\ +0.440 \\ -1.785 \end{bmatrix}.$$

Since $\boldsymbol{\tau} = \boldsymbol{\gamma} - \boldsymbol{\delta}$ and $\boldsymbol{\tau}_0 = \boldsymbol{\gamma} + 2\boldsymbol{\delta}$,

$$\hat{\boldsymbol{\gamma}} = \begin{bmatrix} +1.758 \\ -0.439 \\ -0.384 \\ -0.935 \end{bmatrix} \quad \text{and} \quad \hat{\boldsymbol{\delta}} = \begin{bmatrix} +0.644 \\ -0.631 \\ +0.412 \\ -0.425 \end{bmatrix}.$$

When it comes to testing significances, the high value of σ_0^2, associated as it is with only three degrees of freedom, makes difficult any effective examination of local and remote effects. It shows that the inter-block should be precise and have adequate replication. If all twelve trees had been available comparisons would have been much better, though still not good enough. As it is the covariance matrix of $\hat{\boldsymbol{\tau}} = (\hat{\boldsymbol{\gamma}} - \hat{\boldsymbol{\delta}})$ is $0.5173\mathbf{\Omega} = \mathbf{X}$, say, while that of $\hat{\boldsymbol{\tau}}_0 = (\hat{\boldsymbol{\gamma}} + 2\hat{\boldsymbol{\delta}})$ is $7.9164\mathbf{\Omega}_0 = \mathbf{X}_0$, say. Hence, the covariance matrices of $\hat{\boldsymbol{\gamma}}$ and $\hat{\boldsymbol{\delta}}$ are $(4\mathbf{X} + \mathbf{X}_0)$ and $(\mathbf{X} + \mathbf{X}_0)$ respectively. Such variances are much too high to show anything significant. The whole analysis demonstrates the need for a sound inter-block determination of $\boldsymbol{\tau}_0$. Here there was too much variation between trees; also, the inter-block efficiencies were too low. In so far as the experiment shows anything, it confirms the belief of practical farmers that Cox's Orange Pippin does produce larger fruit when pollinated with King of the Pippins.

C. The following data have been used for purposes of demonstration before (Pearce, 1953; Cox, 1958, Bliss, 1970). They come from an experiment intended to study the best way of forming ground cover in an apple plantation. Treatment O represents the usual method used in England, i.e. keeping the land clean cultivated during the growing season but letting the weeds grow up towards the end, ready to be turned in when cultivation is resumed in the spring. Treatments A, B, C, D and E represented the growing of various permanent crops, e.g. grass, clover, etc. under the trees.

There were four randomized blocks. The variate, y, was the crop in pounds over a four-year period following the initiation of treatments. It is to be adjusted by x, the crop in bushels over the preceding four years. In all pairs of data, x precedes y. (1 pound = 453.6 grams, 1 bushel = 36.4 litres).

Treatments	Blocks I		II		III		IV	
A	8.2	287	9.4	290	7.7	254	8.5	307
B	8.2	271	6.0	209	9.1	243	10.1	348
C	6.8	234	7.0	210	9.7	286	9.9	371
D	5.7	189	5.5	205	10.2	312	10.3	375
E	6.1	210	7.0	276	8.7	279	8.1	344
O	7.6	222	10.1	301	9.0	238	10.5	357

Carry out the necessary covariance adjustments on the lines suggested in Section 9.5.

Note: The trees were very old. They occupied a field between the East Malling Research Station and the railway, and had become virtually derelict, a fact that incited passing travellers to ungenerous comment. In fact, the Station purchased them almost to save its good name. Four years were spent rehabilitating them, during which time volume records were taken on the fruit. Then someone had the idea of using them for an experiment on ground cover, that being a fashionable topic of study when applied to aged trees. Objection was raised that they were much too variable to be used in an experiment, but the statisticians assured everyone that, when trees are old, they have established cropping patterns and so error can be reduced by covariance on previous cropping records. In the event the adjustment served another purpose as well.

The variety was Barnack Beauty. Plots were long and narrow and comprised several trees.

Answer: The design being orthogonal, the method of summation terms can be used. For the products x- and y-values are multiplied together instead of being squared separately. Terms are:

	x^2	xy	y^2
Total	1 713.94	56 649.5	1 896 948
Blocks	1 682.31	55 939.4	1 872 766
Treatments	1 664.08	55 006.4	1 825 663
Correction	1 656.68	54 984.6	1 824 914

That leads to the analyses of variance and covariance, i.e.

Source	*d.f.*	x^2	xy	y^2
Blocks	3	25.63	954.8	47 852
Treatments	5	7.40	21.8	749
Error	15	24.23	688.3	23 433
Total	23	57.26	1 664.9	72 034

It does not appear that the treatments are having any effect, the F-value for y

being only 0.10. There is, however, evidence that y is closely related to x, the correlation coefficient calculated from the error line being $+0.913$. Here U in (2.3.3) $= 688.3$ and $V = 24.23$. Hence, the error sum of squared deviations for y after adjustment by x is $23\,433 - (688.3)^2/24.23 = 3881$ with 14 degrees of freedom. When the treatment line is merged with error, the corresponding figure is $24\,182 - (710.1)^2/31.63 = 8240$ with 19 degrees of freedom. That gives this revised analysis of variance:

Source	*d.f.*	*s.s.*	*m.s.*	*F*
Treatments	5	4359	871.8	3.51
Error	14	3881	277.2	
Total	19	8240		

The impression now is very different. The analysis of covariance, as expected, has usefully reduced the error variance.

There is, however, a second advantage. Examination of the treatment means shows that the randomization process has assigned the control, O, to particularly good plots, thus:

	A	B	C	D	E	O
x	8.45	8.35	8.35	7.93	7.48	9.30
y	285	268	275	270	277	280

The regression coefficient is $V/U = 688.3/24.23 = 28.41$. That is to say, every increase of one bushel in the calibrating period implies an increase of 28.4 pounds in the experiment itself. Changing all values of x to 8.31 gives adjusted means of y, as follows

A	B	C	D	E	O
281	267	274	281	301	252

It is now possible to enquire how far any of the treatments have improved on the control. To take Treatment E first, there is an advantage of 49. The standard error is

$$\sqrt{\left[277.2\left(\frac{1}{4}+\frac{1}{4}+\frac{1.82^2}{24.23}\right)\right]} = 13.3.$$

The figure of 1.82 represents the difference in x-values between the two treatments, i.e. $7.48 - 9.30$, the amount by which adjustment has been made. Clearly the two treatments do differ very significantly. To go to the other extreme, let Treatment B be studied instead. The difference is 15 with a standard error of

$$\sqrt{\left[277.2\left(\frac{1}{4}+\frac{1}{4}+\frac{0.95^2}{24.23}\right)\right]} = 12.2,$$

so here there is no evidence of any gain from using the cover crop. The other treatments can be studied in the same way.

Chapter 10

The people involved

10.1 THE PLANNING OF RESEARCH

Anyone who sits down with pencil and paper to design an experiment should see the task as the culmination of a long process. It started perhaps in a government office when someone decided to allot more funds to agricultural research, there being some defined reason for doing so, e.g., to raise local nutritional levels, to reduce imports or to earn foreign currency. Following that original decision someone decided what crops needed to be developed, which in turn led to a decision how to increase yields, perhaps by initiating a breeding programme, by seeking control measures for a damaging disease or in some other way. By that time there was probably a specific research institute designated to undertake the work and its staff would be involved in the discussions, so continuing the chain of decisions, each made in furtherance of earlier ones, until at last an experiment has been decided upon and someone has to design it. The choices to be made, like everything else, must depend upon what has gone before in pursuance of the aims defined.

Some research institutes are very jealous of their autonomy and may protest that no one tells them what to do. They really mean that no one is so tactless as to give them direct orders, but their funds come from somewhere, whether from an aid-giving agency, the local farming community or a charitable trust, and the flow of money would cease if there were no satisfaction with their work. Other institutes have a commercial basis and there again work must ultimately be directed to defined objectives.

In preceding chapters all the emphasis has been on the specification of contrasts of interest and the best way of estimating them. Further, a warning was given in Section 1.6 about putting a lot of treatments together and using some statistical technique like multiple comparisons to sort them out. It was objected that such a procedure indicates poor scientific thinking. It may be added that it is also an inadequate way of pursuing a defined objective.

There is, however, something peculiar about this long chain of decisions. The first step is made largely on a statistical basis after consultation with economists, marketing experts, sociologists and perhaps many others. The last, the designing of an experiment, also has a large statistical content, but what of the steps in between? Here there is often little quantitative thinking at all, yet it is needed as much in the middle as at the ends.

One reason is the complexity of what is involved in relation to the information available. The minister who wants to know the amount spent on imported fertilizer or the average weight of the potato crop can expect an answer. The biometrician who asks about the uniformity of soil over the experimental area can be told, but the questions raised at an intermediate level may call for quite a lot of research. To take an example, one region of a country is growing a crop that is quite valuable but at the limit of its range so yields are unreliable. Consequently in bad seasons supplies are imported from neighbouring countries and the traders are difficult to dislodge when local supplies are good. A committee meets to decide the allocation of research funds voted from higher up and one suggestion is the initiation of a breeding programme to develop more reliable varieties. Such a decision should be based on quantitative information. If the alternative to better varieties is the eventual extinction of the crop in that region, what alternatives could be grown? How much labour will each alternative need in relation to area of land and will any of them lead to rural unemployment? If so, is there anything else for the displaced farm workers to do? Would increased production of the alternatives lead to a collapse in prices? Questions of that sort could well have been raised at an earlier stage. When they are asked by important people, someone makes the effort to find the answers, but it is by no means certain that similar efforts will be made to help a subordinate committee. In the event the decision may well be made with no quantitative basis at all. Perhaps the politicians yield to pressure from the farming community or perhaps the plant breeders see scientific difficulties in doing what would be asked of them and contrive to kill the project. Either way a wrong decision may result.

Agronomists who find themselves in the unfamiliar environment of a planning committee are apt to assert that everything should be decided on 'technical', i.e. agronomic, grounds, but they should look wider. Suppose, for example, that a valuable crop is being grown only in a certain limited area. Farmers elsewhere may ask for research so that they can grow it too. The agronomic contribution may be limited to confirmation that such research is feasible. If it is approved, there could well be complaints from the area that had the initial monopoly and the ultimate decision will probably lie with the politicians and not with the technical people at all.

At a lower level it may be decided that a crop must be helped but differences arise whether better fertilization or pest control provides the better way forward. The government statisticians could have helped in the last example; here the biometricians could offer guidance by assessing how much land, how much plant material, how much farm labour and how much recording would be needed for each approach. It is true that a lot will depend upon other people's assessment of the relative chances of success, but the decision needs to be made in the light of the biometrician's comments also. The writer recalls sitting with growing horror through a lecture given by a distinguished pathologist. The speaker made the legitimate point that a widespread disease that does little damage can be more serious than a disease that is rare but devastating when it attacks. He then went on to say that 'all that was needed' was a simple experiment that would take the main commercial varieties of a species and compare disease-free stocks with ordinary stocks in order to detect for each the minute differences in crop that he designated as important. The writer's horror came from the realization that if the idea were taken up, he would have to design the experiment and he foresaw difficulties. The species indicated is notoriously variable. In a study of those diseases wide buffers are needed to protect disease-free plots from infection. Since detailed symptom records were proposed, special observers would have to be trained and so it went on. When the writer got home he sketched out what would be needed—at least 50 ha of land and so much labour as to reduce drastically the number of other experiments elsewhere in the institute. Nevertheless the speaker genuinely believed that he was making a modest proposal.

Sometimes the difficulty lies the other way. Some momentous decision has to be made, e.g. a committee has to choose between two quite diverse approaches, and no one is certain. Then a 'preliminary' experiment is suggested to decide which is the more promising. If four plots were used, two to represent each approach, the committee would 'know'. (Biometricians develop violent reactions to the word 'preliminary'. They know it as a synonym for 'slovenly'.) The odd thing is that the same committee, once it has decided its approach, may insist upon excessive replication 'to be certain', yet the complete abandonment of a possible line of approach surely requires a higher level of certainty than the temporary adoption of a practice that is less than optimal.

All this is leading to a plea for statisticians to be involved at all stages of planning, not just the first and the last. In particular, at a research institute there is need of biometrical thinking at the top. It is absurd to allocate land and order plant material before someone with experience of design has had a look at the problem. That person also needs to have a good idea how much labour will be involved in the application of treatments and in the recording of crops and should be aware of expected difficulties. For example, women workers may be hard to obtain during school holidays. If there is a committee that sanctions experiments, the biometrician should be an active member of it. If files are circulated, the biometrician's name should stand high on the list so that warnings can be given early. King Louis XIV of France is reputed to have expressed regrets that he was not present on the Day of Creation, because there were a few improvements he could have suggested. At a well-run research institute biometricians are more fortunate.

10.2 THE PRINCIPAL EXECUTIVES

At this point something needs to be said about the organization of field experiments and the people who have to carry out the work. The whole process may be likened to a dramatic production in which there are several parts to be played. It does not matter a lot if one actor has to play more than one part—

indeed, with a small company that may be inevitable—but all parts must be filled or the result will be absurd. It is so with an experiment, where the cast comprises four main characters.

First, there is the scientist who wants the experiment carried out. It is to be supposed that he has read the relevant papers and has attended conferences of other workers in the area of study so he can give a coherent account of why he wants to do that experiment and no other. But can he? The writer seems to have been plagued all his life by people who say that they are studying, say, 'the fertilization of maize in Fantasia'. Asked what they want to know, they reply that the subject is a most important one and any information would be helpful. If objectives are vague, no satisfactory experiment will result unless by analogy with monkeys endlessly tapping typewriters and eventually by chance producing a Shakespearean sonnet. The scientist is the leading actor; his thinking and intentions must dominate and inspire all else. If he fails, no one else will do much of value. In particular, he should learn that other people are depending upon him for a lead. Sometimes it is not forthcoming as he retreats from the bewildering practicalities of the experimental field to his laboratory, where he has everything under control. Further he needs to value other people's skills and get them to work for him. If the writer has been plagued by bad experimenters, he has also been greatly expanded in his thinking by the good ones, whose clearly formulated needs in experimental design and data analysis have provided a continual inspiration and stimulus.

The next part to be played may be called that of the farm supervisor, who has to prepare the land, perhaps after choosing it in the first place, order the plants or seeds, arrange for planting, control the weeds, apply fertilizer, keep an eye on pests and diseases and eventually harvest and market the crop. All that would be difficult enough without the complications caused by small plots and differential treatments. With those complications it is a wonder that anyone succeeds at all in the part. As a body farm supervisors are the unsung heroes of agricultural research. They also tend to be autocratic, which is perhaps inevitable with people who provide an unobtrusive service that receives too little appreciation. The test comes when someone does notice them and asks for something different. One scientist who asked for wider plots was told that the farm machinery allowed only one width and he must accept that or go without, there being no thought that scientific considerations might count as well. The writer has some sympathy with the research institute that established its experimental farm some twelve kilometres from the laboratories on the grounds, first, that scientists were not much interested in their experiments anyway and, secondly, that, when they did take an interest, their suggestions and requests were so unpractical as to be a nuisance. A better solution is to encourage the scientist and the farm supervisor to know one another.

The third part in the play is that of the recorder, who has to keep the experiment under review and measure what needs to be measured. More than the others this part is often combined with another, the work being undertaken perhaps by the farm supervisor or perhaps by the scientist and his assistants. At a large institute there may be a separate group who perform such duties and they may be virtually autonomous or part of a wider organization, so the possibilities are many. The most important record is probably that of crop yield but at harvest it may be important to know about quality and moisture content as well. Usually records are required at other times, when blossoms break or pests appear or mineral deficiencies develop. A certain, if limited, autonomy is essential. The farm supervisor may want to get the crop quickly to market but it has to be weighed first and mistakes will occur if the recorders are unduly pressed. Also, someone has to talk sense to the scientist who wants detailed records of blossom development but does not appreciate how much work is involved and how long it will take at a time when changes are taking place fast. Not least of a recorder's responsibilities is that of keeping his eyes open, because in the regular surveillance of growth, the taking of leaf samples, etc., he will encounter abnormalities of development that others may miss. One research institute used to have a standard recording sheet on which was written at the bottom

Notes of special features

. . . .

Well, write something! You have eyes, haven't you?

The language sounds rather peremptory but the idea was a good one.

The last part to be played is that of the biometrician. It has deliberately been left to the end because this is a book written by a biometrician principally for his own kind. If he wishes and if the organization of the institute allows him, he can be a catalyst. He should be involved in the problems of experimental design and that means first finding out what the scientist wants to know, but it may also involve consulting the farm supervisor and the chief recorder about what they can actually do. Also, he is the final recipient of the data and, more than most others, he can notice unsuspected variability in the site or variates that are so wildly uncontrolled as to be meaningless. Also, he may have to help the scientist to understand the statistical analyses and, if they reveal little or nothing, he can initiate discussion on what went wrong. Too often the biometrician does none of those things, sometimes because he is content to do the minimum and sometimes because he is told to mind his own business, which is considered to be the pushing of data into a computer that someone else has programmed.

Standards of experimentation vary a lot. There are quite small and ill-equipped institutes that perform excellent experiments and some much better funded ones whose work is terrible indeed. A recent paper by Preece (1982) sets out how terrible it can be. It is not just a matter of money or official support. It should not be possible for a lazy or incompetent scientist to set in motion a process by which the biometrician designs a trial, which the farm supervisor carries out and from which the recorders produce data, which the biometrician analyses to produce a report, which the scientist puts into a publication. (Perhaps he tells his friends how good it is to work at an institute where he has time to 'think'.) Of course, good scientists are not like that. They involve themselves in every stage of the experiment and inspire everybody to give of their best. Unfortunately, not all attain such heights.

10.3 THE ROLE OF THE BIOMETRICIAN

As has been said, the word biometrician in this text refers to a job. At a large institute it may be done by a qualified statistician; at a small one the person concerned may have had little special training. Another distinction and one that is perhaps more important concerns the degree of compulsion about the advice given. Some professional biometricians speak eloquently about the advantages of giving advice that may be freely rejected. They speak of themselves as a village priest might depict himself, as guiding the faithful yet still a friend to those of no religion, who come to him in time of need. That all sounds very delightful, but if the biometrician has been involved at the planning stage there can be no question of his agreeing to something different when it comes to design. Furthermore, if an experiment has been approved subject to certain conditions he must insist upon compliance. For example, difficulties may have been foreseen about excessive recording at a time when labour will be short. The biometrician himself may have urged approval on the grounds that there is land available, so it would be feasible to use fewer replicates than usual but with larger plots. If the scientist talks too much about past replication and so on, the biometrician may have to say bluntly that the experiment goes ahead as approved or it does not go ahead at all, explaining that more but smaller plots may indeed save land and plant material but will cause difficulties at harvest that cannot be faced.

The author, who has had quite a lot of experience in statistical consultancy, will perhaps be forgiven if he interpolates some comments on the subject. It has, incidentally, quite a large literature (e.g. Finney, 1956, 1968; Cox, 1968; Sprent, 1970) and an older generation will recall a facetious article on the subject (Anon, 1949) that pointedly expressed some home truths the biometricians should never forget or they will become pompous. First of all, even if a biometrician does effectively have powers of veto and can prevent a badly designed experiment being planted, it is better to persuade. For one thing it makes relations easier next time; for another, persuasion is a form of education. Reasoned objections to a course of action, if well presented, can lead to useful thought. Next, in a staff of biologists the biometrician's contribution is numeracy and he will be better regarded if his skills appear outside his work. If his views on the new pension scheme or the financial position of the tennis club are hailed as revelations, people will be more ready to seek his opinion about their data. They will also come more readily if they know him as a person. It follows that recluses do not make good biometricians. Finally, his most important safeguard is a reputation for integrity. If he tells the director one thing and the scientist another and the farm supervisor something else, he may put off one awkward moment but he will have many more later when they come to distrust him. Also, if the idea gets around that he will agree to anything if he is shouted at, no one will address him in ordinary tones. In any case, falsehood is self-corrupting, but strict truthfulness is urged here for a much less exalted reason as being part of a biometrician's survival kit. Without a reputation for integrity he will be subject to continual and disconcerting pressure.

There is another quality required in a biometrician though it is not always recognized. It is width of interest. When the author graduated he went to the East Malling Research Station, where the Head of Statistics was T. N. Hoblyn, a former associate of the great R. A. Fisher and himself something of a genius. One day Hoblyn had to entertain a distinguished visitor and over lunch conversation turned to the Pilgrims Way, which runs near East Malling. It came out that the visitor had never heard of the mediaeval pilgrimages to Canterbury, let alone of Geoffrey Chaucer, the first poet of the English language, who wrote about them. Hoblyn remarked afterwards that the visitor was obviously very clever but much too ignorant to make a statistician. The point was well made. It is not that planning committees want discourses on fourteenth century poetry or that colleagues wish to be informed about mediaeval religious practices. The point is that their interests range far and they do not express themselves in statistical language. Unless the biometrician readily takes a genuine interest in problems as such, there will be little understanding on either side. The narrow specialist, who can explain matters only by using algebraic formulae and who takes no interest in other people's doings, will be quite lost.

As has been said, the designer needs a clear appreciation of what is to be established. Here scientists vary a lot. At the top are a small minority who know and can express it in statistical terminology. At the bottom are another minority who do not know and appear to think that precise objectives do not matter. In between lie the majority who can talk about the experiment in their own terms but need prompting. Getting them to talk easily is one of the chief skills of a statistical consultant and it is here that width of interest comes in. A botanist talking about adventitious roots can strike someone trained in mathematics as boring. By a great effort the consultant might sort out the main points, but the best results are achieved by someone who can catch the enthusiasm and respond to it by questions which, though perhaps uncomprehending, are nonetheless prompted by genuine interest. Of course, it helps to recognize that scientist and biometrician are not really interested in quite the same aspects of the problem. The scientist is concerned with the substance of the conclusions sought. A negative result could destroy a line of thought while a positive one might bring a successful innovation within reach. The biometrician, on the other hand, is concerned with the form of the problem, e.g. the aim is to establish whether there is or is not an interaction. Emotional involvement is better avoided, one conclusion being as acceptable as the other. It should be clear what contrasts are important and what extraneous factors, like soil variability or texture of soil, will have to be taken into account.

The other information that should come out of the consultation concerns resources, e.g. how much land will be needed and what will have to be recorded. Here other people may be involved and such matters can bring out the biometrician's role as a catalyst. A tactful suggestion that the farm supervisor or the chief recorder should be consulted can avoid trouble later. The biometrician at this point sometimes has to be firm. If, as he sees it, the experiment needs 0.5 ha and the scientist can manage only 0.2 ha, there must be no evasion of the impracticability of proceeding. If tact fails, one has to be blunt.

Finally, a consultation can help to sort out someone's thoughts and the biometrician should not think that a discussion has failed because he himself has said very little. Many years ago when the author was a young man, he was asked to advise a visitor about the design of experiments in a certain country. It so happened that he had recently heard a lecture about agriculture in that country, which turned out to be very fortunate. The visitor explained briefly that he was interested in fertilizer trials on a certain crop but did not seem disposed to enlarge on the problems, so the writer asked if he worked in a high rainfall area. That started a long explanation about the difficulties of conducting fertilizer trials in conditions where nutrients are speedily leached out. When that came to an end, the writer tried a second question. Were the experiments to be conducted on a laterite soil? That provoked an explanation of the manifold difficulties when nutrient elements are applied but become locked up in complex molecules. At the end the visitor shook the writer's hand and thanked him enthusiastically for a most helpful consultation. Later he wrote the director a charming letter. Not only, it said, had Dr. Pearce disposed of all the difficulties but he had explained everything so clearly. Dr. Pearce has only one comment. If his visitor really discovered how to conduct a fertilizer trial on a laterite soil with heavy rainfall, he might have explained. The other party to the consultation is still trying to find out! The point of the story is, however, quite serious. It helps to talk to a sympathetic listener who makes the right noises from time to time. If a consultation clarifies matters, it is immaterial who does the talking.

10.4 POST-MORTEMS ON CONCLUDED EXPERIMENTS

There is one important function of a resident biometrician that is often neglected. It is that of looking back over completed experiments to see how things have turned out, not from the point of view of agronomic results but of biometrical efficacy.

The first quantity to look at is the error variance. Any research institute should have a good idea how much error to expect for a given plot size of a given species when recording a given quantity. There need be no mystery about such matters because the information often lies in past analyses of variance. Better than looking back, however, is the keeping of a register of error variances as data are analysed. It may reveal surprising things quite apart from its main purpose in suggesting an expected value that can be used in future. For example, it may show a steady change in the level of error over the years, possibly due to changes in the control of a dominant disease or in cultural practices or for some other reason, which needs to be identified. Also, the occasional 'rogue' experiments may reveal a pattern. If they always occur in a certain field or when a certain variety is used the conclusion will be obvious, but in all instances they need to be noted and followed up as a matter of course to find out what went wrong. (A difficulty here lies in the time that often elapses between events in the field and the completion of analyses of data. It points the need for a good diary for each field trial complete with locations and dates for all important happenings.) It need hardly be added that different measurements will have different variances so the size of an experiment should depend upon the principal variates for analysis.

There is another important use of past data, namely, to resolve doubts remaining at the inception of an investigation. There may, for example, have been a difference of opinion whether to use blocks or rows and columns. A decision had to be made, but a re-examination of the data using the Kuiper–Corsten iteration will show the sort of error variances to be expected had the decision been made the other way (Pearce, 1976c). That can be a useful guide for the future. (It is true that the randomization will have accorded with the blocks, etc., actually used. That must lead to some bias in the error, though it is likely to be small relative to the gain given by a correct blocking system compared with an incorrect one.) Where a rogue experiment has been conducted, it is similarly possible to enquire whether some other configuration of blocks would have eliminated the large variation encountered. If a positive answer is reached, something will have been learnt for the future. It would, of course, be quite wrong to analyse and re-analyse a body of data until a desired conclusion was reached, but that is not proposed. The intention is to obtain information about experimental and statistical techniques so as to avoid making the same mistake a second time. Example 10A takes a body of data and shows how such a study might proceed.

In fact, a post-mortem program that will put in blocks, etc., as instructed and find the corresponding error variance is as desirable an adjunct as a catalogue of past variances. One sometimes hears statements like 'we always use row-and-column designs' or 'we never use split-plots'. If they are based on well-tried lore appropriate to the conditions, such assertions deserve respect and could provide a basis of practice for other people similarly placed, but they must not merely be prejudices.

No less, past data can be used to check whether or not the best plot size has been adopted. If there is reason to suspect that plots could advantageously have been smaller, it is necessary to take steps during the life of the experiment by recording, say, half-plots instead of whole ones. Simple addition will give the data that would otherwise have been obtained, while the Kuiper–Corsten iteration will discover the variance of those half-plots within suitably chosen half-blocks. If, on the other hand, it looks as if larger plots could have been used, the existing plots can be merged in pairs and the variance of those double plots within blocks of double size also can be found using the Kuiper–Corsten iteration, possibly with Worthington's extension (Section 6.3). The method has been examined in more detail elsewhere (Pearce, 1976a). The approach obviates any need for the old-fashioned 'uniformity trial', in which an area without differential treatments was harvested in minute plots, which were then grouped into plots of various shapes and sizes to find the variance appropriate to each. In any event, that method provided information only for one field in one season whereas a series of post-mortems can do more.

Just as a biometrician should have a good idea what error variances he can expect in different circumstances, so he should have soundly based ideas on the reasonableness of assuming that the effects of blocks and treatments can be added. The subject receives curiously little attention. Biological thinking goes directly against any such assumption, while facile assertions that it is confirmed by general experience must often be discounted. In fact, the assumption lies deep in statistical thinking and discarding it would demolish too much. Like normality of distribution and equality of variance of all observations it is convenient to assume and also difficult to test. Consequently the statistician is loath to question it.

That is not to say that the assumption is wrong. Mostly, at least, it holds well enough. It will do so, for example, if there are no effects of blocks. If it is true that blocks are mostly chosen with little thought and so remove little or no variation, their non-existent effect can hardly interact with anything else. Again there may be no effect of treatments. Also, if, as is quite likely, there is a moderate effect of both blocks and treatments, whatever happens the outcome cannot be far from an additive relationship, just as a segment from a smooth curve can be approximated by a straight line if the arc is short enough. The danger comes when one or two blocks are markedly different from the rest and the treatments have a large effect. In such a case an assumption of additivity could be absurd.

Also, of course, some treatments are more likely to interact with site differences than others. Thus, varieties are often recommended for their tolerance of specified adverse conditions, like strong winds or drought. Given a site in which there are areas exposed to wind while others are sheltered or in which there is a range of moisture, no one should expect varietal differences to be the same over the whole experiment. Every case should be considered separately because there can be no general conclusion. Nevertheless, a body of lore based on past experiences can usefully be built up. Too often an institute has only a few studies based on data that raised doubts; that can be misleading. A serious study of typical data, on the other hand, could be most valuable.

Similar comments apply to the interaction of rows and columns, though here the problems are much simpler. If a research institute conducts its experiments on terrain that does not lend itself to row-and-column designs, a few post-mortems will soon reveal the fact. Again, a little thought will sometimes suggest that a particular site should be used in some other way. The writer recalls an experiment in the West Indies. The land sloped towards a stream and parallel drains had been dug to take away the rain, which could be heavy. The site appeared to be ideal for rows and columns, so a Latin square was laid down, the rows corresponding to altitude and the columns to the spaces between the drains. The only difficulty came with a corner plot, which lay partly on the lower slopes of a hill behind the site. In the event two things went wrong. The corner plot had a large residual. Also, one of the drains became blocked and in consequence the upper parts of two adjacent columns did not fit the additive hypothesis of rows and columns. The institute is one that considers these matters sensibly and two conclusions were reached. One was to be more suspicious of the hills in future. The other was to pay special attention to keeping drains clear where a row-and-column design is in use. It would be a good thing if all institutes were so thoughtful.

The matters mentioned in this Section are practical in nature and call for practical solutions. Biometricians who work in a context of agricultural research are surrounded by people learning from experience, who put established methods to the test and seek something better where results are disappointing. It is here suggested that biometry should be more of an experimental science itself, each research institute building up a body of lore about its own land, its plant material and its techniques and aware of the degree of error that results from those various sources. Of course, there are institutes where the biometricians set themselves very high standards in these matters, but there are others where they stay close to the computer without ever putting on their boots and joining their agronomic colleagues outside to see exactly how treatments are applied, exactly how plant material is selected and distributed in the field, exactly how records are taken and diseases assessed. If they did, they would be in less danger of solving unrealistic or non-existent problems, a danger pointed out with some vigour by Federer (1976). Also, if they did those things they would find a wealth of problems that called for attention and they would no longer have to complain, as some do, that they have few openings for research. It is true that directors and those who allot funds do not always encourage their biometricians to be so active mentally, but they might change if they became aware of the economies, financial and otherwise, that can be effected by making good use of them at every stage. If such a change of attitude is to come about, however, some biometricians themselves will have to change. There are ample rewards for those who engage in the activities around them with minds that are alert and critical and also constructive.

10.5 DATA RECORDING

In writing a book the data are necessarily presented as if they appeared from nowhere; anyone who has worked at a research institute knows that they are obtained with great labour and sometimes with great difficulty. This is a matter where the biometrician can help a lot provided he goes out into the field and learns the procedures at first hand. Skill at detecting outliers in data can be acquired in an office; it is much better to know how rogue values occur and so be able to obviate them. Here again, with experience and the comparing of instances it is possible to build a body of lore that will provide guidance for the future and that is much better than dealing with each crisis in isolation as it arises.

Broadly speaking there are two kinds of records: those, such as harvesting, that involve farm workers, and those that call for specialists, like keeping track of the development of a disease. The distinction is not completely clear, but it will serve. Where people are involved who have had little scientific training it is necessary to lay down precise procedures and insist that they are observed. The point needs special emphasis in a developing country where workers may even be illiterate. It is no good placing red markers round a plot and red labels on a set of containers and supplying the harvesters with red armbands if no one objects when they wander off to work alongside friends on the blue plot. In fact, simple people are often cheerfully complaisant about observing regulations that they can understand even if they regard the whole exercise as slightly absurd. Also, crops often have to be gathered speedily and harvesters, if they are country folk, understand that. There must nonetheless be pressures to prevent them leaving small amounts that a commercial farm would not bother about, its collection being uneconomic, whereas on a research farm it must be recorded. Town folk who come out to earn a little at harvest time can be difficult and may need a different set of instructions, not only to hurry them up and to obviate wrong procedures but to divert them from the damage they sometimes cause. A tactful field supervisor is most important; crop recording to scientific standards is a human problem as well as a technical one.

The chief difficulty is often the demarcation of plots. Sometimes a discard area or an alley is sufficient; sometimes a wire has to be strung on canes to show where the division comes. When a plot boundary is reached, there must be clear instructions what to do. Either the harvester takes the crop to be weighed or the containers must be left in a designated place. They can then either be picked up and taken to a weighing centre or be recorded in the field—if the former, they must be clearly and consistently labelled. Since mistakes are bound to occur from time to time, it can be a help if all containers available are given a number and allocated to plots as may be convenient. Then if someone goes round and notes the numbers beforehand and if at weighing the numbers are noted again by someone else, a useful check will have been obtained.

Recording should not stop at noting harvest weights. Some argue that the aim of an agricultural process is to produce crop and as much of it as possible, to which sentiment no one need object. They therefore conclude that supplementary records are irrelevant. In a scientific context, however, it is necessary to know how crop is obtained, so records of germination, plant height, pests and diseases and much else may be required, nor should such records be confined to what is apparent. The development of roots and the presence of soil fungi or colonies of nematodes could all be important. The danger here is of institutionalizing records—'We carry out chemical analyses as a matter of course' or 'We always record leaf colour' sound fine—as if all contingencies were provided for in some comprehensive scheme, but does such inflexibility lead to a wise allocation of resources? Given an outbreak of caterpillar, does it really help to know the chemical content of the leaves they devour? Or their colour? Perhaps it does, but such matters require thought.

The records now being considered are not the gross records of crop weight but rather the measurements required for scientific assessment. Here the biometrician has a large part to play. First of all, there is the allocation of effort between various stages of the recording procedure. A chemist may go out into the field, take a few leaves with little thought as to their age or position on the plant, bring them back to his laboratory and then demand a platinum crucible so that his determinations shall be precise; nor are chemists the only offenders. Everyone thinks that his stage of the procedure is the important one. In general, random errors combine according to Pythagoras's rule

$$\sigma_{12}^2 = \sigma_1^2 + \sigma_2^2$$

where σ_1 and σ_2 are the standard errors introduced in Stages 1 and 2 and σ_{12} measures their combined effect. If σ_2, say, is less than one third of σ_1, σ_{12} will lie between σ_1 and $1.055\sigma_1$ and the total elimination of error at the second stage would do little good. The aim must always be to reduce the larger component. In the case of the leaf samples, the need would be to sample better in the field. Until that problem was settled, quite simple methods of chemical analysis would suffice. If it is objected that such less elaborate analyses might introduce bias, it would still be necessary to ask how far that would matter. Where only comparative figures are needed it can be ignored, just as anyone who wants to measure the lapse of time does not have to enquire if the clock is fast or slow. (As has been pointed out in Chapter 3, field experiments usually measure contrasts anyway, not means.)

Another problem for the biometrician is degree of precision. In general three significant places are enough, e.g. 25.4, 0.0173, and often two are satisfactory. At the least, if the analysis of variance is to be used, there should be ten possible values of the variate. (Data confined, say, to the values, 0, 1, and 2 cannot be regarded as continuous.) A danger here is the spurious precision introduced by a change of scale. Thus, crop weights might start with

6350, 7711, 7258, . . . grams

and the first reaction might be to accuse the recorder of wasteful meticulousness. When a few more data are seen, however, repeated values begin to arise and it becomes obvious that the recorder has been measuring

14, 17, 16, . . . pounds

and multiplying by 453.6. In fact, his precision is scarcely good enough. It is better to keep as close as possible to the original data and to make any conversions at the end. It is then much easier to pick up any quirks in the data.

Another problem arises when a recorder fails to make up his mind and introduces half gradations or the like to avoid problems. He thus produces data like

$$52, 61, 59, 49, 50\tfrac{1}{2}, 52, \ldots.$$

(That situation arises with the data in Example 2A.) Some clear-cut convention is needed about values that lie between two gradations. For that matter, reading 'to the nearest tenth' also produces some odd results. Many people do in fact produce data that end mostly in 2 or 3, 5, 7 or 8, though others become self-conscious and avoid 5. The papers of Yule (1927) and Preece (1981) provide reading that is entertaining as well as instructive.

Grading causes even more problems than measuring. If there is no objective background, it is impossible for most people to carry more than five grades in their head. To ask someone to grade leaf colour with 20 gradations is to invite trouble; most people will settle down to using about five of the marks available. Further, the uncritical, who think that everything is either marvellously good or astonishingly bad, will slash down extreme values and outweigh the cautious, who will stick to a few middle values, though it is the latter who are probably the better judges. The writer's liking is for five grades, A, B^+, B, B^-, C, the recorder being told that A and C are to be used only for what is really outstanding, leaving B^+, B and B^- respectively for 'above average', 'about average', and 'less than average'. Even that is to court oddities like B−−?. In that case the recorder must be dealt with severely, as when he produces an excess of A's or C's. Of course, with so few grades the data are discontinuous; either they must be accepted as such or several recorders must be used and their results aggregated. (With such grades recorders do not usually agree very well and matters are improved by gathering the impressions of as many people as can conveniently be recruited.) Where the recorder has no objective way of deciding his grades, he cannot produce objective continuous data. He can do little more than formalize his impressions and that limitation must be faced.

Where there are objective background records to provide a check, more grades can be used, but they should be reasonable ones. The biometrician may sympathize with the entomologist who wants a scale like

0	No insects present
1	One insect found
2	Two insects found
3	A few insects found
4	Many insects found.

Nevertheless, he should point out the difficulties it raises. Having got such data he can try transforming them, but the whole situation is less than satisfactory. Anyway, before he complains too much, he should ask himself whether he has anything better to suggest. One possibility in the above case is to recommend that the sample be made a lot larger andthat the recorder should note only presence and absence. For a Poisson distribution the natural logarithm of the zero count (i.e. the proportion of units that have no insects) provides an estimate of the mean, but is a Poisson distribution likely? And what happens if most units are infested to some extent? Such problems do not admit of a good general solution.

Nevertheless, even if there is no perfect solution, as in many other cases the biometrician has a part to play. He should not stay in his office and wait for the data. He should know where the figures come from and what they represent. He should know how a certain quantity has been measured in the past and should have ideas how it could be measured in the future if changes are called for. Also, his suggestions should be practical if he is to retain the confidence of his colleagues.

Recorders should be aware of the benefits that come from a good understanding with the biometricians. The one produces the data that the other analyses, so they have a shared interest. Nevertheless what is done in the field must basically be under the control of those who have to work in it. Elaborate forms intended to facilitate summarization, either by the scientist or the biometrician, are a mistake. A recorder must go into the field with the plots written down on his sheet in the order in which he will come to them; the process of derandomization can be carried out later. (In any case, the less he is reminded of the treatments the better.)

There is, however, another reason why the recorder and the biometrician should work together, namely, the statistical background to much of a recorder's work. For example, if only some plots can be measured or if some are to be measured today and some tomorrow, the biometrician will know about block boundaries and experimental design. Again, many records have to be taken by sampling and that is quite a specialist subject in itself. Not least, people are sometimes sent out into the field to do silly things and may need support if they are to protest. For example, people cannot put yields of trees into grades allowing for tree size. There is a certain irony there. If a biometrician were to make proposals for recording yield or tree size without actually measuring anything he would be told not to be pretentious. If he were to enunciate a relationship between the two that was supposed to be of general application regardless of treatment, he would be told loftily that he did not understand botany, but some see nothing wrong in asking a recorder to perform such a task.

10.6 COMPUTING SERVICES

These days no agricultural research institute will get far without a good computing service to store the data in a form, e.g. on magnetic tape or discs, accessible to computers, so that summaries and analyses can readily be obtained. Further, computer science is a highly technological subject and those who have attained proficiency in it possess skills different from and complementary to those of the agronomist and the biometrician. Consequently they should be heard with respect when they speak about their own subject, but by the same argument they are in no position to tell the staff of an institute which calculations are needed. In approaching a computer scientist for help the same ground rules apply as for the consultation with any other expert, i.e. the client explains what is wanted and the expert advises how to obtain it.

There are several features of data from field experiments that should be borne in mind when considering what computer facilities are required. One is that a large experiment generates a lot of data but not all at the same time. Also, figures collected later may have to be related to those collected earlier so there must be some uniformity of format. Here it should be emphasized that the worst standard format is one in which the data are de-randomized into a neat $b \times v$ table, the blocks one way and treatments the other. Whatever the design and whatever the conditions, it interposes an additional and avoidable copying process between the field data and the computer store and one that is productive of errors. In any case, such a format may suit an orthogonal design but it causes difficulty with an incomplete block design ($k < v$) and chaos with one in extended blocks ($k > v$). Also, if the computer is required to print out residuals according to the field plan, it has to know the randomization anyway. A much better practice is to number the plots according to some convenient sequence for recording them and to use it consistently, starting with the blank sheets taken to the field and continuing through storage in the computer to input of the analysis program. If the person who wrote that program saved himself trouble by a bland statement that data should be derandomized before input, it is advisable to ignore his efforts and find something better.

The next feature to look for is variate control. As has been said, data have to be combined. A crop like tomatoes might give 50 or 60 harvests in the year and tea is picked at intervals for years. Even if there is only one harvest, there could still be a need to relate yields to water content or number of stems. Hence there should be ample facilities for making new variates from those measured in the field, e.g., the agronomist might want the percentage of crop in the first three harvests and the biometrician may add that it should be analysed in angular transformation. (Perhaps a complication arises because one plot was lost after the fourth harvest!) If those facilities are not present, the computer installation is not going to be much use as it stands. (Of course, people can always write additional programs for it.)

Next, there should be programs able to deal with incomplete data and with covariance adjustments. Again, they can be written or acquired from elsewhere, but the whole suite of programs is either going to need quite a large computer or else a good deal of skill and co-operation in reading intermediate results in a form available to a succeeding program or what a computer specialist calls 'overlay'.

At this point someone accustomed to the conditions of a developed country may protest that he has all the facilities required and a console in every laboratory. Others are not so lucky. The fact is that the average agricultural research institute is not large enough to justify so large a computer installation for its exclusive use. In a technologically developed country that is not necessarily a difficulty. It is possible to have access to a distant computer with all the needed facilities but there can still be problems. If the computer is in some government department that regards the paying of pensions as its real function, the intrusion of a lot of analyses of variance will not be welcomed, even if satisfactory programs are found and, what is unlikely, the hardware is suitable. Also, since people who do not get their pensions are more of a nuisance than scientists who do not get their analyses, the priorities will not favour the scientists. Also, it is unlikely that the computer staff will be very knowledgeable if the programs start producing nonsense. The idea is one that appeals to administrators, but it has its disadvantages. A university computer is better because its operation is geared to a multitude of diverse tasks. Also, a group of scientists could well relate better to its staff. The trouble with universities is their tendency to let students learn from their mistakes. That can be awkward for people who make theirs with a long stretch of telephone wire between them and the computer.

Anyhow, even if all that is available in a developed country, it is not so in a poor one, where the telephone system is probably overextended and transmission is of poor quality. Here the preferred solution is often to provide a computer for the institute with the idea that it will work out wages, keep track of the stores and, incidentally, serve all scientific purposes as well. That creates a nearly hopeless situation. Adequate computer programs cannot be written, especially if the installation is unsuitable, and alternatives like sending the data to the nearest city will be rejected by the administrators. They will point out that there is no need, because the institute has a perfectly good computer of its own. (They may add that they can vouch for its excellence. Only rarely does it make a mistake with the wages.)

Fortunately the advent of microcomputers gives hope of improvement. New ones have surprisingly large cores and the use of terse computer languages, like APL, extends their range of usefulness. It is too early to say that they provide a solution but at least they give hope of one.

However, in the tropics the difficulties are not finished. Whatever the hardware, line voltage is likely to be very variable, humidity may go to extremes, while sand and dust may pervade everything mechanical. No doubt the experts are right when they say that such problems can readily be solved, but it is fair to comment that the solutions profferred are not always successful.

However, assuming that a computer is available and that everyone is on speaking terms, a matter of some importance, it becomes necessary to choose the programs. At one extreme there are the standard packages, which have become standard because people have used them, liked them and recommended them to others, so all must have their advantages. One stands out as having been developed in an agricultural context, namely, GENSTAT, and certainly it has a lot to commend it. Its facilities for data control are extensive and it can cover a range of designs in different strata so split-plots are no problem to it. Its strength and also its weakness is its close adherence to the concept of general balance (see Section 4.8). It uses the method of sweeping set out in Sections 6.2 and 6.3, combining data from strata as explained in Section 6.4. However, other programs have their adherents also and it is advisable to survey the field before deciding. Whatever package is chosen, there will be things it cannot do. For example, many packages give trouble with a non-factorial partition of the treatment sum of squared deviations. That is not intended as a condemnation of packages but rather as a warning that a computer is not there to write a report but to provide the figures upon which a report can be based. There will always be a need for supplementary calculations.

At the other extreme there is a lot to be said for occasionally writing a little program that will do just what is wanted and no more. It can be a lot easier than using an extensive program that was not really intended for jobs of that particular sort and then adapting the output. Incidentally, when writing programs oneself or when trying to discover what the unknown writer of a package did in some contentious and difficult situation, there is much to be said for analysing pseudo-variates, such as those in Sections 7.4, 7.5 and 7.7, to make sure that all is correct or to find out exactly what the program does do in those circumstances.

However, there will always be instances where mishaps have caused such chaos that any program devised for ordinary designs is going to be at a loss. It is therefore useful to have some facility for minimizing on to a given set of parameters. In principle any analysis of variance can be carried out given two minimizations. If a post-mortem is required of the sort described in Section 10.4, only the error variance is needed and one minimization will suffice. If such a program has to be written locally there is a lot to be said for using the Kuiper–Corsten iteration (Section 3.5), possibly with Worthington's extension (Section 6.2).

However, this chapter is primarily about people. Many of the questions raised in this section find a solution if the various experts involved recognize one another's skills and needs. As has been said, the computer scientist must be allowed to run the installation in his own way. Equally he must be told exactly what is wanted or he may well produce something else. A difficult area is the writing of programs. The computer expert may justly claim to know more about the economical use of space and difficulties with rounding errors etc. but it is unlikely that he knows much about the calculations. Usually it is best for the person who wants the program to take the lead but to work in close association with the computer scientists. In the last resort it is better to have a program that produces useful results wastefully than one that generates irrelevancies in an economical manner, but given good sense and co-operation both faults can be avoided.

A modern hazard comes from the ability of someone who knows nothing about data analysis to sit at a console and produce output that is assumed to mean something. Cautious biometricians regard it rather as medical men regard self-medication. They fear a series of patients who have poisoned themselves with proprietary drugs. In fact the medical profession has lived for a long time with the problem and biometricians will no doubt come to terms with it too. If people use techniques they do not understand they will make mistakes and the biometricians can fairly decline responsibility. Nevertheless, they could try educating their colleagues. It is also a wise provision for directors who are asked to approve a report for publication to enquire if the biometrician has seen it. That could stop some nonsense going further.

In fact, there are two chief dangers. One is that of misunderstanding. The experimenter may not appreciate, say, the effect that split-plots will have on the analysis and does everything in one stratum instead of two. Another fault is taking a variate that is confined to a few values, say, 0, 1, and 2, and treating it as if it were continuous. Such matters can be picked up afterwards by someone more knowledgeable. The other danger is that of data snooping, which is nearly impossible to detect afterwards unless it leads to absurd results. The owner of the data sits down and analyses everything he has in all the transformations anyone ever heard of. If he tries hard enough, the one-in-twenty chance will come off several times and there will be something to report with that recognized 'mark of approval', $P < 0.05$. If it appears that the three-factor interaction is significant for the fifth harvest but not for the others and that one of the two-factor interactions shows up for the mean width of nodes divided by the number of aphides, no one will take much notice of the results anyway, but there are dangers nonetheless. The fortuitous results might make misleading sense.

10.7 CONCLUSION

This book has covered a range of topics, some apparently theoretical and others apparently practical, but the distinction is unreal. Theory should be directed to practical ends and practice should be soundly based, so the two are not opposed but complementary.

One danger comes from obsession with techniques, whether statistical or otherwise. Good experiments do need good organization and sound procedures, but that is not enough. However magnificent the equipment and however impressive the range of experts consulted, little good will be achieved unless the scientific thinking is clear and directed to useful ends. Some quite humble little institutes have achieved great things and some large ones have failed to justify the funds invested in them, so whatever conduces to success it is not size or money as such.

Two ingredients of success can, however, be discerned. One is sound planning in which the needs of the farmers are related to scientific knowledge and objectives are set out, which are clear and defined and such that everyone can work to them, biologists, field workers and biometricians alike. That is not an easy task for those who have to perform it. The need is for clear, sharp, and disciplined thinking. It is always difficult to secure precise agreement among any group of people, no obscure corners being tactfully left unilluminated lest clarity destroy the illusion of unity. It is especially difficult when the group consists of diverse experts—perhaps a botanist, a rural economist, a practically minded farm supervisor, a biometrician and a marketing expert. Nevertheless genuine unity has to be achieved.

The second essential is really anterior to the first. The clarity of mind that leads to good experimentation depends upon concord between all those diverse experts, who must respect one another's skills and recognize that no one person's knowledge is enough. Given that state of mind they will consult one another as a matter of course and their combined wisdom can produce something good. If, on the other hand, they all retreat into their respective offices or laboratories, each declaring to anyone who will listen that the others just do not understand, it is unlikely that the outcome will do much for the people who really matter, namely, the farming community. The resulting experiment may be adequate in each of its parts, but the whole can be ineffective. It is true that the scientist must play the leading part and true also that the planning objectives are paramount, but an experiment must also be conducted economically or projects with lower priorities will be deferred. The design should please the farm supervisor, who has to prepare the land and ensure good husbandry, the people who have to apply the differential treatments and those who have to record everything, and the biometrician, who has to report on the contrasts of interest and to do so with a clear conscience that the figures can be relied upon.

All that calls for willing co-operation of a high order, which will be found only if those concerned think of themselves as a team united by a common purpose. Napoleon once said that in making war three-quarters depends upon the state of mind of the troops. Much the same can be said of field experimentation. These days everything is so technical and technological that good equipment is essential but of itself it does not ensure a good outcome. That needs people too.

Example 10

A. Turning back to the data of Example 2A, can anything be learnt from them to guide the design of future similar experiments?

Answer: The experimental design made no attempt to control local variation, but the calculations in Examples 2B and 2C show that the land was far from uniform, so some kind of post-mortem examination is indicated. Those examples do not constitute a post-mortem on their own account because they represent alternative ways of analysing the data and give no guidance as to what designs would be better. They do, however, show what can be done without any improvements and therefore set standards.

Since there are four replicates of most treatments and eight of the other, an obvious suggestion is to use four blocks, with each treatment except O having one plot in each block and O having two. Further, on the general grounds that blocks should be compact there is a case for each to cover two columns. Introducing such blocks and using the Kuiper–Corsten iteration gives an error variance of 21.80, which is a great improvement on 44.92. Further, it compares quite favourably with the values of 19.26 and 30.28 found from the methods used respectively in Examples 2B and 2C. If anyone proposed making blocks out of rows, he would still have blocks of eight plots each, but this time the Kuiper–Corsten iteration gives a variance of 44.54, which is no real improvement. (Such a block structure might well be effective if outside effects dominated, but that is scarcely to be expected here where there is a patchy incidence of fungus.)

Since double columns are so effective as blocks there is a case for exploring the usefulness of single columns as well. They would contain only four plots and would therefore require a non-orthogonal design, but that possibility need not be rejected if a substantial improvement in error variance could be obtained. Introducing such blocks gives a figure of 9.37, which is so satisfactory that non-orthogonality does look eminently worthwhile. Further, now that the column effects have been removed, it could be that a row-and-column design has possibilities. It is true that rows did little good by themselves, but they were then cutting across the contours of infectivity, and it was fortunate that they did not do positive harm. However, imposing both rows and columns on the plots and using the Kuiper–Corsten iteration in the appropriate forms gives an error variance of 15.23, so it shows no advantage, being less good than single columns alone.

Where does all that lead? First of all, of course, it does nothing to improve the completed experiment. (In any case, data should be analysed in the way intended at the inception of the investigation. To do anything else, i.e. to analyse now this way and then that way until something agreeable emerges, is just 'data-snooping' in an especially offensive form, because the elaboration of techniques can disguise what is going on.) Further, too much should not be built on a single example. The next one might suggest something different; the aim is to build up a body of knowledge. (Mark Twain had a little anecdote to show that it is possible to learn too much from experience. It relates how a cat that once jumped on to a hot stove would never thereafter jump on to a cold one either.) In fact, the post-mortem just carried out confirms conventional wisdom. It suggests that blocks are advantageous, that they need to be compact and are especially effective if they can be made to be along contours, always assuming that the direction of the contours is known. Since the incidence of soil fungus can be expected to form patches of the sort assumed by the conventionally wise, the conclusions need occasion no astonishment. They do nonetheless provide useful guidance for the design of similar experiments in the future.

Actually, post-mortems are especially useful when it is possible to examine the site afterwards. If an area of poor growth or cropping can be related to soil texture or exposure to wind or any other feature, something valuable will have been learned. The difficulty about practical experimental design is often that there is so much that could be taken into account—soil maps, previous land usage, altitude, exposure, etc. etc.—that a choice has to be made. It is not possible to allow for everything, so experience of what has mattered in the past can provide valuable guidance.

The study just made was perhaps helped by the lack of blocks, but it could still have been carried out if there had been an attempt to control local variation. Actual blocks could have been ignored, because the aim is only to calculate an error variance. The restriction on randomization implied by actual blocks is unlikely to be serious and, anyway, no one is going to build much on one post-mortem.

Appendix on Matrix Algebra

In this text extensive use has been made of matrices and matrix algebra despite the difficulties that may be caused to those whose mathematics has not been taken so far. The decision was not made carelessly. Ordinary algebra deals only with single quantities, but in statistical theory the argument concerns groups of quantities, like data or treatment replications, and it requires a calculus able to deal with aggregations of numbers. In fact, matrix algebra is not very difficult to anyone who can understand ordinary mathematical notation and it has seemed better to use a language appropriate to the problems in hand and to explain it in this appendix rather than to assay the formidable complications of writing in a more familiar but quite unsuitable medium. This appendix sets out the main concepts introduced in the text. It makes no claims to mathematical rigour. To a complete beginner it may seem like offering a dictionary to someone called upon to read a book in an unfamiiiar language. Even so, a dictionary would be a help and there are those who know a little about the language and could usefully know more. This appendix should in any case help the reader to work through the examples and that is as much as some will require.

Ordinary algebra deals with single quantities called 'scalars'. Traditionally they are represented by italic letters, like x or A, or by Greek letters, like θ or Γ. It is true that they can be put into groups by the use of suffices. For example, the data could be called $y_1, y_2, \ldots$, etc. Also, scalar algebra does make some provision for dealing with such a group. Thus Σ means 'the sum of', so Σy_i means the sum of quantities like y_i, i.e. the sum of the data. It is more correct to specify the suffix over which summation is taking place, i.e. $\Sigma_i y_i$, but sometimes that is so obvious that it can be omitted. Sometimes too it is necessary to set limits on the summation. Thus, if there were n data, $\Sigma_{i \neq 1} y_i$ or $\Sigma_{i=2}^{n} y_i$ would both mean the sum of all data except the first. The symbol Π similarly means 'the product of'.

A matrix can be defined as a set of quantities arranged in rows and columns, the rows and columns being meaningful. For example, the rows of a matrix might represent crops and its columns years, the 'elements' being the national production of each crop in each year. In fact, what a mathematician calls a matrix is what everyone else calls a table, though there is one difference. A table may contain voids. Thus, if no one knows how much mustard was produced in 1976 it is permissible in a table to put a dash, but a matrix must be complete. Where a matrix has only one row or only one column it is called a 'vector'. In this text, as in some others, the convention will be adopted of regarding all vectors as having a single column unless otherwise specified. Matrices are represented by bold-face letters, e.g. **x** or **A**. Where Greek letters are used, a heavy type face will be chosen, e.g. $\boldsymbol{\theta}$ or $\boldsymbol{\Gamma}$. There are some who like matrices to be represented by capital letters and vectors by lower case ones. They have a sound point but adoption of the practice in the present text would have upset too much traditional notation. For that reason it has not been followed.

Where a matrix has been introduced, specific elements can be indicated by the use of suffices. Thus, if there is a matrix **R**, then R_{ij} means the element in the ith row and the jth column. For a vector, a single suffix is enough.

After that introduction the main part of the appendix can begin.

(1) *Addition and subtraction.* Two matrices are added simply by adding corresponding elements. Supposing that $\mathbf{A}+\mathbf{B}=\mathbf{C}$, then $C_{ij}=A_{ij}+B_{ij}$ for all i and j. Thus, if **M** represents the number of male students reading various subjects at a range of universities and if **F** is a similar matrix for female students, someone might want to add them to find **S**, referring to students as a whole. The two matrices must be 'conformable', i.e. both must have the same number of rows and the same number of columns. Further, to make sense, corresponding rows and columns should have the same meaning. Subtraction follows the same rule as addition.

(2) *Multiplication of a matrix by a scalar.* Each element of the matrix is multiplied by the scalar, i.e. if $\mathbf{X}=\alpha\mathbf{Y}$, then $X_{ij}=\alpha Y_{ij}$ for all i and j. To take an example, a matrix gives mean height of plants in inches, that being the local unit of measurement, but before publication all elements are multiplied by 2.54 to convert inches to centimetres.

(3) *Transposition of a matrix.* A matrix is transposed when its rows and columns are interchanged. Hence, if $\mathbf{P}=\mathbf{Q}'$, $P_{ij}=Q_{ji}$. Transposition similarly changes a row vector into a column vector and *vice versa.* (It is this ability to transpose that enables all vectors to be defined in a single column. If a row vector is needed, it is sufficient to add a prime, $'$.)

(4) *Inner product of two vectors.* If two vectors are taken and corresponding elements are multiplied, the sum of products is called the 'inner product'. Thus, if the vectors are

$$\mathbf{a}=\begin{bmatrix}3\\-1\\4\end{bmatrix} \quad \text{and} \quad \mathbf{b}=\begin{bmatrix}2\\2\\-1\end{bmatrix},$$

their inner product is $6-2-4=0$. It is written $\mathbf{a}'\mathbf{b}$, which is the same as $\mathbf{b}'\mathbf{a}$. (The first vector is written as a row vector and the second as a column vector.) The process is sometimes called 'multiplying out' the two vectors. In this context conformability means only that the two vectors have the same number of elements. If they do not, the operation is impossible. The calculation is often required. For example, if **a** represents the number of leaves in various segments of a plant and if **b** represents the mean area of leaves within each segment as determined by sampling, **a**′**b** gives an estimate of the total leaf area for the plant.

(5) *Multiplication of matrices.* Here an important difference is found from scalar algebra. There $xy = yx$ but in general **XY** is not the same as **YX**, though in special cases it may be. Accordingly a distinction must be made between the first matrix of a product, which is said to 'pre-multiply', and the second, which is said to 'post-multiply'. To be conformable for purposes of multiplication, the number of columns in the first matrix must equal the number of rows in the second. Then if $\mathbf{Z} = \mathbf{XY}$, Z_{ij} is found by multiplying out the ith row of **X** and the jth row of **Y**. If

$$\mathbf{X} = \begin{bmatrix} 4 & 3 & 1 \\ 2 & 3 & 0 \end{bmatrix} \quad \text{and} \quad \mathbf{Y} = \begin{bmatrix} 2 & 2 & 5 \\ 1 & 3 & -1 \\ 2 & 4 & 3 \end{bmatrix}, \quad \text{then}$$

$$\mathbf{Z} = \mathbf{XY} = \begin{bmatrix} 8+3+2 & 8+9+4 & 20-3+3 \\ 4+3+0 & 4+9+0 & 10-3+0 \end{bmatrix} = \begin{bmatrix} 13 & 21 & 20 \\ 7 & 13 & 7 \end{bmatrix}$$

On the other hand, **YX** does not exist because the two matrices are not conformable for that product.

The definition just given may appear arbitrary but it corresponds to a number of necessary calculations, as an example will show. Let a matrix, **S**, set out the yields of various species given by a range of cropping systems. The rows represent systems and the columns the species, say, millet, sorghum, groundnut, soyabean and a number of others. Let **C** be a conversion matrix. Its elements give conversion rates for the several species concerned, setting out how much energy, protein, vitamin C, etc., are given by one kilogram of crop. The rows represent species in the same order as before and the columns represent the dietary components. Then $\mathbf{A} = \mathbf{SC}$ has the same rows as **S** and the same columns as **C** and its elements give an assessment of the various cropping systems in terms of contributions to diet. Other forms of multiplication can be devised and are sometimes useful, but, unless otherwise stated, matrix multiplication implies the operation defined above.

(6) *Extended products.* In an extended product the order of the multiplications does not matter. Thus, if **PQR** is required, calculating **PQ** and using it to pre-multiply **R** comes to the same thing as calculating **QR** and using it to post-multiply **P**. On the other hand, **QPR**, if it exists at all, could be quite different.

(7) *Transposing a product.* Note that $(\mathbf{AB})' = \mathbf{B}'\mathbf{A}'$.

(8) *Multiplication with vectors.* Multiplication with vectors follows the same rules as with matrices. The product of two vectors can be a little surprising. Thus, **a**′**b** is a scalar, but **ab**′ is a matrix. Going back to (4)

$$\mathbf{ab}' = \begin{bmatrix} 3 \\ -1 \\ 4 \end{bmatrix} [2,\ 2,\ -1] = \begin{bmatrix} 6 & 6 & -3 \\ -2 & -2 & 1 \\ 8 & 8 & -4 \end{bmatrix}.$$

In general any extended product that starts with a row vector and ends with a column vector is a scalar. Thus, in (3.3.9) it is shown that the error sum of squared deviations (a scalar) always equals $\mathbf{y}'\boldsymbol{\psi}\mathbf{y}$, where $\mathbf{y}$ is the vector of data and $\boldsymbol{\psi}$ is a matrix that depends upon the design.

(9) *Eigenvectors and eigenvalues.* If $\mathbf{M}$ is a square matrix and $\mathbf{u}$ is a vector, such that $\mathbf{Mu} = \lambda\mathbf{u}$, where λ is some scalar, then $\mathbf{u}$ is said to be an eigenvector of $\mathbf{M}$ with eigenvalue λ. For example, if

$$\mathbf{M} = \begin{bmatrix} 2 & 2 & 8 \\ 0 & 3 & -2 \\ 1 & -2 & -2 \end{bmatrix} \quad \text{and} \quad \mathbf{u} = \begin{bmatrix} 2 \\ -2 \\ 1 \end{bmatrix}, \quad \text{then}$$

$\mathbf{Mu} = 4\mathbf{u}$. In fact, if $\mathbf{u}$ had been $(4, -4, 2)'$ or $(1, -1, \frac{1}{2})'$ the relationship would still have held, so it is useful to have some convention as to scaling. It is often convenient, though not always essential, to scale an eigenvector, $\mathbf{u}$, so that $\mathbf{u}'\mathbf{u} = 1$. Thus, the present example could well be written $(\frac{2}{3}, -\frac{2}{3}, \frac{1}{3})'$.

It may be noted that $\mathbf{M}$ has three eigenvectors in all, namely

$$\mathbf{u}_1 = \tfrac{1}{3}\begin{bmatrix} 2 \\ -2 \\ 1 \end{bmatrix}, \quad \mathbf{u}_2 = \frac{1}{\sqrt{5}}\begin{bmatrix} 2 \\ 1 \\ 0 \end{bmatrix} \quad \mathbf{u}_3 = \frac{1}{\sqrt{153}}\begin{bmatrix} 10 \\ -2 \\ -7 \end{bmatrix}.$$

with eigenvalues respectively equal to 4, 3 and -4.

(10) *Directrices.* If a matrix is square with n rows and n columns, then any set of n elements are said to form a directrix if none of them have a row or column in common. Thus in the matrix

$$\mathbf{X} = \begin{bmatrix} a & b & c \\ d & e & f \\ g & h & i \end{bmatrix},$$

$b\,f\,g$ and $c\,d\,h$ are both directrices. The directrix $a\,e\,i$ has a special role and is called the 'leading diagonal'.

(11) *Idempotent numbers.* In scalar arithmetic there are two numbers with special properties, namely, 0 and 1. They are called 'idempotent' because they equal their own powers, i.e.

$$0 = 0^2 = 0^3 = \ldots \text{etc.}, \; 1 = 1^2 = 1^3 = \ldots \text{etc.}$$

They are differentiated by their behaviour in multiplication, because any number multiplied by 0 gives 0, whereas any number multiplied by 1 is unchanged. Further, any number added to 0 also remains the same.

In matrix arithmetic the corresponding concepts are the 'null-matrix' or the 'null-vector', in which all elements equal zero, and the 'identity matrix', in which elements on the leading diagonal equal one and all others are zero. For example

$$\begin{bmatrix} 0 & 0 \\ 0 & 0 \\ 0 & 0 \end{bmatrix} = \mathbf{0} \quad \begin{bmatrix} 1 & 0 & 0 \\ 0 & 1 & 0 \\ 0 & 0 & 1 \end{bmatrix} = \mathbf{I}_3$$

It will be seen that identity matrices are necessarily square. In this text the symbol **I** will usually carry a suffix to show the size of the matrix, but that will not be done for **0**. For one thing, it has two dimensions, which would need two suffices and that can be cumbersome. For another, once a matrix has been shown to be null, it is mostly not used further and its dimensions are not going to matter much.

Somewhat similar is a vector in which all elements equal one. It will here be written $\mathbf{1}_p$, where the suffix gives the number of elements. Post-multiplication by it sums the rows of a matrix; pre-multiplication by such a vector transposed sums its columns.

(12) *Trace of a matrix*. The sum of the elements in the leading diagonal is called the 'trace' of the matrix. Taking the matrix in (10), tr $(\mathbf{X}) = a+e+i$.

Special interest attaches to the trace of a matrix like $\mathbf{B} = \mathbf{AA}'$. A diagonal element, B_{ii}, equals $\Sigma_j A_{ij}^2$. Hence tr $(\mathbf{AA}')$ equals the sum of the elements of **A** after each has been squared. If it equals zero, **A** must be a null-matrix.

(13) *Determinant of a matrix*. If all elements of a directrix are multiplied together and the products added, the sum is called the 'permanent' of the matrix. Thus, in (10) the permanent of **X** is

$$a\,e\,i+a\,f\,h+b\,d\,i+b\,f\,g+c\,d\,h+c\,e\,g.$$

Examination of the six products shows that one, $a\,e\,i$, is derived from the leading diagonal, while three, $a\,f\,h$, $b\,d\,i$, and $c\,e\,g$, come from directrices that can be made into the leading diagonal by a single interchange of rows or columns. The others, $b\,f\,g$ and $c\,d\,h$, would require two such interchanges. Giving a positive sign to those products that require an even number of interchanges and a negative one to those that require an odd number, leads to the 'determinant'. Here

$$\det(\mathbf{X}) = a\,e\,i-a\,f\,h-b\,d\,i+b\,f\,g+c\,d\,h-c\,e\,g.$$

(14) *Some properties of determinants*. If two columns (or two rows) of a matrix are interchanged, the sign of the determinant is reversed. Obviously that is so, because the directrices remain the same but each now requires one more or one less interchange to make it into the leading diagonal; hence, the sign of its product is altered. As a corollary, if two columns (or rows) are the same, the determinant must be zero, because interchanging the similar columns (or rows) must reverse the sign of the determinant without changing its value. The same applies if the columns (or rows) are not equal but one can be obtained from the other by multiplying throughout by a constant, because that merely multiplies the determinant by the same amount. As a further corollary, if a matrix is altered by taking a column (or row), multiplying it by a constant and adding it to another column (or row), the determinant remains the same. Thus,

$$\det\begin{bmatrix} a & b & c+\theta a \\ d & e & f+\theta d \\ g & h & i+\theta g \end{bmatrix} = \det\begin{bmatrix} a & b & c \\ d & e & f \\ g & h & i \end{bmatrix} + \det\begin{bmatrix} a & b & \theta a \\ d & e & \theta d \\ g & h & \theta g \end{bmatrix}$$

and the second determinant equals zero. It follows that if a column (or row) is a linear combination of the others, the determinant is zero, i.e.

$$\det\begin{bmatrix} a & b & \theta a+\phi b \\ d & e & \theta d+\phi e \\ g & h & \theta g+\phi h \end{bmatrix} = \det\begin{bmatrix} a & b & \phi b \\ d & e & \phi e \\ g & h & \phi h \end{bmatrix} = \det\begin{bmatrix} a & b & 0 \\ d & e & 0 \\ g & h & 0 \end{bmatrix}$$

which is zero because all directrices must contain a zero element. Where that property holds, the columns (or rows) are said to be 'dependent' and the determinant equals zero. Also, there is some eigenvector with a zero eigenvalue. Thus,

$$\begin{bmatrix} a & b & \theta a+\phi b \\ d & e & \theta d+\phi e \\ g & h & \theta g+\phi h \end{bmatrix}\begin{bmatrix} \theta \\ \phi \\ -1 \end{bmatrix} = \begin{bmatrix} 0 \\ 0 \\ 0 \end{bmatrix} = 0\begin{bmatrix} \theta \\ \phi \\ -1 \end{bmatrix}$$

Conversely, if the determinant is zero there must be dependence between the columns (or rows), but the proof is beyond the scope of the present text.

(15) *Latent roots and characteristic vectors.* It is an old problem to consider the square matrix, $(\mathbf{M}-\lambda\mathbf{I})$ and to enquire what values of λ will make its determinant equal zero. Plainly, if the matrix has p rows and p columns, the product of elements on the leading diagonal will give a term in λ^p so the equation

$$\det(\mathbf{M}-\lambda\mathbf{I}) = 0$$

must have p solutions, or 'roots' as they are called. Further, for each such root there must be dependence between the columns of $(\mathbf{M}-\lambda\mathbf{I})$, i.e. there must be a vector, $\mathbf{u}$, such that

$$(\mathbf{M}-\lambda\mathbf{I})\mathbf{u} = \mathbf{M}\mathbf{u}-\lambda\mathbf{u} = 0.$$

In fact, the solution, λ, which is called 'a latent root' and the vector, $\mathbf{u}$, which is called a 'characteristic vector' are an eigenvalue and an eigenvector under other names.

The current approach does, however, have some notable strengths. First of all, it establishes that there are p eigenvalues. Accordingly, it is no coincidence that three were found for the matrix in (9).

In what follows it will be convenient to reverse signs in the determinantal equation and write

$$\det(\lambda\mathbf{I}-\mathbf{M}) = 0.$$

It is a polynomial of order p in λ. Here a short examination of polynomials may help. Such an expression can always be written in terms of its roots. Thus, scaling it so that the coefficient of x^2 is 1, a second-order polynomial may be written

$$(x-x_1)(x-x_2) = x^2-(x_1+x_2)x+x_1x_2$$

and a third-order thus,

$$(x-x_1)(x-x_2)(x-x_3) = x^3-(x_1+x_2+x_3)x^2 + (x_2x_3+x_1x_3+x_1x_2)x-x_1x_2x_3.$$

However much the order is increased, the expression will start with a term, x^p, and then there will follow a term in x^{p-1} with coefficient $-\Sigma x_i$. The last term will be a constant, $(-1)^p \Pi x_i$.

Returning to the determinantal equation in its new form, the first term is indeed λ^p. A term in λ^{p-1} can arise only as a product that has $(p-1)$ elements from the leading diagonal of $\lambda\mathbf{I}$ and one from the leading diagonal of $\mathbf{M}$. The coefficient of λ^{p-1} is therefore $-\Sigma M_{ii} = -\mathrm{tr}(\mathbf{M})$. It follows then that the sum of elements on the leading diagonal of $\mathbf{M}$ is also the sum of solutions for λ, i.e.

$$\Sigma\lambda_i = \mathrm{tr}(\mathbf{M}).$$

Reference to (9) will show that the trace of that matrix, namely $2+3-2=3$, is indeed the sum of its eigenvalues, $4+3-4=3$.

The constant term of a polynomial is found by putting the variable equal to 0. In $\det(\mathbf{M}-\lambda\mathbf{I})$ that gives $\det(\mathbf{M})$ but in $\det(\lambda\mathbf{I}-\mathbf{M})$ it gives $(-1)^p \det(\mathbf{M})$. Considering polynomials in general, it has already been noted that the constant term is $(-1)^p \Pi\lambda_i$. Hence

$$\Pi\lambda_i = \det(\mathbf{M}).$$

Again, reference to (9) provides confirmation, because $\det(\mathbf{X}) = -48$, which is the product of the eigenvalues, i.e. $4 \times 3 \times (-4) = -48$.

(16) *Inverse matrices.* If an element, M_{ij}, is selected in a square matrix $\mathbf{M}$ and all elements in its row and column are removed, what is left is said to be its minor. Thus, if

$$\mathbf{M} = \begin{bmatrix} a & b & c \\ d & e & f \\ g & h & i \end{bmatrix},$$

the minor of a is

$$\begin{bmatrix} e & f \\ h & i \end{bmatrix} \quad \text{and of } f \quad \begin{bmatrix} a & b \\ g & h \end{bmatrix}, \quad \text{and so on.}$$

If now, $\mathbf{M}$ is transposed and each element is replaced by the determinant of its minor, the resultant matrix, $\mathbf{M}^*$, has some interesting and useful properties. (Elements on the leading diagonal take a positive sign, those on adjoining diagonals take a negative one and so on alternately.) It appears that

$$\mathbf{M}^* = \begin{bmatrix} ei-fh & ch-bi & bf-ce \\ fg-di & ai-cg & cd-af \\ dh-eg & bg-ah & ae-bd \end{bmatrix}.$$

If now $\mathbf{M}^*\mathbf{M}$ or $\mathbf{M}\mathbf{M}^*$ are found, each diagonal element of the product will equal $\det(\mathbf{M})$. Thus, to multiply out the first row of $\mathbf{M}^*$ and the first column of $\mathbf{M}$, the result is

$$a(ei-fh)+d(ch-bi)+g(bf-ce).$$

That is to say, ei and fh are products of elements on directrices through a and they need a for completion. Similarly, ch and bi need d and so on. If, however, the first row of $\mathbf{M}^*$ is multiplied out by the second column of $\mathbf{M}$, the result is

$$b(ei-fh)+e(ch-bi)+h(ef-ce)=0.$$

The reason is that the determinant now formed is that of

$$\begin{bmatrix} b & b & c \\ e & e & f \\ h & h & i \end{bmatrix},$$

which has two identical columns. It follows that

$$\mathbf{MM}^* = \mathbf{M}^*\mathbf{M} = \det(\mathbf{M})\mathbf{I}.$$

The matrix, $\mathbf{M}^{-1} = \mathbf{M}^*/\det(\mathbf{M})$, is called the 'inverse' of $\mathbf{M}$. In matrix algebra it corresponds to the reciprocal, $1/m$, of a scalar, m, which likewise is sometimes written m^{-1}. In general, multiplication by the inverse of a matrix, whether pre-multiplication or post-multiplication, has many resemblances to division. For example, in (2.3.5) the reduction in sum of squared deviations brought about by allowing for a concomitant variable is shown to be U^2/V. When there are several such variables, U becomes a vector, $\mathbf{U}$, and V becomes a matrix, $\mathbf{V}$. It appears from (2.3.11) that the reduction, still a scalar, equals $\mathbf{U}'\mathbf{V}^{-1}\mathbf{U}$.

If $\det(\mathbf{M}) = 0$, no inverse exists and $\mathbf{M}$ is then said to be 'singular'. Since $\det(\mathbf{M}) = \Pi\lambda_i$, one or more zero eigenvalues make a matrix singular.

Although inverses can be found by the method given, other easier methods exist. In practice almost every computer has a sub-routine for inverting a matrix, so nowadays inversions are rarely calculated 'by hand', though some care may be needed in choosing the best sub-routine for the particular matrix.

A simple expression exists for the inversion of a product, namely $(\mathbf{AB})^{-1} = \mathbf{B}^{-1}\mathbf{A}^{-1}$, because $(\mathbf{B}^{-1}\mathbf{A}^{-1})(\mathbf{AB}) = \mathbf{B}^{-1}\mathbf{IB} = \mathbf{B}^{-1}\mathbf{B} = \mathbf{I}$, so $\mathbf{B}^{-1}\mathbf{A}^{-1}$ is the inverse of $\mathbf{AB}$.

To take an example of inversion,

$$\begin{bmatrix} 4 & 2 & -1 & 1 \\ 0 & 6 & 2 & 3 \\ 2 & 1 & 3 & 1 \\ 2 & 0 & -1 & 4 \end{bmatrix}^{-1} = \frac{1}{338}\begin{bmatrix} 67 & 29 & 40 & 5 \\ 36 & 50 & 34 & -38 \\ -42 & 2 & 96 & -12 \\ -44 & 14 & 44 & 84 \end{bmatrix}.$$

(17) *The eigenvectors of a symmetric matrix.* A matrix is said to be symmetric if it equals its own transpose, i.e. if $\mathbf{S}' = \mathbf{S}$. Such a matrix is necessarily square and its eigenvectors have some interesting properties. For example, provided $\lambda_i \neq \lambda_j$, $\mathbf{u}_i'\mathbf{u}_j = 0$ because

$$\mathbf{u}_i'\mathbf{S}\mathbf{u}_j = \mathbf{u}_i'(\lambda_j\mathbf{u}_j) = \lambda_j\mathbf{u}_i'\mathbf{u}_j.$$

Also

$$\mathbf{u}_i'\mathbf{S}\mathbf{u}_j = (\mathbf{u}_i'\mathbf{S}')\mathbf{u}_j = (\lambda_i\mathbf{u}_i')\mathbf{u}_j = \lambda_i\mathbf{u}_i'\mathbf{u}_j.$$

Since $\lambda_i \neq \lambda_j$, $\mathbf{u}_i'\mathbf{u}_j$ necessarily equals zero.

It has already been provided that $\mathbf{u}_i'\mathbf{u}_i = 1$. Accordingly, if a matrix, $\mathbf{U}$, is formed such that its ith column is the ith eigenvector, i.e. '$\mathbf{U}$ holds the eigenvectors in its columns', then $\mathbf{U}'\mathbf{U} = \mathbf{I}$. That is to say, the transpose of $\mathbf{U}$ is also its inverse, $\mathbf{U}' = \mathbf{U}^{-1}$. It follows that $\mathbf{U}\mathbf{U}'$ also equals $\mathbf{I}$.

The reader may ask what does happen if $\lambda_i = \lambda_j$. The difficulty is merely that the eigenvectors are not uniquely defined, for if $\mathbf{S}\mathbf{u}_i = \lambda\mathbf{u}_i$ and $\mathbf{S}\mathbf{u}_j = \lambda\mathbf{u}_j$, then $\mathbf{S}(\theta\mathbf{u}_i + \phi\mathbf{u}_j) = \lambda(\theta\mathbf{u}_i + \phi\mathbf{u}_j)$, where θ and ϕ are any scalars. Nevertheless, acceptable solutions can always be found. If $\mathbf{u}_i'\mathbf{u}_j = \alpha \neq 0$, then alternatives are

$$\mathbf{u}_{ai} = \frac{\mathbf{u}_i + \mathbf{u}_j}{\sqrt{2(1+\alpha)}}, \qquad \mathbf{u}_{aj} = \frac{\mathbf{u}_i - \mathbf{u}_j}{\sqrt{2(1-\alpha)}}.$$

It will be seen that $\mathbf{u}_{ai}'\mathbf{u}_{ai} = 1 = \mathbf{u}_{aj}'\mathbf{u}_{aj}$ and that $\mathbf{u}_{ai}'\mathbf{u}_{aj} = 0$. Further, assuming that $\mathbf{u}_i'\mathbf{u}_k = 0 = \mathbf{u}_j'\mathbf{u}_k$, where $\mathbf{u}_k$ is any other eigenvector, then $\mathbf{u}_{ai}'\mathbf{u}_k = 0 = \mathbf{u}_{aj}'\mathbf{u}_k$ also.

Returning to the expression, $\mathbf{U}'\mathbf{U} = \mathbf{I}$, the element of $\mathbf{U}'\mathbf{U}$ in the ith row and the jth column is $\Sigma_k U_{ki} U_{kj}$, which value could have been obtained equally well from the matrix, $\Sigma_k(\mathbf{u}_k\mathbf{u}_k')$. In fact, for a symmetric matric,

$$\sum_k (\mathbf{u}_k\mathbf{u}_k') = \mathbf{I}$$

For example, the matrix

$$\begin{bmatrix} 2 & -1 & -1 \\ -1 & 2 & -1 \\ -1 & -1 & 2 \end{bmatrix}$$

has eigenvectors

$$\frac{1}{\sqrt{3}}\begin{bmatrix} 1 \\ 1 \\ 1 \end{bmatrix}, \quad \frac{1}{\sqrt{6}}\begin{bmatrix} 2 \\ -1 \\ -1 \end{bmatrix}, \quad \text{and} \quad \frac{1}{\sqrt{2}}\begin{bmatrix} 0 \\ 1 \\ -1 \end{bmatrix},$$

so

$$\frac{1}{3}\begin{bmatrix} 1 & 1 & 1 \\ 1 & 1 & 1 \\ 1 & 1 & 1 \end{bmatrix} + \frac{1}{6}\begin{bmatrix} 4 & -2 & -2 \\ -2 & 1 & 1 \\ -2 & 1 & 1 \end{bmatrix} + \frac{1}{2}\begin{bmatrix} 0 & 0 & 0 \\ 0 & 1 & -1 \\ 0 & -1 & 1 \end{bmatrix} = \begin{bmatrix} 1 & 0 & 0 \\ 0 & 1 & 0 \\ 0 & 0 & 1 \end{bmatrix}.$$

Most computers are equipped with sub-routines for finding the eigenvalues and eigenvectors of a given matrix, so there need be no problems of calculation. However, some procedures break down if the eigenvalues are not all distinct and, unfortunately, in experimental design equality of eigenvalues of certain matrices is a sought-after property. Nevertheless, a suitable sub-routine can almost always be acquired.

(18) *Expression of a general vector in terms of eigenvectors.* If $\mathbf{x}$ is any vector with p elements and if $\mathbf{u}_1, \mathbf{u}_2, \ldots, \mathbf{u}_p$ are eigenvectors of a symmetric matrix of order p, then it is always possible to express $\mathbf{x}$ as a linear combination of those eigenvectors, i.e. $\mathbf{x} = \Sigma(\alpha_i\mathbf{u}_i)$. The solution is given by putting $\alpha_i = \mathbf{x}'\mathbf{u}_i$. Then

$$\sum(\alpha_i\mathbf{u}_i') = \sum(\mathbf{x}'\mathbf{u}_i\mathbf{u}_i') = \mathbf{x}'\sum(\mathbf{u}_i\mathbf{u}_i') = \mathbf{x}'\mathbf{I} = \mathbf{x}'.$$

The result has an important consequence. If two symmetric matrices, **X** and **Y**, have the same eigenvalues and eigenvectors, they must be identical. Clearly $(\mathbf{X}-\mathbf{Y})\mathbf{c} = \mathbf{0}$ if **c** is an eigenvector, but the same relationship will hold if **c** is a linear combination of eigenvectors. Hence $(\mathbf{X}-\mathbf{Y})$ gives a null-vector whatever vector post-multiplies it. It must therefore itself be a null-matrix.

(19) *Spectral decomposition.* Let **S** be a symmetric matrix with eigenvalues $\lambda_1, \lambda_2, \ldots$, corresponding respectively to eigenvectors $\mathbf{u}_1, \mathbf{u}_2, \ldots$, and let

$$\mathbf{X} = \sum(\lambda_k \mathbf{u}_k \mathbf{u}_k').$$

Then, $\mathbf{X}\mathbf{u}_j = \lambda_j \mathbf{u}_j$, because $\mathbf{u}_k'\mathbf{u}_j = 0$ if $k \neq j$ and $\mathbf{u}_k'\mathbf{u}_j = 1$ if $k = j$. It follows that **X** has the same eigenvectors and eigenvalues as **S** and so $\mathbf{X} = \mathbf{S}$. If

$$\mathbf{Y} = \sum_k(\mathbf{u}_k\mathbf{u}_k'/\lambda_k),$$

then $\mathbf{XY} = \mathbf{I} = \mathbf{YX}$, again because of the known result for multiplying out eigenvectors, a quantity like $(\lambda_i\mathbf{u}_i\mathbf{u}_i')(\mathbf{u}_j\mathbf{y}_j'/\lambda_j)$ being zero if $i \neq j$ and $\mathbf{u}_i\mathbf{u}_i'$ if $i = j$. Hence, an inverse of **S** can be found by replacing all its eigenvalues by their reciprocals, i.e.,

$$\mathbf{S}^{-1} = \sum(\lambda_i^{-1}\mathbf{u}_i\mathbf{u}_i') \quad \text{if} \quad \mathbf{S} = \sum(\lambda_i\mathbf{u}_i\mathbf{u}_i'),$$

provided there are no zero eigenvalues of **S**, i.e. provided no $\lambda_i = 0$.

Similarly

$$\mathbf{S}^2 = \mathbf{SS} = \sum(\lambda_i^2\mathbf{u}_i\mathbf{u}_i') \qquad \text{and in general} \qquad \mathbf{S}^n = \sum(\lambda_i^n\mathbf{u}_i\mathbf{u}_i').$$

It follows that if a symmetric matrix is idempotent, i.e. if $\mathbf{S}^n = \mathbf{S}$ for all n, then its eigenvalues also must be idempotent, i.e. all must equal 0 or 1.

(20) *Rank of a matrix.* For purposes of the present text the 'rank' of a matrix will be defined as the number of its non-zero eigenvalues. (The definition is a little unconventional but it does no violence to the usual meaning, because only symmetric matrices are to be considered here.) There are several points to note. Thus, if **M** is square and of rank r so is any matrix, $\mathbf{XMX}^{-1}$, where **X** is non-singular, because if $\mathbf{Mu} = \lambda\mathbf{u}$ then

$$\mathbf{XMX}^{-1}(\mathbf{Xu}) = \mathbf{XMu} = \lambda(\mathbf{Xu})$$

and the eigenvalues are the same, though not the eigenvectors.

Also a matrix is of the same rank as its transpose because $\det(\mathbf{X}) = \det(\mathbf{X}')$, the directrices being the same. Hence $\det(\mathbf{M}-\lambda\mathbf{I}) = \det(\mathbf{M}'-\lambda\mathbf{I})$ and the latent roots of $\mathbf{M}'$, i.e., its eigenvalues, are the same as those of **M**.

It follows that if **A** and **B** are both square, then their product cannot be of higher rank than that of either **A** or **B**, for if $\mathbf{Bv} = \mathbf{0}$, then $\mathbf{ABv} = \mathbf{0}$ also. Similarly, if $\mathbf{A}'\mathbf{u} = \mathbf{0}$, then $(\mathbf{AB})'\mathbf{u} = \mathbf{B}'\mathbf{A}'\mathbf{u} = \mathbf{0}$.

It may also be noted that the trace of an idempotent matrix equals the rank. (For such a matrix all eigenvalues that are not zero equal one.)

(21) *Partitioned matrices.* It is sometimes convenient to think of a matrix as made up of sub-matrices. Thus, if the rows and columns of a matrix indicate

treatments, it may be important to distinguish between those that have been involved in some mishap and those that have not. First, the treatments must be ordered so that one set stands first and then it is possible to write

$$\mathbf{P} = \begin{bmatrix} \mathbf{X} & \mathbf{Y} \\ \mathbf{Y}' & \mathbf{Z} \end{bmatrix}$$

or some similar expression. The multiplication of partitioned matrices follows the usual rules. Thus

$$\begin{bmatrix} \mathbf{A} & \mathbf{B} \\ \mathbf{C} & \mathbf{D} \end{bmatrix} \begin{bmatrix} \mathbf{W} & \mathbf{X} \\ \mathbf{Y} & \mathbf{Z} \end{bmatrix} = \begin{bmatrix} \mathbf{AW}+\mathbf{BY} & \mathbf{AX}+\mathbf{BZ} \\ \mathbf{CW}+\mathbf{DY} & \mathbf{CX}+\mathbf{DZ} \end{bmatrix}$$

supposing all sub-matrices to be conformable where required.

(22) *Differentiating by a matrix.* When a scalar is differentiated by a matrix, it is differentiated by each element in turn and the results are set out in the same form as in the original matrix. Thus, if $a = x^3 + xz + wy^2 + 3y$ and it is to be differentiated by

$$\mathbf{M} = \begin{bmatrix} w & x \\ y & z \end{bmatrix},$$

then

$$\frac{\mathrm{d}a}{\mathrm{d}\mathbf{M}} = \begin{bmatrix} y^2 & 2x^2 + z \\ 2wy + 3 & x \end{bmatrix}.$$

Particular interest attaches to the differentiation of a quantity like $A = \mathbf{p}'\mathbf{X}\mathbf{q}$, where $\mathbf{p}$ and $\mathbf{q}$ are vectors. Here

$$\frac{\mathrm{d}A}{\mathrm{d}\mathbf{p}} = \mathbf{Xq}, \qquad \frac{\mathrm{d}A}{\mathrm{d}\mathbf{q}} = \mathbf{X}'\mathbf{p}.$$

If $B = \mathbf{p}'\mathbf{X}\mathbf{p}$, then $\mathrm{d}B/\mathrm{d}\mathbf{p} = 2\mathbf{Xp}$. The above results are readily derived because $A = \Sigma\Sigma_{ij}(X_{ij}p_j q_i)$ and $B = \Sigma\Sigma_{ij}(X_{ij}p_i p_j)$.

(23) *Quadratic forms.* A scalar like $\mathbf{y}'\mathbf{A}\mathbf{y}$, where A is a symmetric matrix, is said to be a 'quadratic form'. Expressing A as its spectral decomposition, as in (19),

$$\mathbf{y}'\mathbf{A}\mathbf{y} = \sum(\lambda_i \alpha_i^2) \qquad \text{where} \qquad \alpha_i = \mathbf{y}'\mathbf{u}_i.$$

Hence, if no λ_i is negative it is not possible for $\mathbf{y}'\mathbf{A}\mathbf{y}$ to be negative. Also, if all λ_i are positive, then the quadratic form must be positive also.

(24) *Diagonal matrices.* If all off-diagonal elements of a square matrix are zero, the matrix is said to be 'diagonal', e.g.

$$\begin{bmatrix} 4 & 0 & 0 \\ 0 & -1 & 0 \\ 0 & 0 & 3 \end{bmatrix}.$$

In the study of experimental design the same set of quantities often appears both as a vector and as a diagonal matrix. A special notation has accordingly been evolved. If $\mathbf{x}$ is a vector, $\mathbf{x}^{\alpha\delta}$ indicates a diagonal matrix having the same elements as $\mathbf{x}$, each raised to the power α.

Thus, if $\mathbf{x} = \begin{bmatrix} 4 \\ 9 \\ 1 \end{bmatrix}$, then $\mathbf{x}^{\delta} = \begin{bmatrix} 4 & 0 & 0 \\ 0 & 9 & 0 \\ 0 & 0 & 1 \end{bmatrix}$, $\mathbf{x}^{\frac{1}{2}\delta} = \begin{bmatrix} 2 & 0 & 0 \\ 0 & 3 & 0 \\ 0 & 0 & 1 \end{bmatrix}$,

$$\mathbf{x}^{-\delta} = \begin{bmatrix} \frac{1}{4} & 0 & 0 \\ 0 & \frac{1}{9} & 0 \\ 0 & 0 & 1 \end{bmatrix} = \frac{1}{36}\begin{bmatrix} 9 & 0 & 0 \\ 0 & 4 & 0 \\ 0 & 0 & 36 \end{bmatrix} \text{ and}$$

$$\mathbf{x}^{-\frac{1}{2}\delta} = \begin{bmatrix} \frac{1}{2} & 0 & 0 \\ 0 & \frac{1}{3} & 0 \\ 0 & 0 & 1 \end{bmatrix} = \frac{1}{6}\begin{bmatrix} 3 & 0 & 0 \\ 0 & 2 & 0 \\ 0 & 0 & 6 \end{bmatrix} \text{ and so on.}$$

Obviously $\mathbf{x}^{\alpha\delta}\mathbf{x}^{\beta\delta} = \mathbf{x}^{(\alpha+\beta)\delta}$ and $\mathbf{x}^{0\delta} = \mathbf{I}$.

References

(The number in square brackets at the end of each reference is the section or example in which the reference is made)

Abraham, J. K. (1960). On an alternative method of computing Tukey's statistic for the Latin square model. *Biometrics*, **16**, 686–91. [8.9]

Allen, F. E. and Wishart, J. (1931). A method of estimating the yield of a missing plot in field experimental work. *J. agric. Sci.* (Cambridge), **20**, 399–406. [7.4]

Andrews, D. J. (1972). Plots of high dimensional data. *Biometrics*, **28**, 125–36. [9.8]

Andrews, D. J. and Herzberg, A. M. (1980). *Data: a collection of problems and data from many fields for the student and research worker in statistics* (Imperial College, London). [8B]

Anon (1949). Notes and News. *Biometrics*, **5**, 256–7. [10.3]

Anscombe, F. J. (1948). The transformation of Poisson, binomial and negative binomial data. *Biometrika*, **35**, 246–54. [9.12]

Atkinson, A. C. (1969). The use of residuals as a concomitant variable. *Biometrika*, **56**, 33–41. [1.5]

Bailey, N. T. J. (1959). *Statistical Methods in Biology*. English Universities Press, London. [1.1]

Bailey, R. A. (1981). A unified approach to design of experiments. *J. Royal. statist. Soc.*, **A144**, 214–23. [4.9]

Bartlett, M. S. (1937). Some examples of statistical methods of research in agriculture and applied biology (with discussion) *J. Royal statist. Soc.*, Suppl. **4**, 137–83. [1.1, 7.5]

Bartlett, M. S. (1938). The approximate recovery of information from field experiments with large blocks. *J. agric. Sci.* (Cambridge), **28**, 418–27. [2.4]

Bartlett, M. S. (1978). Nearest neighbour models in the analysis of field experiments (with discussion). *J. Royal statist. Soc.*, **B40**, 147–74. [2.4 twice]

Berry, G. (1967). A mathematical model relating plant yield and arrangement for regularly spaced crops. *Biometrics*, **23**, 505–15. [9.1]

Besag, J. (1972). Nearest neighbour systems and the auto-logistic model for binary data. *J. Royal statist. Soc.*, **B34**, 75–83. [1.5]

Bleasdale, J. K. A. (1967a). The relationship between the weight of a plant part and total weight as affected by plant density. *J. hort. Sci.*, **42**, 51–8. [9.1]

Bleasdale, J. K. A. (1967b). Systematic designs for spacing experiments. *Exptl. Agric.*, **3**, 73–86. [9.1 twice]

Bleasdale, J. K. A. and Nelder, J. A. (1960). Plant population and crop yield. *Nature* (London), **188**, 342. [9.1]

Bliss, C. I. (1967). *Statistics in Biology*, Vol. 1. McGraw-Hill, New York. [1.1]

Bliss, C. I. (1970). *Statistics in Biology*, Vol. 2. McGraw-Hill, New York. [1.1]
Bose, R. C., Clatworthy, W. H., and Shrikande, S. S. (1954). Tables of partially balanced designs with two associate classes. *N. Carolina Agric. Expt. Sta., Tech. Bull.* **107**. [5.8 three times]
Bose, R. C. and Nair, K. R. (1939). Partially balanced incomplete block designs. *Sankhya*, **4**, 337–42. [5.8]
Box, G. E. P. and Cox, D. R. (1964). An analysis of transformations. *J. Royal statist. Soc.*, **B26**, 211–43. [9.12]
Bradley, R. A., Walpole, R. E., and Kramer, C. Y. (1960). Intra- and inter-block analysis for factorials in incomplete block designs. *Biometrics*, **16**, 566–81. [5.8]
Burns, W. D., Charlton, P. J., and Li, W. K. (1978). Comparison of missing plot techniques. *Biom. J.*, **20**, 737–42. [7.3]
Caliński, T. (1971). On some desirable patterns in block design (with discussion). *Biometrics*, **27**, 275–92. [3.5 twice, 3.6]
Caliński, T. (1977). On the notion of balance in block designs. In *Recent Developments in Statistics* (ed. J. R. Barra *et al.*). North-Holland, Amsterdam. [3.6, 5.1]
Caliński, R. and Ceranka, B. (1974). Supplement block designs. *Biom. J.*, **16**, 299–305. [5.3]
Campbell, R. C. (1967). *Statistics for Biologists*. Cambridge Univ. Press, Cambridge. [1.1, 4A]
Catchpole, E. A. (1981). On the convergence of Jones' iteration. *Biom. J.*, **23**, 495–500. [3.4]
Chakrabati, M. C. (1962). *Mathematics of Design and Analysis of Experiments*. Asia Publishing House, Bombay. [3.1]
Clarke, G. M. (1963). A second set of treatments in a Youden square design. *Biometrics*, **19**, 98–104. [9.7]
Clatworthy, W. H. (1973). Tables of two associate-class partially balanced designs. *Appl. Math. Ser.*, **63**, U.S. National Bur. Standards, Washington. [5.8]
Cleaver, T. J., Greenwood, D. J., and Wood, J. T. (1970). Systematically arranged fertilizer experiments. *J. hort. Sci.*, **45**, 457–69. [9.1]
Cochran, W. G. (1937). Problems arising in the analysis of a series of similar experiments. *J. Royal statist. Soc.*, Suppl. **4**, 102–18. [1.1]
Cochran, W. G. and Cox, G. M. (1950). Experimental designs. Wiley, New York (Second Edition, 1957). [1.1, 2A, 6.7, 9.9]
Conniffe, D. and Stone, J. (1974). The efficiency factor of a class of incomplete block designs. *Biometrika*, **61**, 633–6. [5.1]
Conniffe, D. and Stone, J. (1975). Some incomplete block designs of maximum efficiency. *Biometrika*, **62**, 685–6. [5.1]
Corsten, L. C. A. (1958). Vectors: a tool in statistical regression theory. *Meded. Landbouwhogeschool*, Wageningen, **58**, 1–92. [3.5]
Corsten, L. C. A. (1976). Canonical correlation in incomplete blocks. In *Essays in probability and statistics* (ed. S. Ikeda *et al.*). Shinka Tsusho, Tokyo. [4.6]
Cox, C. P. (1968). Some observations on the teaching of statistical consultancy. *Biometrics*, **24**, 789–802. [10.3]
Cox, D. R. (1958). *Planning of Experiments*. Wiley, New York. [1.1, 9C]
Crowther, E. M. (1947). *Fertilizer experiments in colonial agriculture. Memoranda on colonial fertilizer experiments*, II. H.M.S.O. (London), Colonial **214**. [1.1]
Curnow, R. N. (1961). Optimal programmes for varietal selection (with discussion). *J. Royal. statist. Soc.*, **B23**, 282–318. [1.2]
Dagnelie, P. (1981). *Principes d'expérimentation*. Les presses agronomiques de Gembloux, Belgium. [1.1]
Das, M. N. (1958). On reinforced incomplete block designs. *J. Ind. Soc. agric. Statist.*, **10**, 73–7. [5.3 twice]

Dempster, A. P., Laird, N. M., and Rubin, D. B. (1977). Maximum likelihood from incomplete data via the EM algorithm (with discussion). *J. Royal statist. Soc.*, **B39**, 1–38. [7.4]

Duby, C., Guyon, X., and Prum, B. (1977). The precision of different experimental designs for a random field. *Biometrika*, **64**, 59–67. [1.5]

Duncan, D. B. (1955). Multiple range and multiple *F* tests. *Biometrics*, **11**, 1–42. [2A]

Dyke, G. V. (1974). Comparative experiments with field crops. Butterworth, London. [1.1]

Dyke, G. V. and Shelley, C. F. (1976). Serial designs balanced for effects of neighbours on both sides. *J. agric. Sci.* (Cambridge), **87**, 303–5. [9.2]

Dyson, W. G. and Freeman, G. H. (1968). Seed orchard designs for sites with a constant prevailing wind. *Silvae Genet.*, **17**, 12–5. [9.2]

Eccleston, J. A. and Jones, B. (1980). Exchange and interchange procedures to search for optimal row-and-column designs. *J. Royal statist. Soc.*, **B42**, 372–6. [9.6]

Euler, L. (1782). Recherches sur une nouvelle espèce de quarrés magiques. *Verhandelingen Zeeuwsch Genootschap Wetenscaffen*, Vlissingen, **9**, 85–239. [1.1]

Farazdagh, H. and Harris, P. M. (1968). Plant competition and crop yield. *Nature* (London), **217**, 289–90. [9.1]

Federer, W. T. (1955). *Experimental Design: Theory and Application.* Macmillan Company, New York. [1.1, 5.8, 8.2, 8.9, 8F]

Federer, W. T. (1956). Augmented (or Hoonuiaku) designs *Hawaiian Planters Record*, **55**, 191–207. [4.4]

Federer, W. T. (1976). Sampling, blocking and model considerations for the completely randomized, randomized complete block and incomplete block experimental designs. *Biom. J.*, **7**, 511–25. [10.4]

Federer, W. T. (1979). Statistical designs and response models for mixtures of cultivars. *Agron. J.*, **71**, 701–6. [9.8]

Federer, W. T. (1980). Some recent results in experimental design with a bibliography, I. *Internat. statist. Rev.*, **48**, 357–68. [5.1]

Federer, W. T. (1981a). Some recent results in experimental design, II. Bibliography, A–K. *Internat. statist. Rev.*, **49**, 95–109. [5.1]

Federer, W. T. (1981b). Some recent results in experimental design, III. Bibliography, L–Z. *Internat. statist. Rev.*, **49**, 185–97. [5.1]

Federer, W. T. and Balaam, L. N. (1972). *Bibliography on Experiment and Treatment Design: Pre-1968.* Oliver and Boyd, Edinburgh (for the Internat. Statist. Inst.). [5.1]

Federer, W. T., Hedayat, A., Lowe, C. C., and Raghavarao, D. (1976). Application of statistical design theory to crop estimation with special reference to legumes and mixtures of cultivars. *Agron. J.*, **68**, 914–9. [9.8]

Federer, W. T. and Schottefeld, C. S. (1954). The use of covariance to control gradients in experiments. *Biometrics*, **10**, 282–9. [2.3]

Finch, S., Skinner, G., and Freeman, G. H. (1976). The effect of plant density on populations of the cabbage root fly on four cruciferous crops. *Ann. appl. Biol.*, **83**, 191–7. [Fig. 4]

Finney, D. J. (1945a). Some orthogonal properties of the 4 × 4 and 6 × 6 Latin squares. *Ann. Eugenics*, **12**, 213–19. [9.7]

Finney, D. J. (1945b). The fractional replication of factorial experiments. *Ann. Eugenics*, **12**, 291–301. [8.6]

Finney, D. J. (1946a). Orthogonal partitions of the 5 × 5 Latin squares. *Ann. Eugenics*, **13**, 1–3. [9.7]

Finney, D. J. (1946b). Orthogonal partitions of the 6 × 6 Latin squares. *Ann. Eugenics*, **13**, 184–96. [9.7]

Finney, D. J. (1946c). Recent developments in the design of field experiments, III. Fractional replication. *J. agric. Sci.* (Cambridge), **36**, 184–91. [8.6]

Finney, D. J. (1950). The fractional replication of factorial experiments—a correction. *Ann. Eugenics*, **15**, 276. [8.6]

Finney, D. J. (1953). Response curves and the planning of experiments. *Ind. J. agric. Sci.*, **23**, 167–86. [9.1]

Finney, D. J. (1955). *Experimental Design and its Statistical Basis*. Cambridge Univ. Press, Cambridge. [1.1]

Finney, D. J. (1956). The statistician and the planning of experiments (with discussion). *J. Royal statist. Soc.*, **A119**, 1–27. [1.1, 10.3]

Finney, D. J. (1960). *An Introduction to the Theory of Experimental Design*. Univ. Chicago Press, Chicago. [1.1]

Finney, D. J. (1962). *An Introduction to Statistical Science in Agriculture*. Oliver and Boyd, Edinburgh and Munksgaard, Copenhagen. [1.1]

Finney, D. J. (1968). Teaching biometry in the university. *Biometrics*, **24**, 1–12. [10.3]

Fisher, R. A. (1925). *Statistical Methods for Research Workers*. Oliver and Boyd, Edinburgh. [1.1]

Fisher, R. A. (1934). Fourth edition of Fisher (1925) with certain additions. [1.1]

Fisher, R. A. (1935). *The Design of Experiments*. Oliver and Boyd, Edinburgh. [1.1]

Fisher, R. A. and Mackenzie, W. A. (1923). Studies in crop variation, II. The manurial response of different potato varieties. *J. agric. Sci.* (Cambridge), **13**, 311–20. [1.1]

Fisher, R. A. and Wishart, J. (1930). The arrangement of field experiments and the statistical reduction of the results. *Imp. Bur. Soil Sci., Tech. Commun.*, **10**. [1.1]

Fisher, R. A. and Yates, F. (1938). *Statistical Tables for Biological, Agricultural and Medical Research*. Oliver and Boyd, Edinburgh. [1.6, 5.2]

Freeman, G. H. (1957a). Some experimental designs of use in changing from one set of treatments to another. Part I. *J. Royal statist. Soc.*, **B19**, 154–62. [2.6, 6.6, 9.7]

Freeman, G. H. (1957b). Some experimental designs of use in changing from one set of treatments to another. Part II. *J. Royal statist. Soc.*, **B19**, 163–65. [2.6, 6.6, 9.7]

Freeman, G. H. (1958). Families of designs for two successive experiments. *Ann. Math. Statist.*, **29**, 1063–78. [9.7]

Freeman, G. H. (1959). The use of the same experimental material for more than one set of treatments. *Appl. Statist.*, **8**, 13–20. [9.7]

Freeman, G. H. (1963). The combined effect of environmental and plant variation. *Biometrics*, **19**, 273–7. [1.5]

Freeman, G. H. (1964a). The addition of further treatments to Latin square designs. *Biometrics*, **20**, 713–29. [9.7]

Freeman, G. H. (1964b). The use of a systematic design for a spacing trial with a tropical field crop. *Biometrics*, **20**, 200–3. [9.1]

Freeman, G. H. (1966). Some non-orthogonal partitions of 4×4, 5×5 and 6×6 Latin squares. *Ann. Math. Statist.*, **37**, 666–81. [9.7]

Freeman, G. H. (1967). The use of cyclic balanced incomplete block designs for directional seed orchards. *Biometrics*, **23**, 761–78. [9.2]

Freeman, G. H. (1969). The use of cyclic balanced incomplete block designs for non-directional seed orchards. *Biometrics*, **25**, 561–71. [9.2]

Freeman, G. H. (1973). Experimental designs with unequal concurrences for estimating direct and remote effects of treatments. *Biometrika*, **60**, 559–63. [9.4]

Freeman, G. H. (1975). Row-and-column designs with two groups of treatments having different replications. *J. Royal statist. Soc.*, **B37**, 114–28. [6.6]

Freeman, G. H. (1976a). On the selection of designs for comparative experiments. *Biometrics*, **32**, 195–9. [9.6]

Freeman, G. H. (1976b). A cyclic method of constructing regular group divisible incomplete block designs. *Biometrika*, **63**, 555–8. [5.7]

Freeman, G. H. (1979a). Some two-dimensional designs balanced for nearest neighbours. *J. Royal statist. Soc.*, **B41**, 88–95. [9.2]

Freeman, G. H. (1979b). Complete Latin squares and related experimental designs. *J. Royal statist. Soc.*, **B41**, 253–62. [9.2]
Freeman, G. H. (1981). Further results on quasi-complete Latin squares. *J. Royal statist. Soc.*, **B43**, 314–20. [9.2]
Freyman, S. and Dolman, D. (1971). A simple systematic design of planting density experiments with set row widths. *Canad. J. Plant Sci.*, **51**, 340–2. [9.1]
Gnot, S. (1976). The mean efficiency of block designs. *Math. Operationsvorsuch Statist.*, **7**, 75–84. [5.1]
Gomez, K. A. and Gomez, A. A. (1976). *Statistical procedures for agricultural research with emphasis on rice.* The International Rice Research Institute, Manila, Philippines. [1.1]
Goodchild, N. A. (1971). Designs having fruit trees as blocks and incorporating analysis of covariance. *Biometrics*, **27**, 313–23. [9.4]
Goode, J. E. and Marchant, W. T. B. (1962). Use of mechanical barriers instead of guard rows in field experiments. *Nature* (London), **195**, 410. [1.3, 9.1]
Gosset, W. S. (1936). Cooperation in large-scale experiments: A discussion. *J. Royal Statist. Soc.*, Suppl. **3**, 114–36. [1.1]
Goulden, C. H. (1952). *Methods of Statistical Analysis.* Wiley, New York. [1.1]
Greenham, D. W. P. (1965). A long-term manurial trial on dessert apple trees. *J. hort. Sci.*, **40**, 213–35. [9.11]
Grundy, P. M. and Healy, M. J. R. (1950). Restricted randomization and quasi-Latin squares. *J. Royal statist. Soc.*, **B12**, 286–91. [4.9]
Hall, W. B. and Williams, E. R. (1973). Cyclic superimposed designs. *Biometrika*, **60**, 47–53. [9.7]
Harris, J. A. (1920). Practical universality of field heterogeneity as a factor influencing plot yields. *J. agric. Sci.*, (Cambridge), **19**, 279–313. [1.1]
Hartley, H. O. (1956). Programming analysis of variance for general purpose computers. *Biometrics*, **12**, 110–22. [7.4]
Healy, M. J. R. and Westmacott, M. H. (1956). Missing values in experiments analysed on automatic computers. *Appl. Statist.*, **5**, 203–6. [7.4]
Hedayat, A. (1973). Self orthogonal Latin square designs and their importance. *Biometrics*, **29**, 393–6. [9.7]
Hedayat, A. (1975). Self orthogonal Latin square designs and their importance, II. *Biometrics*, **31**, 755–9. [9.7]
Hedayat, A., Parker, E. T., and Federer, W. T. (1970). The existence and construction of two families of designs for two successive experiments. *Biometrika*, **57**, 351–5. [9.7]
Hedayat, A., Parker, E. T., and Federer, W. T. (1971). Amendment to the above paper. *Biometrika*, **58**, 687. [9.7]
Hedayat, A. and Seiden, E. (1970). *F*-squares and orthogonal *F*-squares design: a generalization of Latin square and orthogonal Latin square design. *Ann. Math. Statist.*, **41**, 2035–44. [9.7]
Hoblyn, T. N. (1931). Field experiments in horticulture. *Tech. Commun.* **2**, Bur. Fruit Prod., East Malling. [1.1]
Hoblyn, T. N. (1941). Manurial trials with apple trees at East Malling, 1920–39. *J. Pomol. hort. Sci.*, **18**, 325–43. [9.11]
Hoblyn, T. N., Pearce, S. C., and Freeman, G. H. (1954). Some considerations in the design of successive experiments in fruit plantations. *Biometrics*, **10**, 503–15. [2.6, 5.3, 9.7]
Holliday, R. (1960). Plant population and crop yield. *Nature* (London), **186**, 22–4. [9.1]
Hoyle, M. H. (1973). Transformations—an introduction and a bibliography. *Internat. statist. Rev.*, **41**, 203–23. [9.12]
Irwin, J. O. (1934). On the independence of the constituent items in the analysis of variance. *J. Royal. statist. Soc.*, Suppl. **1**, 236–51. [1.1]

Jaccottet, M. and Tomassone, R. (1976). Méthodes d'analyse factorielle en théorie des plans d'experience. *Ann. Inst. Henri Poricaré*, **12**, 233–56. [4.7]

Jarrett, R. G. (1977). Bounds for the efficiency factor of block designs. *Biometrika*, **64**, 67–72. [5.1]

Jarrett, R. G. (1978). The analysis of designed experiments with missing observations. *Appl. Statist.*, **27**, 38–46. [7.4]

Jarrett, R. G. and Hall, W. B. (1978). Generalized cyclic incomplete block designs. *Biometrika*, **65**, 397–401. [5.7]

John, J. A. (1973). Generalized cyclic designs in factorial experiments. *Biometrika*, **60**, 55–63, [5.7]

John, J. A. and Lewis, S. M. (1976). Mixed-up values in experiments. *Appl. Statist.*, **25**, 61–3. [7.7]

John, J. A. and Prescott, P. (1975). Estimating missing values in experiments. *Appl. Statist.*, **24**, 190–2. [7.5]

John, J. A., Wolock, F. W., and David, H. A. (1972). Cyclic designs. *Appl. math. Ser.*, **62**, U.S. Nat. Bur. Standards. [5.7]

Jones, B. (1976). An algorithm for deriving optimal block designs. *Technometrics*, **18**, 451–8. [3.4, 9.6]

Jones, B. (1979). Algorithms to search for optimal row-and-column designs. *J. Royal statist. Soc.*, **B41**, 210–6. [6G, 9.6]

Jones, B. and Eccleston, J. A. (1980). Exchange and interchange procedures to search for optimal designs. *J. Royal statist. Soc.*, **B42**, 238–43. [9.6]

Kempthorne, O. (1952). The design and analysis of experiments. Wiley, New York. [1.1]

Kempton, R. A. and Howes, C. W. (1981). The use of neighbouring plot values in the analysis of variety trials. *Appl. Statist.*, **30**, 59–70. [2.4]

Kramer, C. Y. and Bradley, R. A. (1957). Examples of intra-block analysis for factorials in group divisible, partially balanced, incomplete block designs. *Biometrics*, **13**, 197–224. [5.8]

Kuiper, N. H. (1952). Variantie-analyse. *Statistica*, **6**, 149–94. [3.5 twice]

Kurkijan, B. and Zelen, M. (1963). Application of the calculus of factorial arrangements, I. Block and direct product designs. *Biometrika*, **50**, 63–73. [5.5]

Li, J. C. R. and Keller, K. R. (1951). An application of serial correlation in field experiments. *Agron. J.*, **43**, 201–3. [1.5]

Lockwood, G. (1979). Improving precision of cocoa progeny trials using calibrator trees. *Exptl. Agric.*, **15**, 209–15. [9.5]

Lockwood, G. (1980). Adjustment by neighbouring plots in progeny trials with cocoa. *Exptl. Agric.*, **16**, 81–9. [2.4]

Lockwood, G. and Martin, K. J. (1976). The use of whole plot data from progeny trials with cocoa (*Theobrama cacao* L), in Ghana. *J. hort. Sci.*, **51**, 353–8. [1.3]

Mandel, J. (1971). A new analysis of variance model for non-additive data. *Technometrics*, **13**, 1–18. [8.1]

Marchant, W. T. B. and Boa, W. (1962). An implement for laying plastic foil root barriers, *J. agric. Engng. Res.*, **7**, 258–9. [1.3, 9.1]

Martin, F. B. (1973). Beehive designs for observing variety competition. *Biometrics*, **29**, 397–402. [9.1]

Martin, K. J. (1980). A partition of a two-factor interaction, with an agricultural example. *Appl. Statist.*, **29**, 149–55. [8.8]

Martin, K. J. and Lockwood, G. (1979). Edge effects and interaction with environment in cocoa. *Exptl. Agric.*, **15**, 225–39. [1.3]

Matérn, B. (1972). Performance of various designs of field experiments when applied to random fields. *Proc. Third Conf. Advis. Group Forest Statist.*, Jouy-en-Josas, I.N.R.A., Paris, 119–28. [1.5 twice]

Mead, R. (1966). A relationship between individual plant-spacing and yield. *Ann. Bot.*, **30**, 301–9. [9.1]

Mead, R. (1967). A mathematical model for the estimation of interplant competition. *Biometrics*, **23**, 189–205. [9.1]

Mead, R. (1979). Competition experiments. *Biometrics*, **35**, 41–54. [9.1]

Mead, R. and Riley, J. (1981). A review of statistical ideas relevant to intercropping (with discussion). *J. Royal statist. Soc.*, **A144**, 462–509. [9.8]

Mead, R. and Stern, R. D. (1980). Designing experiments for intercropping research. *Exptl. Agric.*, **16**, 329–42. [9.8]

Mead, R. and Willey, R. W. (1980). The concept of a 'Land equivalent ratio' and advantages in yields from intercropping. *Exptl. Agric.*, **16**, 217–28. [9.8]

Mendez, I. (1971). Refinamiento a la technica de seleccion masal moderna. *Agrociencia*, **A6**, 87–91. [2.4]

Mercer, W. B. and Hall, A. D. (1911). The experimental error of field trials. *J. agric. Sci.*, **4**, 107–32. [1.1]

Mitscherlich, E. A. (1934). Die Verarbeitung landwirtschaftlicher und anderer biologischer Versuchsergebnisse. *Schr. Konigsberger Gelehrten Ges.*, **7**, 199–229. [9.1]

Mukerjee, R. (1981). Construction of effect-wise orthogonal factorial designs. *J. Statist. Plan. Infer.*, **5**, 221–9. [5.5]

Myers, R. H. (1971). *Response Surface Methodology*. Allyn and Bacon, Boston. [7.4]

Nelder, J. A. (1962). New kinds of systematic designs for spacing experiments. *Biometrics*, **18**, 283–307. [9.1]

Nelder, J. A. (1965a). The analysis of randomized experiments with orthogonal block structure: Block structure and the null analysis of variance. *Proc. Royal Soc.*, **A283**, 147–62. [4.8, 6.1 twice]

Nelder, J. A. (1965b). The analysis of randomized experiments with orthogonal block stucture: Treatment structure and the general analysis of variance. *Proc. Royal Soc.*, **A283**, 163–78. [4.8, 6.2]

Nelder, J. A. (1968). The combination of information in generally balanced designs. *J. Royal statist. Soc.*, **B30**, 303–9. [6.4]

Nelder, J. A. (1977). A reformulation of linear models (with discussion). *J. Royal statist. Soc.*, **A140**, 48–76. [8.1]

Neyman, J. (1935). Statistical problems in agricultural experimentation (with discussion). *J. Royal statist. Soc.*, Suppl. **2**, 107–80. [1.1]

Nigam, A. K. (1976a). On some balanced row-and-column designs. *Sankhya*, **B38**, 87–91. [6.6]

Nigam, A. K. (1976b). Nearly balanced incomplete block designs. *Ind. J. Statist.*, **B38**, 195–8. [5.8]

Okigbo, B. N. (1966). The use of covariance in the adjustment for a fertility gradient in a cassava pre-planting cultivations experiment. *Nigerian J. Sci.*, **1**, 55–64. [2.4]

Outhwaite, A. D. and Rutherford, A. A. (1955). Covariance adjustment as an alternative to stratification in the control of gradients. *Biometrics*, **11**, 431–40. [2.3]

Palluel, C. de (1788). Sur les avantages et l'économie que procurent les racines employées à l'engrais des moutons à l'étable. *Mem. agric. écon. rurale domest.* Soc. royale Agric., Paris, 17–23. [2.6]

Papadakis, J. (1937). Méthode statistique pour des expériences sur champ. *Bull. Inst. Amél. Plantes, Salonique (Grèce)*, **23**. [2.4]

Papadakis, J. (1940). Comparaison de différentes méthodes d'expérimentation phytotechnique. *Rev. Argentina Agron.*, **7**, 297–362. [2.4]

Patterson, H. D. (1964). Theory of cyclic rotation experiments (with discussion). *J. Royal statist. Soc.*, **B26**, 1–45. [9.10]

Patterson, H. D. and Thompson, R. (1971). Recovery of inter-block information when block sizes are unequal. *Biometrika*, **58**, 545–54. [3.8]

Patterson, H. D. and Williams, E. R. (1976). A new class of resolvable incomplete block designs. *Biometrika*, **63**, 83–92. [5.7]

Patterson, H. D., Williams, E. R., and Hunter, E. A. (1978). Block designs for variety trials. *J. agric. Sci.* (Cambridge), **90**, 395–400. [5.8]

Pearce, S. C. (1948). Randomized blocks with interchanged and substituted plots. *J. Royal statist. Soc.*, **B10**, 252–6. [5.3]

Pearce, S. C. (1953). Field experimentation with fruit trees and other perennial plants. Commonwealth Agric. Bur., Farnham Royal, Bucks, England. *Tech. Commun.*, **23**. [1.1, 2.3, 3A, 6.7, 8A, 9.11, 9C]

Pearce, S. C. (1957). Experimenting with organisms as blocks. *Biometrika*, **44**, 263–71. [9.4 twice, 9B]

Pearce, S. C. (1960a). Supplemented balance. *Biometrika*, **47**, 263–71. [5.3]

Pearce, S. C. (1960b). A method of studying manner of growth. *Biometrics*, **16**, 1–6. [9.7]

Pearce, S. C. (1963). The use and classification of non-orthogonal designs (with discussion). *J. Royal statist. Soc.*, **B126**, 353–77. [5.3, 5.6, 5D, 6.6 twice, 6H]

Pearce, S. C. (1964). Experimenting with blocks of natural size. *Biometrics*, **20**, 699–706. [5.2]

Pearce, S. C. (1965). *Biological Statistics: An introduction*. McGraw-Hill, New York. [1.1, 5.3]

Pearce, S. C. (1968). The mean efficiency of equi-replicate designs. *Biometrika*, **55**, 251–3. [5.1]

Pearce, S. C. (1970a). The efficiency of block designs in general. *Biometrika*, **57**, 339–46. [4.2, 5.1]

Pearce, S. C. (1970b). A comment on totally confounded designs. *Biometrics*, **26**, 329–32. [8.3]

Pearce, S. C. (1975). Row-and-column designs. *Appl. Statist.*, **24**, 60–74. [6H, 6I]

Pearce, S. C. (1976a). Second edition of Pearce (1953). [1.1, 2.3, 2.6, 9.1, 9.11]

Pearce, S. C. (1976b). An examination of Fairfield Smith's law of environmental variation. *J. agric. Sci.* (Cambridge), **87**, 21–4. [1.5]

Pearce, S. C. (1976c). The conduct of 'post-mortems' on concluded field trials. *Exptl. Agric.*, **12**, 151–62. [10.4 twice]

Pearce, S. C. (1976d). Concurrences and quasi-replications: An alternative approach to precision in designed experiments. *Biom. J.*, **18**, 105–12. [3.4]

Pearce, S. C. (1978). The control of environmental variation in some West Indian maize experiments. *Trop. Agric.* (Trinidad), **55**, 97–106. [2.4, 2.6]

Pearce, S. C. (1980). Randomized blocks and some alternatives: a study in tropical conditions. *Trop. Agric.* (Trinidad), **57**, 1–10. [2.4, 2.6]

Pearce, S. C. and Brown, A. H. F. (1960). Improving fruit tree experiments by a preliminary study of the trees. *J. hort. Sci.*, **35**, 56–65. [9.11]

Pearce, S. C., Caliński, T., and Marshall, T. F. de C. (1974). The basic contrasts of an experimental design with special reference to the analysis of data. *Biometrika*, **61**, 449–60. [3.6, 6.3]

Pearce, S. C. and Gilliver, B. (1978). The statistical analysis of data from intercropping experiments. *J. agric. Sci.* (Cambridge), **91**, 625–32. [9.3, 9.8]

Pearce, S. C. and Gilliver, B. (1979). Graphical assessment of intercropping methods. *J. agric. Sci.* (Cambridge), **93**, 51–8. [9.3, 9.8]

Pearce, S. C. and Jeffers, J. R. N. (1971). Block designs and missing data. *J. Royal statist. Soc.*, **B33**, 131–6. [7.4 twice]

Pearce, S. C. and Moore, C. S. (1976). Reduction of error in perennial crops, using adjustment by neighbouring crops. *Exptl. Agric.*, **12**, 267–72. [2.4]

Pearce, S. C. and Taylor, J. (1948). The changing of treatments in a long-term trial. *J. agric. Sci.* (Cambridge), **38**, 402–10. [6G, 9.7]

Pimental-Gomes, F. (1953). The use of Mitscherlich's regression law in the analysis of experiments with fertilizers. *Biometrics*, **9**, 498–516. [9.1]

Preece, D. A. (1966). Some row and column designs for two sets of treatments. *Biometrics*, **22**, 1–25. [9.7 twice]

Preece, D. A. (1971). Iterative procedures for missing values in experiments. *Technometrics*, **13**, 742–53. [7.4]

Preece, D. A. (1975). Bibliography of designs for experiments in three dimensions. *Austral. J. Statist.*, **17**, 51–5. [2.6]

Preece, D. A. (1977). Orthogonality and designs: A terminological muddle. *Utilitas math.*, **12**, 201–23. [4.1]

Preece, D. A. (1979). Supplementary bibliography of designs for experiments in three dimensions. *Austral. J. Statist.*, **21**, 170–2. [2.6]

Preece, D. A. (1980). Covariance analysis: factorial experiments and marginality. *The Statistician*, **29**, 97–122. [2.3, 9.5]

Preece, D. A. (1981). Distribution of final digits in data. *The Statistician*, **30**, 31–60. [10.5]

Preece, D. A. (1983). The design and analysis of experiments—What has gone wrong? *Utilitas math.*, **21A**, 201–44. [10.2]

Preece, D. A., Bailey, R. A., and Patterson, H. D. (1978). A randomization problem in farming designs with superimposed treatments. *Austral. J. Statist.*, **20**, 111–25. [2.1, 4.9, 9.7]

Preece, D. A. and Gower, J. C. (1974). An iterative computer procedure for mixed up values in experiments. *Appl. Statist.*, **23**, 73–4. [7.7]

Preece, D. A., Pearce, S. C., and Kerr, J. R. (1973). Orthogonal designs for three-dimensional experiments, *Biometrika*, **60**, 349–58. [2.6]

Puri, P. D. and Nigam, A. K. (1975a). On patterns of efficiency-balanced designs. *J. Royal statist. Soc.*, **B37**, 457–8. [5.1]

Puri, P. D. and Nigam, A. K. (1975b). A note on efficiency balanced designs. *Sankhya*, **B37**, 457–60. [5.1]

Puri, P. D. and Nigam, A. K. (1976). Balanced factorial experiments, I. *Communications Statist.*, **A5**, 599–619. [5.1]

Puri, P. D. and Nigam, A. K. (1977a). Partially efficiency balanced designs. *Communications Statist.*, **A6**, 753–71. [5.1]

Puri, P. D. and Nigam, A. K. (1977b). Balanced block designs. *Communications Statist.*, **A6**, 1171–9. [5.1]

Puri, P. D. and Nigam, A. K. (1978). Balanced factorial experiments *Communications Statist.*, **A7**, 591–605. [5.1]

Puri, P. D., Nigam, A. K., and Narain, P. (1977). Supplemented block designs. *Sankhya*, **B39**, 189–95. [5.1]

Quenouille, M. H. (1953). *The Design and Analysis of Experiments*. Griffin, London. [1.1]

Ragavarao, D. (1971). *Construction and Combinatorial Problems in Designs of Experiments*. Wiley, New York. [5.2, 5.8]

Rao, C. R. (1952a). *Advanced Statistical Methods in Biometric Research*. Wiley, New York. [9.3]

Rao, C. R. (1952b). General methods of analysis of incomplete block designs. *J. Amer. statist. Assoc.*, **42**, 541–61. [5.2]

Rayner, A. A. (1969). *A First Course in Biometry for Agriculture Students*. University of Natal Press, Pietermaritzburg. [1.1, 7A, 8D]

Rives, M. (1969). Description de l'hétérogéneité des champs d'expérience. *Biom. J.*, **11**, 113–22. [1.5]

Rogers, I. S. (1972). Practical considerations in the use of systematic spacing designs. *Austral. J. exptl., Agric. Anim. Husb.*, **12**, 306–9. [9.1]

Rubin, D. B. (1972). A non-iterative algorithm for least squares estimation of missing values in any analysis of variance design. *Appl. Statist.*, **21**, 136–41. [7.4]

Russell, E. J. (1966). *A History of Agricultural Science in Great Britain, 1620–1952*. Allen and Unwin, London. [1.1]

Satterthwaite, F. E. (1946). An approximate distribution of estimates of variance components. *Biometrics Bull.*, **2**, 110–4. [6.5]

Scheffé, H. (1959). *The analysis of variance*. Wiley, New York. [2A]

Singh, M. and Dey, A. (1978). Two-way elimination of variation. *J. Royal statist. Soc.*, **B40**, 58–63. [6.6]

Singh, M. and Dey, A. (1979). Block designs with nested rows and columns. *Biometrika*, **66**, 497–502. [6.6]

Singh, M., Dey, A., and Nigam, A. K. (1979). Two-way elimination of heterogeneity. *Sankhya*, **B41**, 1–9. [6.6]

Smith, H. F. (1938). An empirical law, describing heterogeneity in the yields of agricultural crops. *J. agric. Sci.* (Cambridge), **28**, 1–23. [1.5]

Smith, P. L. (1981). The use of analysis of covariance to analyse data from designed experiments with missing or mixed-up values. *Appl. Statist.*, **30**, 1–8. [7.5]

Snedecor, G. W. (1956). *Statistical Methods*. The Collegiate Press, Menaska, Wisconsin. [1.1]

Sprent, P. (1970). Some problems of statistical consultancy (with discussion). *J. Royal statist. Soc.*, **A133**, 139–64. [10.3]

Stewart, A. B. (1947). *Planning and Conduct of Fertilizer Trials*, I. H.M.S.O. (London), Colonial **214**. [1.1]

Student, (1908). The probable error of a mean. *Biometrika*, **6**, 1–24. [1.1]

Tocher, K. D. (1952). The design and analysis of block experiments (with discussion). *J. Royal statist. Soc.*, **B14**, 45–100. [3.4]

Trail, S. M. and Weeks, D. L. (1973). Extended complete block designs generated by BIBD. *Biometrics*, **29**, 565–78. [8.9]

Tukey, J. W. (1949). One degree of freedom for non-additivity. *Biometrics*, **5**, 232–42. [8.9]

Tukey, J. W. (1955). Reply to Query 113. *Biometrics*, **11**, 111–3. [8.9]

Veevers, A. and Boffey, T. B. (1975). On the existence of levelled beehive designs. *Biometrics*, **31**, 963–8. [9.1]

Veevers, A. and Boffey, T. B. (1979). Designs for balanced observation of plant competition. *J. statist. Planning Infer.*, **3**, 325–32. [9.1]

Vyvyan, M. C. (1955). Interrelation of scion and rootstock in fruit trees. I. Weights and relative weights of young trees formed by the reciprocal unions, as scion and rootstock, of three apple rootstock varieties, M.IX, M.IV and M.XII. *Ann. Bot.*, **19**, 401–23. [8.8]

Wilkinson, G. N. (1958a). Estimation of missing values for the analysis of incomplete data. *Biometrics*, **14**, 257–86. [7.4]

Wilkinson, G. N. (1958b). The analysis of variance and derivation of standard errors for incomplete data. *Biometrics*, **14**, 360–84. [7.4]

Wilkinson, G. N. (1970). A general recursive procedure for analysis of variance. *Biometrika*, **57**, 19–46. [6.3]

Wilks, S. S. (1932). Certain generalizations in the analysis of variance. *Biometrika*, **24**, 471–94. [9.3]

Willey, R. W. (1979a). Intercropping—its importance and research needs. 1. Competition and yield advantages. *Field Crop Abstracts*, **32**, 1–10. [1.7, 9.8]

Willey, R. W. (1979b). Intercropping—its importance and research needs. 2. Agronomy and research approaches. *Field Crop Abstracts*, **32**, 73–85. [1.7, 9.8]

Williams, E. J. (1952). The interpretation of interactions in factorial experiments. *Biometrika*, **39**, 65–81. [8.1]

Williams, E. R. and Patterson, H. D. (1977). Upper bounds for efficiency factors in block design. *Austral. J. Statist.*, **19**, 194–201. [5.1]

Williams, R. M. (1952). Experimental designs for serially correlated observations. *Biometrika*, **39**, 151–67. [1.5]

Wishart, J. (1934). Statistics in agricultural research. *J. Royal statist. Soc.*, Suppl. 1, 26–61. [1.1]

Wishart, J. (1936). Tests of significance in analysis of covariance. *J. Royal statist. Soc.*, Suppl. **3**, 79–82. [1.1]

Wishart, J. and Sanders, H. G. (1936). *Principles and Practice of Field Experiments*. Empire Cotton Grow. Corp., London, (Reissued 1955 as *Tech. Commun.*, **18**. Commonwealth, Bur. Plant Breeding, Cambridge) [1.1]

Worthington, B. A. (1975). General iterative method for analysis of variance when block structures are orthogonal. *Biometrika*, **62**, 113–20. [6.2]

Yates, F. (1933a). The principles of orthogonality and confounding in replicated experiments. *J. agric. Sci.* (Cambridge), **23**, 108–45. [4.1, 5.2]

Yates, F. (1933b). The analysis of replicated experiments when the field results are incomplete. *Empire J. exptl. Agric.*, **1**, 129–42. [7.4]

Yates, F. (1935). Complex experiments (with discussion). *J. Royal statist. Soc.*, Suppl. **2**, 181–247. [1.1 twice]

Yates, F. (1936a). A new method of arranging variety trials involving a large number of varieties. *J. agric. Sci.* (Cambridge), **26**, 424–55. [2.4, 5.8]

Yates, F. (1936b). Incomplete randomized blocks. *Ann. Eugenics*, **7**, 121–40. [1.1]

Yates, F. (1937). The design and analysis of factorial experiments. *Tech. Commun.*, **35**, Commonwealth Bur. Soil Sci., Rothamsted. [1.1 twice, 5.8, 6.6, 8.7]

Yates, F. (1954). The analysis of experiments containing different crop rotations. *Biometrics*, **10**, 324–46. [9.10]

Yates, F. and Cochran, W. G. (1938). The analysis of groups of experiments. *J. agric. Sci.* (Cambridge), **28**, 556–80. [9.9]

Youden, W. J. (1935). Use of incomplete block replications in estimating tobacco-mosaic virus. *Contrib. Boyce Thompson Inst.*, **9**, 41–8. [6.6]

Yule, G. U. (1927). On reading a scale. *J. Royal statist. Soc.*, **A90**, 570–87. [10.5]

Subject Index

* indicates a passage that contains a definition
† indicates a passage with one or more worked examples